Environmental Science

Series editors: R. Allan · U. Förstner · W. Salomons

Springer-Verlag Berlin Heidelberg GmbH

Sovan Lek · Jean-François Guégan

Artificial Neuronal Networks

Application to Ecology and Evolution

With 99 Figures and 30 Tables

Springer

Editors

Prof. Sovan Lek
C.E.S.A.C., U.M.R. C.N.R.S. 5576 C
Bâtiment IVR3
Université Paul Sabatier Toulouse III
118 route de Narbonne
F-31062 Toulouse cedex 04, France

Dr. Jean-François Guégan
Centre d'Etudes sur le Polymorphisme
des Micro-organismes
C.E.P.M/U.M.R C.N.R.S.-I.R.D. 9926
ORSTOM, 911 Avenue Agropolis, B.P. 5045
F-34032 Montpellier cedex 01, France

Artificial neural networks are inspired from connexions between natural neurons of the brain. In Turing's networks, the neurons are interconnected freely with each other, which roughly simulates the brain functioning. By contrast, modern neural networks as illustrated in the different chapters of this book are connected from layer to layer which implies that the cortex is organized by successive strata. Scientists have always been investigating computers modelled more closely on the human brain, and networks of artificial neurons are thus performative tools to simulate ecological and evolutionary patterns and processes. With permission from Springer Verlag, Berlin.

ISSN 1431-6250
ISBN 978-3-642-63116-0

Die Deutsche Bibliothek – CIP-Einheitsaufnahme

CIP-data applied for
Library of Congress Cataloging-in-Publication Data
Artificial neuronal networks : application to ecology and evolution / Sovan Lek,
Jean-François Guégan (eds). p. cm. -- (Environmental science) Includes bibliographical references.
ISBN 978-3-642-63116-0 ISBN 978-3-642-57030-8 (eBook)
DOI 10.1007/978-3-642-57030-8
1. Ecology--Computer simulation. 2. Evolution (Biology)--Computer simulation. 3. Neural networks (Computer science) I. Lek, Sovan, 1952- II. Guégan, Jean-François. III. Environmental science (Berlin, Germany)
QH541.15.S5 A78 2000 577'.01'13--dc21 00-030795

Originally published by Springer-Verlag Berlin Heidelberg New York in 2000
Softcover reprint of the hardcover 1st edition 2000

Cover Design: Struve & Partner, Heidelberg
Dataconversion: Büro Stasch · Uwe Zimmermann, Bayreuth

SPIN: 10636510 – 3136/xz – 5 4 3 2 1 0 – Printed on acid-free paper

To our wifes, sons and daughters

Preface

By virtue, nature is nonlinear! This cliché embodies the simple fact that natural sciences are irremediably confounded with persistent fuzzy and nonlinear dynamics that any classical statistical methods may not simply solve. The recent developments in computer-aided identification of Artificial Neural Networks (ANNs) provides a way to process nonlinear relationships between variables. This is particularly true in biodiversity research and conservation plans, environmental sciences and applied ecology, or evolutionary ecology, genetics and epidemiology. The question of how ANNs may help to solve some fuzzy problems in natural sciences was explored during a workshop on the Applications of Artificial Neural Networks to Ecological Modelling. The workshop was held in December 1998 at the University of Toulouse III (France) and was co-organized by Sovan Lek (Université Toulouse III, U.M.R. C.N.R.S. 5576) and Jean-François Guégan (U.M.R. C.N.R.S.-I.R.D. 9926, Montpellier). It was jointly funded by the French Centre National de la RechercheScientifique and different sponsoring agencies. The presentations featured different environmental, ecological and evolutionary models, such as landscape ecology and remote sensing, population, community and ecosystem ecology, genetics and evolutionary ecology. The different contributions look at examples of applications of ANNs in a large diversity of research areas, and our intention was to bring together, for the first time, accessible discussions of the use of neural networks. This book probably illustrates some of the most convincing demonstrations of the power of these techniques in natural sciences. Readers who have no special kind of expertise in artificial networks or those who already have a firm grasp on the area will be interested in the different approaches and discussions scattered throughout this book.

The first chapter in this volume was written by Lek, Giraudel and Guégan. Introducing the book, it presents two of the most popular ANN models, one supervised network, the backpropagation algorithm, and one unsupervised network, the Kohonen's Self-Organizing Mapping algorithm (SOM). In particular, it discusses techniques (programmation, algorithms) to generate ANN models, and it then presents some interesting comparisons of ANN results with those obtained from other more conventional statistical methods. The book is then formed of four blocks of papers which correspond to different subheadings, with contributions in landscape ecology and remote sensing forming the first subgroup, then followed by a second subgroup of papers in population to ecosystem ecology, then by a third subgroup in genetics, evolutionary ecology and related fields, and finally by a last one on new perspectives in ANNs.

Kimes, Nelson and Fifer's article, which follows on from Lek, Giraudel and Guégan, begins the second part on *Artificial Neural Networks in Landscape Ecology and Re-*

mote Sensing, and it discusses the use of large data sets of the earth's surface during NASA's Earth Observing System era, and the need to develop efficient algorithms which may incorporate a wide variety of ecosystem data. Much of the discussion is devoted to the relevance of ANNs in the context of specific examples dealing with the extraction of vegetation variables of ecological interest. They round out their paper by discussing some disadvantages of using ANNs with such data, and conclude that ANNs can be more practically used as a variable selection tool in complex natural nonlinear systems to define a set of variables which accurately predict variable(s) of interest.

An interesting result in the same vein can be found in the next article by Foody. Conventional hard statistical classification techniques are often used in mapping vegetation from remotely sensed imagery, but these techniques typically make untenable assumptions, i.e. acceptation of discrete and mutually exclusive vegetation classes which is not really the case in nature, about the remotely sensed data and the vegetation. Here, Foody argues that soft classifications based on ANNs may provide an appropriate representation of both discrete and continuous classes. Such derived soft classifications provide more realistic and appropriate representations of the vegetation, especially for transitional areas and ecotones, mixing of different plant communities or other kinds of vegetation mosaic. Foody's findings reveal that, for instance, a membership of a region to a certain vegetation type, say a woodland, based on conventional mapping may be misclassified, and a soft classification may be more appropriated to reveal the vegetation's mosaic complexity. It is therefore not surprising that it has many applications in the analysis of vegetation profiles.

Another example in which ANNs are applied to obtain results from ground-based photographs is given by Dubois, Cournac, Chave and Riera. They describe how tree diameter distribution may be predicted from ground-based photographs of the landscape. Trained on a set of photographs taken in Equatorial French Guiana, for which *in situ* measurements of tree diameters were available, the authors discussed the main results of their work and perspectives for future application. Indeed, automatic sampling from ground-based photographs may prove to be of real interest for studying ecosystem spatial dynamics. This permits us, for instance, to locate the various stages of a forest during sylvigenetic cycles or the existence of different vegetation types in rain forests, which contribute to the maintenance of plant biodiversity and probably of animal diversity as a whole. Further, as Dubois and collaborators' article discusses, not only does the processing of radar photographs distinguish between vegetation communities, it may also yield efficient tools for detection of the effects of perturbations, e.g. forest fires, and climate modifications, e.g. drier periods, on patterns of plant biodiversity.

The ANN analysis of plant dynamics is not limited to the spatial patterns of plant communities. In their chapter, Tomasel and Paruelo provide an example of the use of ANNs to investigate time-series analysis in plant dynamics. Their results demonstrate that ANN analysis of NDVI (Normalized Difference Vegetation Index) data, which represents a surrogate of Aboveground Net Primary Production (ANPP) of the system, separates between two different phytogeographical areas in Argentina (Patagonia and Monte steppes), and it allows for a satisfactory prediction of the NDVI values up to six months ahead. On the basis of these results, the authors conclude that the combination of ANNs and satellite data offers a very promising approach for the prediction of ANPP and forage availability over extensive rangelands, providing a critical piece of information for ranchers and natural resource managers to devise sustainable systems in arid and semiarid lands.

Manslow, Brown and Nixon's chapter ends the first subheading of this book. This paper presents a novel probabilistic interpretation of area-based fuzzy classifications of mixed pixels, and it may run in counterpoint to the central theme of this book since Manslow and colleagues' contribution concerns the theory of a probabilistic tool, not its direct application. Techniques traditionally used to extract land cover information from remotely sensed images have tended to produce crisp (or hard) classifications of image pixels. This has been criticised since the resulting maps of ground cover consist of grids of pixels of homogeneous class membership, and are hence inherently dissimilar to the true ground cover, which they intend to model. An alternative approach, which is discussed in this paper, is to represent the composition of pixels by the proportions of the sub-pixel area occupied by each cover class, a process sometimes referred to as fuzzy classification. The scope of this chapter obviously is connected to the previous papers, and certainly with that proposed by Foody and Giraudel and collaborators. Probably, much effort has to be made to increase the richness of such pixel based classifications by, for example, relating the probability of class membership of pixels in particular classes to the sub-pixel area occupied by those classes. Indeed, there are some common research interests in the results of Manslow, Brown and Nixon and those which have the same flavour as the results of Foody or Giraudel and his contributors. Manslow and colleagues round out their chapter by discussing new developments in this area.

The third part deals with *Artificial Neural Networks in Population, Community and Ecosystem Ecology*, and it begins with the chapter by Chon, Park and Cha. Patterning temporal dynamics of animal communities is an important topic in ecosystem management, especially in aquatic ecosystems where communities are easily affected by disturbances caused by various natural and anthropogenic agents. Data for community dynamics, however, are complex and difficult to analyse since they consist of many species, varying in nonlinear fashion in spatio-temporal domain. The field data used in this study by Chon and collaborators were monthly densities of selected genera in benthic macroinvertebrate communities collected from urbanized streams in a wide range of organic pollution in the Suyong and Han rivers in Korea. After training the multivariate data, the authors test the feasibility of temporal ANNs in forecasting community changes in a short time period. The authors present some interesting applications of these techniques, including a comparison between macroinvertebrate community changes in the two rivers surveyed. Temporal networks seem to be exciting new methods for time-series analysis and forecast purposes, and a careful reading of this paper by community ecologists will be well regarded.

In the next chapter of Scardi, we see how ANNs may be used to predict the phytoplankton primary production at different, i.e. from global to lowest, spatial scales in the oceans. The role of phytoplankton productivity is crucial in driving the global carbon cycle, and its assessment is a major ecological problem. However, empirical models, which were developed to obtain estimates of the global phytoplankton primary production on the basis of remotely sensed data, may fail to effectively reproduce the local functioning of phytoplankton production at a specific locality. Using two different types of data bases, i.e. a global set which consists of 2 218 phytoplankton biomass, irradiance, temperature and primary production data and a local one with the information collected at a single sampling station in the Gulf of Napoli (Italy) during a five-year cycle of fortnight measurements, Scardi clearly demonstrates in his contribution the major advantage of ANNs for modelling phytoplankton primary production. He also compares the performance of ANNs to conventional models of primary phytoplankton productivity, and he ends his chapter with some new perspectives, i.e. the use of fluorometric data, for instance. It is therefore not surprising that ANNs have many applications in the analysis of phytoplankton productivity.

In their chapter, Boët and Fuhs use results of in situ electrical fish catches covering the whole Seine River basin. These data include the results of over 700 fish catches at 583 sampling stations, which consist of fish species abundance, representing more than 200000 fish belonging to 39 different freshwater species. The 26 most frequent fish species were used, and in addition, for each sampling site, from the 15 covariates describing the habitat conditions, six were retained, which are stream order, river slope and width, water quality, habitat quality, and ecoregion. The determination of the relationships between the habitat characteristics and the presence of fish species was performed using a standard backpropagation algorithm. Such a task may serve to quantify the importance of the environmental variables in the structuration of fish communities, and then it should allow testing the impact of perturbations on fish species composition. This research is of particular importance since there are currently very few predictive models for aquatic ecosystems at the spatial scale of a whole drainage area. Boët and Fuhs have made an interesting study, which has the same flavour as the results of Chon, Park and Cha, and Scardi.

Recknagel and Wilson's article argues that ANNs can be considered as a new generation of inductive models that has not only potential for ecosystem prediction but for ecosystem elucidation as well. The authors of this chapter have chosen two distinct, but aquatic, examples to demonstrate the superiority of neural networks over alternative models (prediction and elucidation of phytoplankton abundance in lakes, and prediction of density of brown trout redds in streams). Both examples demonstrate that ANNs are successful in terms of predictive accuracy and elucidative capacity compared to alternative models in that they overcome previous constraints encountered with traditional techniques and improve predictions. Recknagel and Wilson then present some more interesting developments of these techniques, and they discuss new exciting developments of machine-learning methods in ecological modelling such as genetic algorithms, either in combination with neural networks or as exclusive applications.

In their chapter, Cisneros-Mata, Brey and Jarre-Teichmann compare the performance of regression and artificial neural network models to forecast one year in advance the annual spawning biomass of Pacific sardine (*Sardinops caeruleus*) of the California current. Small pelagic fish species like sardines and anchovies have complex population dynamics as a result of their close relationship to the environmental conditions, short life span and variable recruitment which renders their prediction and management particularly difficult. Nevertheless, forecasting is extremely important for the management of these harvested marine fish populations. First, Cisneros-Mata, Brey and Jarre-Teichmann clearly demonstrate on a 46-year series of sardine catches that past sea temperature is a good indicator of Pacific sardine abundance. Then, the bulk of the discussion is devoted to a comparison of performance between regression and ANN models for fish abundance prediction. The authors report that a crucial step towards using ANNs versus more conventional techniques in time-series analysis may appear to be a threshold length of the series for good performance, which represents a major and serious challenge in population dynamics.

Ball, Palmer-Brown and Mills' article which follows on from Cisneros-Mata and collaborators, concerns a comparative assessment of a range of statistical and ANN modelling techniques to predict the visible ozone injury, characterized by off white chlorotic lesions, in subterranean clover plants during experimental conditions. The data set was generated by exposing plants in acrylic chambers to a range of ozone concentrations and microclimatic conditions during 5 years. The independent factors were ozone dose, leaf age, photosynthetically active radiation, temperature, relative humidity and a random number which represents a variable that had no influence upon the extent of visible injury. The conventional statistical techniques produced a poor performance when modelling the data and were unable to produce accurate predictions on unseen data. On the contrary, ANN models produced the best performance. It is one of the most conclusive of the chapters in the volume since it adopts a comparative view of analyses to demonstrate the superiority of ANNs over more conventional statistical methods.

Next, Giraudel, Aurelle, Berrebi and Lek introduce the following part on *Artificial Neural Networks in Genetics and Evolutionary Ecology*. In this chapter, they discuss more sophisticated techniques for generating Self-Organizing Maps (SOM) with Fuzzy Clustering-Mean (FCM) analysis of genetic data in brown trout (*Salmo trutta*). The SOM procedure, which represents a version of ANNs for visualisation of vectors in a two-dimensional space, is applied to microsatellite loci of French river and domestic fish populations belonging to supposedly different genetic strains. Then, Giraudel and colleagues applying FCM, a procedure which does not assign the membership of one element to exactly one cluster but which classifies the objects by a degree of membership, derive two-dimensional maps which illustrate the genetic structuration of brown trout populations. Giraudel and collaborators' findings support the existence of several wild forms of brown trout in southwestern France, and give some indications about the impacts of fish stocking on natural populations of fish in French rivers. These results show that unsupervised networks and fuzzy clustering algorithms can be successfully applied to complex genetic data, such as microsatellites. Although these results are of more technical nature than most of the other results in this volume, the authors give an excellent illustration of some possible applications of ANNs and soft classifications in population genetics and evolutionary research. A careful reading of this paper by geneticists and evolutionary ecologists will be well rewarded.

Most of Guégan, Thomas, de Meeüs, Lek and Renaud's next article concerns the form of the relationship between presence, or absence, of 15 different infectious and parasitic diseases on a largest scale and a suite of intrinsic and extrinsic factors potentially acting on disease occurrence and their respective spatial distribution. In their contribution, the authors provide a thorough overview of patterns and processes of disease occurrences. The goal of the chapter is to model the parasite species distribution on a global scale using two multivariate models, i.e. logistic regressions and ANNs, from a set of different environmental, demographic and human characteristic descriptors across 153 countries. They discuss the performance of both methods in describing the actual spatial distribution and occurrence of the different parasitic and infectious species at the global level. Indeed, most of the discussion requires some understanding of coevolutionary patterns and processes in host-parasite associations, but the authors have ensured that the chapter is accessible to non-experts.

Teriokhin and Budilova begin their chapter by introducing some of the basic tools (optimal control and ANNs) and knowledge (evolutionary theory of human traits) required in such an analysis. They then focus on fitting these two models to human biodemographic data with a special attention paid to explaining sex distinctions in their respective life histories. The bulk of Teriokhin and Budilova's chapter is devoted to the demonstration that many differences in men's and women's life history traits (later maturity, bigger body size, shorter mean life span, and absence of menopause in men as opposed to women) may be explained as evolutionarily advantageous consequences if only one assumption is made about the difference between the physiology of men and women, namely, if we assume that women, but not men, can accumulate reproductive energy in their offspring. In addition, Teriokhin and Budilova's work may also help to describe the process of distribution of energy in a human organism among its different needs. This area is of particular importance due to the current development of research in human evolutionary ecology.

It seemed appropriate to end this preface with a demonstration that ANNs are evolving in step, and that the major limitation on their use, i.e. they require large amounts of data for training, may be possibly alleviated. Silvert and Baptist begin the last part (*Perspectives*) and the last chapter by exploring several ways of doing this. In most situations the collection of field data is both time consuming and expensive. Since the training and testing of neural networks is very data intensive, this poses serious obstacles to the development of neural network applications in ecology. Silvert and Baptist's chapter is devoted to the topic of using benthic oceanographic data. Further, as the authors discuss, the pre-processing (transformation) of data is a promising solution, but any pre-processing of data in ANNs should be based on ecological knowledge of the system.

We are deeply indebted to many colleagues and friends who generously gave time to the workshop and expert knowledge to review one or two contributions which form this volume: Mr. or Mrs. – Ambroise, Angélibert, Arino, Auger, Auriol, Aussem, Balls, Barbault, Bau, Beacham, Belaud, Bourret, Brabet, Brosse, Canu, Capblancq, Careaux, Cereghino, Cever, Charles, Chau, Chon, Cisneros, Comrie, Cowan, Culverhouse, Dimopoulos, Dreyfus, Dubois, Durand, Ehrman, Flanagan, Foody, Fuhs, Gabas, Galy, Gan, Gascuel, Gedeon, Grandjean, Halls, Hanser, Huntingford, Jarre-Teichman, Jørgensen, Kapetsky, Kaski, Kimes, Kok, Komatsu, Kropp, Lavendier, Lewis, Loot, Manel, Marzban, Masson, Megrey, Melssen, Morimoto, Morlini, Murase, Oberdorff, O'Connor, Park, Pave, Prasher, Puig, Sanchez Manes, Recknagel, Roush, Scardi, Schultz, Shamseldin, Starret, Teriokhin, Thibault, Thiria, Tomasel, Tomassone, Touzet, Vila, Wang, Werner, William, Wong, Yang.

Our special thanks go to the sponsoring agencies: C.N.R.S.-France (Sciences de la Vie), Université Paul Sabatier Toulouse III, Electricité de France, Agence de Bassin d'Adour-Garonne, Ministère des Affaires Etrangères, Région Midi-Pyrénées, Mairie de Toulouse and Oktos-France.

Our gratitude goes to the publishing team of Springer Verlag, particularly Andrea Weber-Knapp and Prof. Dr. Wim Salomons.

Finally, we thank the authors who joined us in the process of publishing this book and who have made this a *Piece of Artificial Life*.

To all of them, our sincere thanks.

Jean-François Guégan, Montpellier
Sovan Lek, Toulouse
Toulouse and Montpellier, 15 March 2000

Contents

Contributors

Didier Aurelle
Laboratoire Génôme et Populations, C.N.R.S. U.P.R. 9060, Cc063
Université Montpellier II
Place Eugène Bataillon
F-34095 Montpellier cedex 05, France
E-mail: aurelle@crit.univ-montp2.fr

Graham R. Ball
Department of Computing, The Nottingham Trent University
Burton Street
Nottingham, NG1 4BU, UK
E-mail: graham.balls@ntu.ac.uk

Martin Baptist
WL Delft Hydraulics, Marine and Coastal Management
P.O. Box 177
NL-2600 MH Delft, The Netherlands
E-mail: martin.baptist@wldelft.nl

Patrick Berrebi
Laboratoire Génôme et Populations, C.N.R.S. U.P.R. 9060, Cc063
Université Montpellier II
Place Eugène Bataillon
F-34095 Montpellier cedex 05, France
E-mail: berrebi@crit.univ-montp2.fr

Philippe Boët
UR qualité et fonctionnement hydrologique des systèmes aquatiques
Cemagref BP44
F-92163 Antony, France
E-mail: philippe.boet@cemagref.fr

Thomas Brey
Alfred Wegener Institute for Polar and Marine Research
Postfach 120161
D-27515 Bremerhaven, Germany

Martin Brown
Image, Speech and Intelligent Systems Research Group
Department of Electronics and Computer Science
University of Southampton
Southampton SO17 1BJ, U.K.
E-mail: m.q.brown@ecs.soton.ac.uk

Elena V. Budilova
Deptartment of Biology, Moscow State University
119899 Moscow, Russia
E-mail: at@ateriokhin.home.biomsu.ru

Eui-Young Cha
Department of Computer Sciences
Pusan National University
Pusan 609-735, Korea

J. Chave
S.P.E.C., C.E.A. Saclay, DSM/DRECAM
Saclay-Orme des Merisiers
F-91191 Gif sur Yvette cedex, France

Tea-Soo Chon
Department of Biology
Pusan National University
Pusan 609-735, Korea
E-mail: ecosys@chollian.net

Miguel A. Cisneros-Mata
Instituto National de la Persa
Calle 20#605 sur Guaymas
Sonora 85400, Mexico
E-mail: cripgym@tetakawi.net.mx

L. Cournac
Laboratoire d'Ecophysiologie de la photosynthèse
D.E.V.M./D.S.V., C.E.A. Cadarache
F-13108 St Paul-lez-Durance cedex, France
E-mail: cournac@dsvcad.cea.fr

Thierry de Meeûs
Centre d'Etudes sur le Polymorphisme des Micro-organismes
C.E.P.M/U.M.R C.N.R.S.-I.R.D. 9926
911 Avenue Agropolis, B.P. 5045
F-34032 Montpellier cedex 01, France
E-mail: demeeus@cepm.mpl.ird.fr

M.A. Dubois
S.P.E.C., C.E.A. Saclay, DSM/DRECAM
Saclay-Orme des Merisiers
F-91191 Gif sur Yvette cedex, France
E-mail: mad@amoco.saclay.cea.fr

Steven T. Fifer
Rathyeon ITSS
4400 Forbes Blvd.
Lanham, MD 20706, USA
E-mail: fifer@forest.gsfc.nasa.gov

Giles M. Foody
Department of Geography
University of Southampton
Highfield
Southampton SO17 1BJ, UK
E-mail: gmf@soton.ac.uk

Thierry Fuhs
Laboratoire d'Ingénierie pour les Systèmes Complexes
Cemagref BP44
F-92163 Antony, France
E-mail: thierry.fuhs@cemagref.fr

Jean-Luc Giraudel
Départment Génie biologique, IUT Périgueux IV
39 rue Paul Mazy
F-24019 Perigueux cedex, France
E-mail: giraudel@montesquieu.u-bordeaux.fr

Jean-François Guégan
Centre d'Etudes sur le Polymorphisme des Micro-organismes
C.E.P.M/U.M.R C.N.R.S.-I.R.D. 9926
911 Avenue Agropolis, B.P. 5045
F-34032 Montpellier cedex 01, France
E-mail: guegan@cepm.mpl.ird.fr

Astrid Jarre-Teichmann
Danish Institute for Fisheries Research, North Sea Centre
P.O. Box 101
DK-9850 Hirtshals, Denmark
E-mail: ajt@dfu.min.dk

Daniel S. Kimes
Biospheric Sciences Branch, Code 923, Laboratory for Terrestrial Physics
NASA Goddard Space Flight Center
Greenbelt
MD 20771, USA
E-mail: dan@pika.gsfc.nasa.gov

Sovan Lek
C.E.S.A.C., U.M.R. C.N.R.S. 5576, Bâtiment IVR3, Université Paul Sabatier
118 route de Narbonne
F-31062 Toulouse cedex 04, France
E-mail: lek@cict.fr

John Manslow
Image, Speech and Intelligent Systems Research Group
Department of Electronics & Computer Science
University of Southampton,
Highfield
Southampton SO17 1BJ, UK
E-mail: jfm96r@ecs.soton.ac.uk

G.E. Mills
Institute of Terrestrial Ecology, University of Wales
Deiniol Road
Bangor LL57 2UP, UK
E-mail: gmi@wpo.nerc.ac.uk

Ross F. Nelson
Biospheric Sciences Branch, Code 923, Laboratory for Terrestrial Physics
NASA Goddard Space Flight Center
Greenbelt
MD 20771, USA
ross@ltpmail.gsfc.nasa.gov

Mark Nixon
Image, Speech and Intelligent Systems Research Group
Department of Electronics & Computer Science
University of Southampton
Highfield
Southampton SO17 1BJ, UK

D. Palmer-Brown
Department of Computing
The Nottingham Trent University
Burton Street
Nottingham NG1 4BU, UK
E-mail: m.q.brown@ecs.soton.ac.uk

Young-Seuk Park
Department of Biology
Pusan National University
Pusan 609-735, Korea

José M. Paruelo
Departamento de Ecología, Facultad de Agronomía, Universidad de Buenos Aires
Av. San Martín 4453
(1417) Buenos Aires, Argentina
E-mail: paruelo@ifeva.edu.ar

Friedrich Recknagel
Department of Soil and Water, University of Adelaide
Glen Osmond
South Australia 5064, Australia
E-mail: frecknag@adelaide.edu.au

François Renaud
Centre d'Etudes sur le Polymorphisme des Micro-organismes
C.E.P.M/U.M.R C.N.R.S.-I.R.D. 9926
911 Avenue Agropolis, B.P. 5045
F-34032 Montpellier cedex 01, France
E-mail: renaud@cepm.mpl.ird.fr

B. Riera
Laboratoire d'Ecologie Générale, M.N.H.N., C.N.R.S., U.R.A. 1183
4, avenue du Petit Château
F-91800 Brunoy, France
E-mail: riera@cimrs1.mnhn.fr

Michele Scardi
Department of Zoology, University of Bari
Via Orabona 4/A
I-70125 Bari, Italy
E-mail: mscardi@mclink.it

William Silvert
Department of Fisheries and Oceans, Bedford Institute of Oceanography
PO Box 1006
Dartmouth, Nova Scotia, Canada B2Y 4A2
E-mail: silvertw@scotia.dfo.ca

Anatoly T. Teriokhin
Deptartment of Biology
Moscow State University
119899 Moscow, Russia
E-mail: at@ateriokhin.home.bio.msu.ru

Frédéric Thomas
Centre d'Etudes sur le Polymorphisme des Micro-organismes
Centre I.R.D. de Montpellier, U.M.R. C.N.R.S.- I.R.D. 9926
911 avenue Agropolis
F-34032 Montpellier cedex 01, France
E-mail: thomas@cepm.mpl.ird.fr

Fernando G. Tomasel
Departamento de Física, Facultad de Ingeniería
Universidad Nacional de Mar del Plata
J.B. Justo 4302
(7600) Mar del Plata, Argentina
E-mail: ftomasel@fi.mdp.edu.ar

Hugh Wilson
Department of Soil and Water, University of Adelaide
Glen Osmond
South Australia 5064, Australia
E-mail: hwilson@tellus.roseworthy.adelaide.edu.au

Part I
Introduction

Neuronal Networks: Algorithms and Architectures for Ecologists and Evolutionary Ecologists

S. Lek · J.L. Giraudel · J.F. Guégan

1.1 Introduction

In ecological research, the processing and interpretation of data play an important role. The ecologist disposes of many methods, ranging from numerical, mathematical, and statistical methods to techniques originating from artificial intelligence (Ackley et al. 1985) like expert systems (Bradshaw et al. 1991; Recknagel et al. 1994), genetic algorithms (d'Angelo et al. 1995; Golikov et al. 1995) and artificial neuronal networks, i.e. ANN (Colasanti 1991; Edwards and Morse 1995).

ANNs were initially developed as models of biological neurons. They are intelligent, thinking machines, working in the same way as the animal brain. They learn from experience in a way that no conventional computer can, and they rapidly solve hard computational problems. With the increasing use of computers, these models could be simulated, and later research was also directed at exploring the possibilities of using and improving these models for performing specific tasks. Research into ANNs has led to the development of various types of neuronal networks, suitable for resolving different types of problems including auto-associative memory, generalization, optimization, data reduction, control and prediction tasks in various scenarios and architectures. Chronologically, we can cite the Perceptron (Rosenblatt 1958), ADALINE, i.e. ADAptive LINear Element (Widrow and Hoff 1960), Hopfield network (Hopfield 1982), Kohonen network (Kohonen 1984), Boltzmann machine (Ackley et al. 1985), and multilayer feed-forward neuronal networks learned by back propagation algorithm (Rumelhart et al. 1986). Descriptions of these methods can be found in various books such as Freeman and Skapura (1992), Gallant (1993), Smith (1994), Bishop (1995), Ripley (1996), etc. The choice of the type of network depends on the type of the problem to be solved. At present, two popular ANNs are multilayer feed-forward neuronal networks, both trained by back propagation algorithm, i.e. back propagation network (BPN) and Kohonen self-organizing mapping, i.e. Kohonen network (SOM). The BPN are the most often used, but other networks are also gaining in popularity nowadays with the emergence of new techniques in various areas of the sciences.

In the last decade research into ANNs has shown explosive growth. They are often applied in physics research like in speech recognition (Rahim et al. 1993; Chu and Bose 1998) or image recognition (Dekruger and Hunt 1994; Cosatto and Graf 1995; Kung and Taur 1995) and in chemical research (Kvasnicka 1990; Wythoff et al. 1990; Smits et al. 1992). In biology most applications of ANNs have been in medicine and molecular biology (Lerner et al. 1994; Albiol et al. 1995; Faraggi and Simon 1995; Lo et al. 1995). Nevertheless, a few applications of this method were reported in ecological and environmental sciences at the beginning of the 1990s. For instance, Colasanti (1991) found similarities between ANNs and ecosystems and recommended the utilization of this tool in ecological modelling. In a review of computer-aided research in biodiversity, Edwards and Morse (1995) underlined that ANNs have an important potential. Relevant examples are found in very different fields in applied ecology, such as modelling the greenhouse effect (Seginer et al. 1994), predicting various parameters in brown trout management (Baran et al. 1996; Lek et al. 1996a,b), modelling spatial dynamics of fish (Giske et al. 1998), predicting phytoplankton production (Scardi 1996; Recknagel et al. 1997), predicting fish diversity (Guégan et al. 1998), predicting the production/ biomass (P/B) ratio of animal populations (Brey et al. 1996), predicting farmer risk preferences (Kastens and Featherstone 1996), etc. Most of these works showed that ANNs performed better than more classical modelling methods.

This book contains working examples of ANN solutions to real ecological problems in various areas. The tasks are as diverse as the neuronal architectures and algorithms themselves, although no attempt has been made to include an example of every shape and form of ANN. We have organized this book so every chapter deals with an interesting example, in which an ANN has been shown to offer a good solution or not. The present textbook is the result of the experiences of leading practitioners in ANN ecological modelling, and we thank them all most warmly.

This book is organized in several chapters. In Chapter 1, two very popular ANN algorithms will be presented: a back propagation neuronal network (BPN) and a Kohonen self-organizing mapping (SOM) network. This chapter offers the reader new to ANN, an introduction with illustrative ecological examples and a comparison to the more classical statistical methods. The following chapters are gathered by ecological theme, from ecosystem studies to evolutionary ecology and even including topics such as remote sensing. The papers show how to obtain solutions for each ecological problem, often making reference to the more classical statistical or mathematical methods like linear or logistic regressions, discriminant analysis, or models based on differential equations, etc.

1.2 Back Propagation Neuronal Network (BPN)

The back propagation neuronal networks, also called multilayer feed-forward neuronal networks or multilayer perceptron, are very popular and are used more than other types of neuronal networks for a wide variety of problems. The BPN is based on the supervised procedure, i.e. the network builds a model based on examples in data with known outputs (Fig. 1.1). It has to extract the input-output relation solely from the examples presented, which together are implicitly assumed to contain the information necessary for this relation. The relationship between problem (input) and solution (output) may be quite general, e.g. the simulation of species richness (where the problem is defined by the characteristic of environment and the solution by the value of species richness) or the abundance of animal expressed by the quality of habitat. A BPN is a powerful system, often capable of modelling complex relationships between variables. It allows one to predict an output object for a given input object.

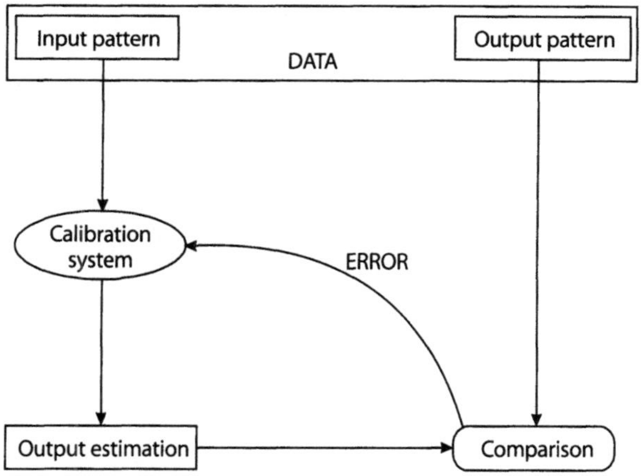

Fig. 1.1. Diagram showing how data are used to establish the model calibration in the supervised modelling procedure. The goal of supervised learning is to find a model, or mapping, that will correctly associate the inputs with the output (or target) data

1.2.1 Structure of BPN

The BPN architecture is a layered feed-forward neuronal network, in which the non-linear elements (neurons) are arranged in successive layers, and the information flows unidirectionally, from input layer to output layer, through the hidden layer(s) (Fig. 1.2). As can be seen in this figure, nodes from one layer are connected (using interconnections or links) to all the nodes in the adjacent layer(s), but no lateral connection between nodes within one layer, or feedback connection(s) are possible. This is in contrast with recurrent networks where feedback connections are also permitted. The number of input and output units depends on the representations of the input and the output objects, respectively. The hidden layer(s) is (are) an important parameter in the network. The BPN with an arbitrary number of hidden units has been shown to be a universal approximate (Cybenko 1989; Hornick et al. 1989) for continuous maps and can therefore be used to implement any function defined in these terms.

1.2.2 BPN Algorithm

BPN is one of the easiest networks to understand. Its learning and update procedure is based on a relatively simple concept: if the network gives the wrong answer, then the weights are corrected so that the error lessens, so future responses of the network are more likely to be correct. The conceptual basis of the back propagation algorithm was first presented in 1974 by Webos, then independently reinvented by Parker (1982), and presented to a wide readership by Rumelhart et al. (1986).

In a training phase, a set of input/target pattern pairs is used for training, which is presented to the network many times. After training is stopped, the performance of the network is tested. The BPN learning algorithm involves a forward-propagating step followed by a backward-propagating step.

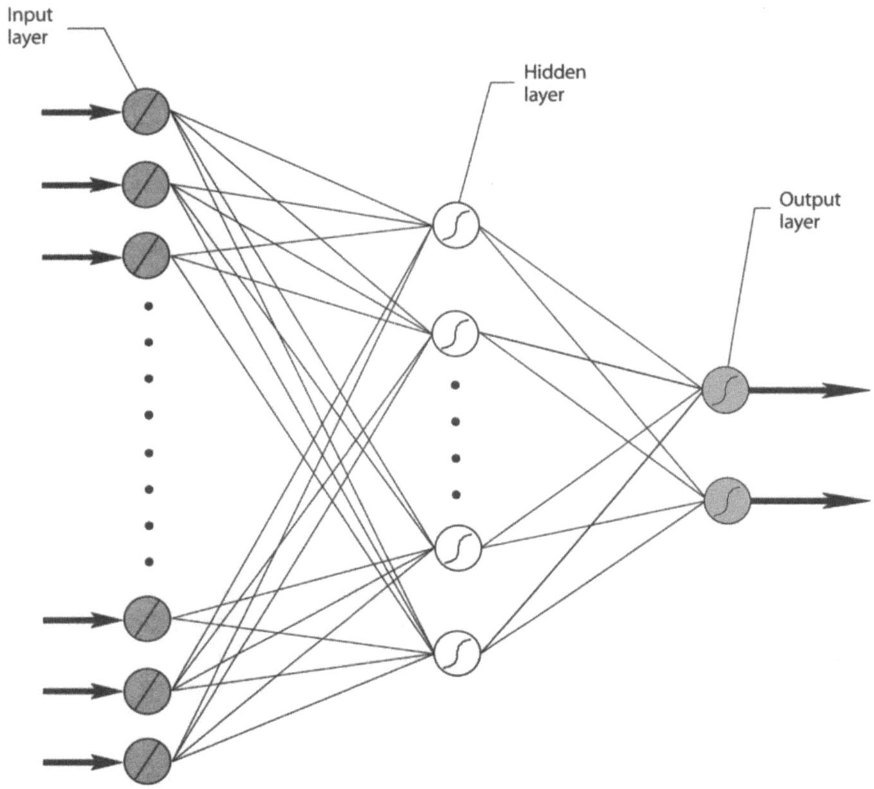

Fig. 1.2. Schematic illustration of a three-layered feed-forward neuronal network, with one input layer, one hidden layer and one output layer

1.2.2.1 *Forward-Propagating Step*

In a natural neuron, the dendrites receive signals from other neurons and send them to the cell body, which elaborates a response. The axon receives response signals from the cell body and carries them away through the synapse to the dendrites of neighbouring neurons. In ANNs, the computational element, i.e. the processing element, is called a neuron (sometimes referred to as node or unit). Figure 1.3 shows the general appearance of a neuron with its connections. Each neuron is numbered; the one in the figure is the *j*th. Like a real neuron, the artificial neuron has many inputs, but only a single output, which can stimulate many other neurons in the network. The input the *j*th neuron receives from the *i*th neurons is indicated as x. Each connection to the *j*th neuron is associated to a quantity called weight or connection strength. The weight on the connection from the *i*th neuron to the *j*th neuron is denoted w_{ji}. An input connection may be excitatory (positive weight) or inhibitory (negative weight). A net input

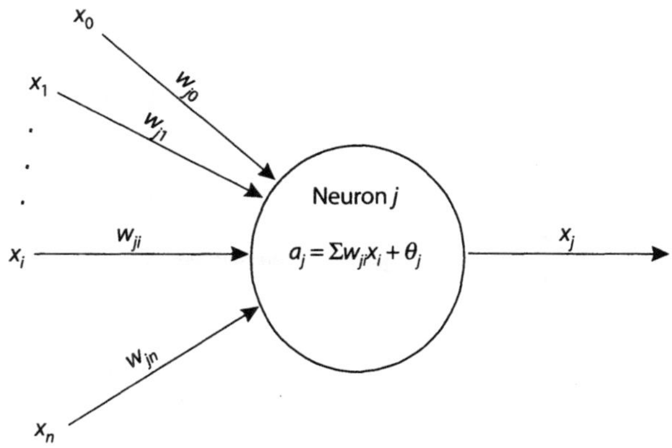

Fig. 1.3. Basic processing element (neuron) in a network. Each input connection value (x_i) is associated with a weight (w_{ji}). The output value can fan out to other units

(called activation) for each neuron is the sum of all its input values multiplied by their corresponding connection weights, expressed by the formula:

$$a_j = \sum_i w_{ji} x_i + \theta_j \tag{1.1}$$

where i is the total number of neurons in the previous layer, θ_j is a bias term, which influences the horizontal offset of the function. The bias θ_j may be treated as the weight from the supplementary input unit, which has a fixed output value of 1. Once the activation of the neuron is calculated, we can determine the output value (i.e. the response) by applying a transfer function:

$$x_j = f(a_j) \tag{1.2}$$

Many transfer functions may be used, e.g. linear function, a threshold function, a sigmoid function, etc. (Fig. 1.4). A sigmoid function is often used. Its formula is:

$$x_j = f(a_j) = \frac{1}{1 + e^{-a_j}} \tag{1.3}$$

The weights play an important role in propagation of the signal in the network. They establish a link between an input pattern and the associated output pattern, i.e. they contain the knowledge of the neuronal network about the problem/solution relationship.

The forward-propagating step begins with the presentation of an input pattern to the input layer, and continues as activation level calculations propagate forward till the output layer through the hidden layer(s). In each successive layer, every neuron sums its inputs and then applies a transfer function to compute its output. The output layer of the network then produces the final response, i.e. the estimated target value.

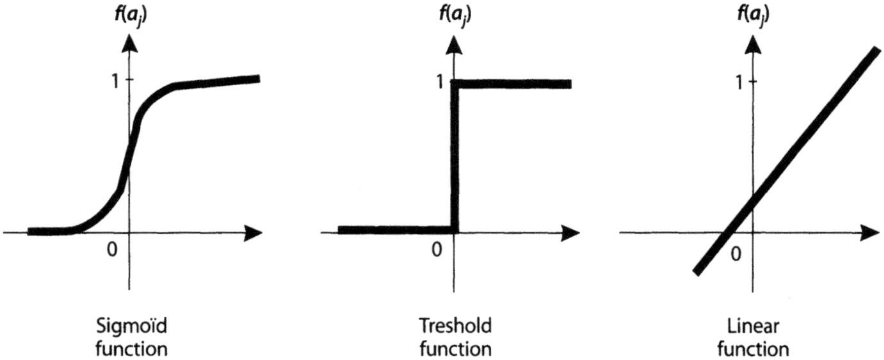

Fig. 1.4. Three types of transfer functions

1.2.2.2
Backward-Propagating Step

The backward-propagating step begins with the comparison of the network output pattern to the target value, when the difference (or error) is calculated. The backward-propagating step then calculates error values and changes the incoming weights, starting with the output layer and moving backward through the successive hidden layers.

The error signal (δ) associated with each processing unit indicates the amount of error associated with that unit. This parameter is used during the weight-correction procedure, while learning is taking place. A large value for δ indicates that a large correction should be made to the incoming weights; its sign reflects the direction in which the weights should be changed.

If the output layer is designated by k, then its error signal is:

$$\delta_k = (t_k - x_k) f'(a_k) \tag{1.4}$$

with t_k: the target value of unit k, x_k: the output value for unit k, f': the derivative of the sigmoid function, a_k: the weighted sum of the input to k, and $(t_k - x_k)$: the amount of error. (The f' part of the term forces a stronger correction when the sum a_k is near the rapid rise in the sigmoid curve.)

For the hidden layer (j), the error signal is computed as:

$$\delta_j = \left[\sum_k \delta_k w_{kj}\right] f'(a_j) \tag{1.5}$$

The adjustment of the connection weights is done using the δ values of the processing unit. Each weight is adjusted by taking into account the δ value of the unit that receives input from that interconnection. The connection weight is adjusted as follows:

$$\Delta w_{kj} = \eta \delta_k x_j \tag{1.6}$$

The adjustment of weight w_{kj}, which goes to unit k from unit j, depends on three factors: δ_k (error value of the target unit), x_j (output value for the source unit) and η. This weight adjustment equation is known as the generalized δ rule (Rumelhart et al. 1986). η is a learning rate, commonly between 0 and 1, chosen by the user, and determines the rate of learning of the network. A very large value of η can lead to instability in the network, and unsatisfactory learning. Too small values of η can lead to excessively slow learning. Sometimes, the learning rate is varied to produce efficient learning of the network during the training procedure. For example, to obtain the best learning performance, the value of η can be high at the beginning of the procedure, and decrease during the learning session.

1.2.3
Training the Network

Before training commences, the connection weights are set to small random values. Values between −0.3 and 0.3 are often used. Next the input patterns are applied to the network, which is allowed to run until an output is produced at each output node. The differences between the output calculations and the targets expected, taken over the entire set of patterns, are used to modify the weights. One complete calculation in the network is called an epoch or iteration of training or learning procedure. This process (epoch) is repeated until a suitable level of error is achieved. The BPN algorithm performs gradient descent on this error surface by modifying each weight in proportion to the gradient of the surface at its location (Fig. 1.5). It is known that gradient descent can sometimes cause networks to get stuck in a depression in the error surface where such a depression exists. These are called "local minima" which correspond to a partial solution for the network in response to the training data. Ideally, we seek a global minimum (lowest error value possible); nevertheless, the local minima are surrounded and the network usually does not leave it by the standard BPN algorithm. Special techniques should be used to get out of a local minimum: changing the learning parameter or the number of hidden units, but notably by the use of momentum term (α) in the algorithm. The momentum term is chosen generally between 0 and 1. Taking this last term into account, the formula for weight modification at epoch $t+1$ is given by:

$$\Delta w_{kj}(t+1) = \eta \delta_k(t+1)\, x_k(t+1) + \alpha \Delta w_{kj}(t) \tag{1.7}$$

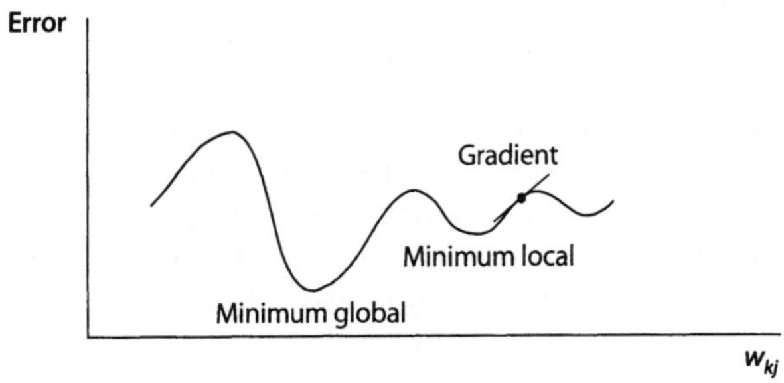

Fig. 1.5. Error surface as function of a weight showing gradient and local and global minima

The learning rate (η) and the momentum term (α) play important roles in the learning process of BPN. If the values of these parameters are wrong, the network can oscillate, or more seriously it can get stuck in a local minimum. In our example (see Section 1.2.7), we obtained good convergence of the networks by initially making $\alpha = 0.7$ and $\eta = 0.01$; then they were modified according to the size of the error by the following algorithm:

```
If present_error > previous_error * 1.04
then η = η * 0.75,
     α = 0,
else η = η * 1.05,
     α = 0.95,
EndIf
```

A training set must have enough examples of data to be representative for the overall problem. However, the training phase can be time-consuming depending on the network structure (number of input and output variables, number of hidden layers and number of nodes in each of them), the number of examples in the training set and the number of iterations.

1.2.4
Testing the Network

Typically the use of a BPN requires both training and test sets. Both sets contain input/output pattern pairs taken from real data. The first is used to train the network, and the second one serves to assess the performance of the network after training is complete. In the testing phase, the input patterns are fed into the network and the desired output patterns compared with those given by the neuronal network. The agreement or the disagreement of these two sets gives an indication of the performance of the neuronal network model.

Another decision that has to be made is the subdivision of the data set into different subsets which are used for training and testing the BPN. The best solution is to have separate data bases, to be able use the first set for training and testing the model, and the second independent set for validation of the model (Mastrorillo et al. 1998). This situation is rarely observed in ecological studies, and partitioning the data set may be applied for testing the validity of the model. We present here two partitioning procedures: *(i)* if enough data sets are available, the data may be divided randomly in two parts to give a training and a test set. The proportion may be 1 : 1, 2 : 1, 3 : 1, etc. for the two sets. However, the training set still has to be large enough to be representative of the problem and the test set has to be large enough to allow correct validation of the network. This procedure of partitioning the data is called k-fold cross-validation, sometimes named the hold-out procedure (Utans and Moody 1991; Efron and Tibshirani 1995; Kohavi and Wolpert 1996; Friedman 1997); *(ii)* if there are not enough examples available to permit the data set to be split into a representative training and test set, other strategies may be used, like cross-validation. In this case, the data set is divided into n parts, usually small, i.e. containing few examples of data. The BPN may now be trained with $n - 1$ parts, and tested with the last part. The same network structure may be repeated to use every test set once in one of the n procedures. The result of these tests together provide the performance of the model. Sometimes, in extreme cases, the test set can have only one example, and this is called the leave-one-out procedure (Efron 1983; Kohavi 1995). The case is often used in ecology when either we have a small database or each observation is a unique piece of information different from the others.

1.2.5
Overtraining or Overfitting the Network

If a network is overfitted (or overtrained), it has a good memory in the detail of data. In such cases, the network will not learn the general features inherently present in the training set, but it will perfectly learn more and more of the specific details of the training set. Thus the network loses its capacity to generalize. Several rules have been developed by many researchers regarding approximate determinations of the required network parameters to avoid overfitting. Three parameters are responsible for this phenomenon: the number of epochs, the number of hidden layers and the number of neurons in each hidden layer. The determination of the appropriate numbers of these elements is the most crucial parameter in BPN modelling. Previously, the optimum size of epochs or hidden layers or hidden nodes were determined by trial and error using training and test sets of data. A typical graph of training and generalization errors versus number of parameters is shown in Fig. 1.6. We can show the errors decrease rapidly as function of parameter complexity. If the error in the training set decreases steadily, the error of the test set can increase after minimal values, i.e. the model is no longer able to generalize. The training procedure must be stopped when the error on the test set is lowest, i.e. the zone corresponding to the best compromise between bias

Fig. 1.6. Criteria of determination of training stop and selection the optimum network architecture

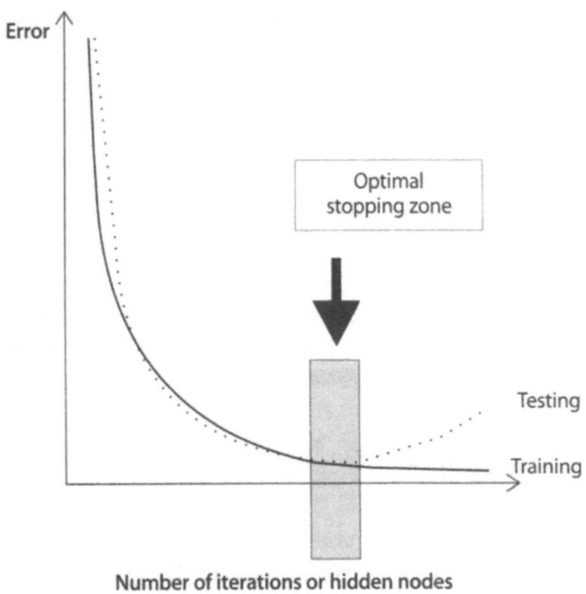

and variance. For an excellent summary of the issues affecting generalization in neuronal networks see Geman et al. (1992).

1.2.6
Use Aspects

In ANN modelling, as mentioned above, many parameters are difficult to grasp and their understanding by the model is often based to some extent on heuristics. We propose to now illustrate this situation through an example taken from fish ecology, prediction of the food consumption by fish relative to their biomass (Q/B) (Palomares and Pauly 1989; Palomares 1991; Lek et al. 1995). Palomares (1991) made a census of and standardized 108 direct evaluations of Q/B involving 65 species and 25 families of fish throughout the world (see data in Appendix). Using the multiple linear regression (MLR) model, we can explain 51% of the variance of Q/B after log transformation of some of the variables:

$$\log(\mathrm{Q/B}) = 0.372 - 0.205\log(W_\infty) + 0.936\log(T) + 0.209\log(A) + 0.529h + 0.425d - 0.019p - 0.165D - 0.477P$$

This model is built with 8 independent variables: the asymptotic weight of the species (W_∞), the morphological ratio A representing the motor activity of the fish, the mean annual temperature (T), three discrete variables defining the diet, herbivorous ($h = 1$), detritivorous ($d = 1$), farmed fish ($p = 1$) and carnivorous ($h = d = p = 0$), and two morphological measurements: D = standard length / height of the body and P = height of the tail / height of the body.

As we can see (Appendix), the variables have different ranges of values. For example, W_∞ has relatively high values and it might dominate or paralyse the model. Scaling of the input variables is then necessary. Different methods may be used, but the best results are often obtained by autoscaling, i.e. centred and reduced variable by this formula:

$$z = \frac{X - \bar{x}}{\sigma_x} \tag{1.8}$$

where z is the scaled value, X is the unscaled value, $\bar{x}$ and σ_x are the mean and standard deviation for the specific variable. The effect of the autoscaling of the variables in the data set is shown in Fig. 1.7.

The output variable(s) also need to be scaled according to the transfer function used. If the sigmoid function is used, the range of variables must be scaled into the interval [0, 1], as given by the formula:

$$z = \frac{x - \min}{\max - \min}(\mathrm{high} - \mathrm{low}) + \mathrm{low} \tag{1.9}$$

where z is the scaled value, x the unscaled value, min and max: the minimum and maximum of the unscaled values for the variable, high and low: lower bound and up-

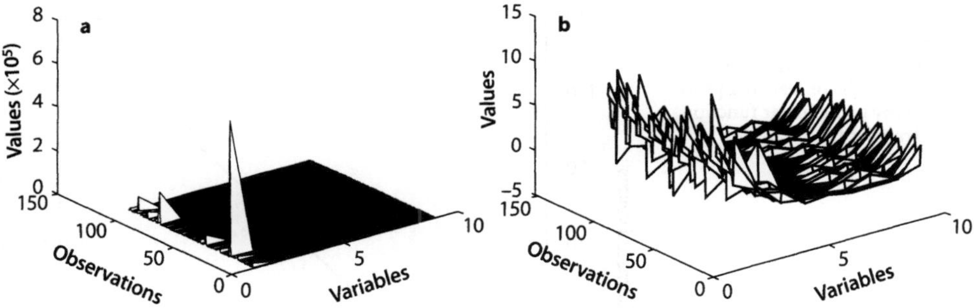

Fig. 1.7. Effect of autoscaling on a data set. The orders of variables (1 to 8) are the same in the table in Appendix; **a** before scaling, only one variable is dominant (W_∞); **b** after scaling, all variables have more or less the same reach

Table 1.1. Values of the synaptic weights linking different independent variables to the dependent variable (Q/B) by using a linear function as the transfer function. The experimentation is repeated 5 times (±*SD*: Standard deviation)

Experiment	W_∞	*T*	*A*	*d*	*p*	*h*	*P*	*D*
1	–0.0131	0.0481	–0.0172	0.0069	–0.0056	0.0540	0.0157	0.0613
2	–0.0096	0.0452	–0.0133	0.0045	–0.0077	0.0538	0.0128	0.0611
3	–0.0138	0.0487	–0.0179	0.0074	–0.0052	0.0540	0.0163	0.0614
4	–0.0040	0.0405	–0.0073	0.0007	–0.0110	0.0536	0.0081	0.0606
5	–0.0039	0.0405	–0.0073	0.0006	–0.0110	0.0536	0.0081	0.0606
Mean	–0.0089	0.0446	–0.0126	0.0040	–0.0081	0.0538	0.0122	0.0610
±*SD*	0.0048	0.0040	0.0051	0.0033	0.0028	0.0002	0.0040	0.0004

per bound of the range over which the data is scaled. Commonly, the bounds used were between 0 and 1, or between 0.1 and 0.9. The latter choice can often speed the rate of convergence.

1.2.7 BPN versus MLR

Our purpose in this section is to compare BPN with MLR, which are largely used in ecological modelling. Other methods are more scarcely applied to ecology and related fields (e.g. General additive models (GAM), Alternating Conditional Expectations (ACE), etc.), but we chose not to include them in this comparative analysis. So, we used the simplest neuronal network and shall now try to see whether any common points can be found with multiple regression.

First, we used a network with no hidden layers and a linear transfer function. To avoid overfitting, the learning procedure was stopped at 1 000 epochs. We repeated this operation 5 times, with different synaptic weights randomly choosing them between

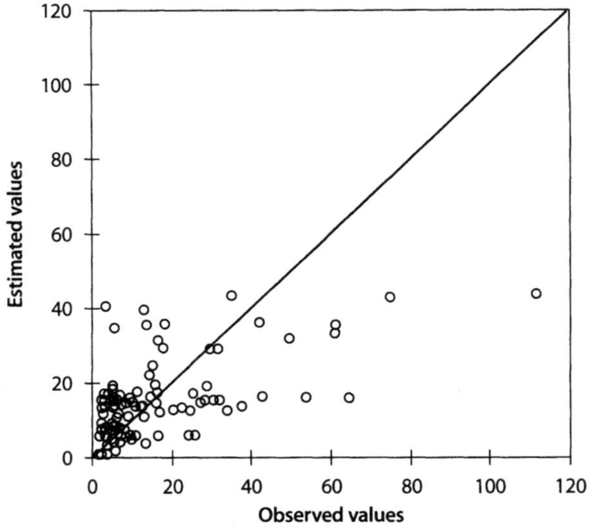

Fig. 1.8. Performance of back propagation neuronal network without hidden layer after 1000 epochs of learning procedure, by using linear function as transfer function

−0.3 and 0.3. In spite of the different starting weights, the final synaptic weights were very close (very low standard deviation), in value and sign (Table 1.1).

Taking into account the mean of the 5 experiments, we can obtain the following model:

$$Q / B = f(-0.0089W_\infty + 0.0446T - 0.0126A + 0.0040d - 0.0081p + 0.0538h + 0.0122P + 0.0610D) \quad (1.10)$$

This equation allows the value of Q/B to be computed with the values of other parameters with f as linear function, in the same way as the MLR model. The correlation coefficient between observed and estimated values is $r = 0.57$ (Fig. 1.8), i.e. the same value obtained by the MLR model performed without transformation of variables.

Table 1.2. Values of the synaptic weights linking different independent variables to the dependent variable (Q/B) by using a sigmoid function as the transfer function. The experimentation is repeated 5 times (*SD*: Standard deviation)

Experiment	W_∞	*T*	*A*	*d*	*p*	*H*	*P*	*D*
1	−6.3935	0.4529	−0.0689	0.1912	−0.2671	0.3030	0.0364	0.4671
2	−6.7578	0.4442	−0.0561	0.1817	−0.2693	0.3091	0.0438	0.4182
3	−6.7647	0.4442	−0.0561	0.1818	−0.2693	0.3090	0.0437	0.4181
4	−6.3932	0.4529	−0.0689	0.1912	−0.2671	0.3030	0.0364	0.4671
5	−6.6569	0.4425	−0.057	0.1806	−0.2718	0.3114	0.0460	0.4068
Mean	−6.5932	0.4473	−0.0614	0.1853	−0.2689	0.3071	0.0413	0.4355
±*SD*	0.1874	0.0051	0.0069	0.0054	0.0019	0.0039	0.0045	0.0293

In a second experiment, by applying the sigmoid function f, we obtained an improvement of the model ($r = 0.67$), given by the following equation (see Table 11.2 for coefficients):

$$Q / B = f(-6.5932W_\infty + 0.4473T - 0.0614A + 0.1853d - 0.2689p + 0.3071h + 0.0413P + 0.4355D) \quad (1.11)$$

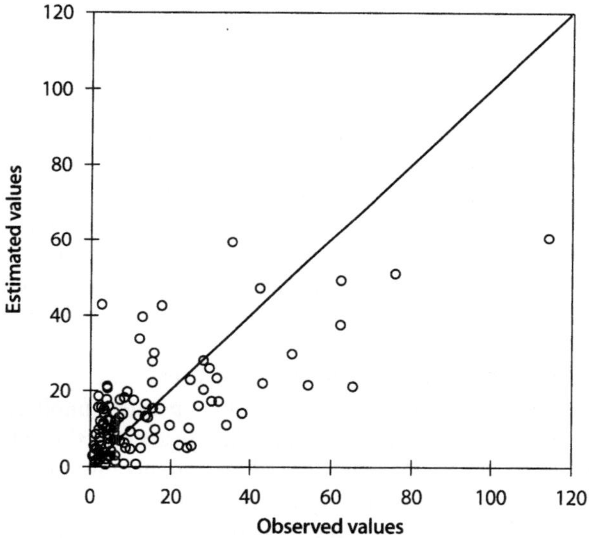

Fig. 1.9. Performance of back propagation neuronal network without hidden layer after 1000 epochs of learning procedure, by using sigmoid function as transfer function

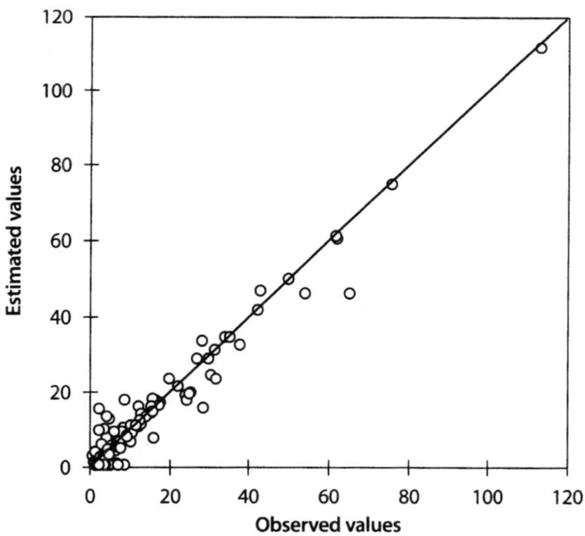

Fig. 1.10. Performance of back propagation neuronal network with 8 neurons in hidden layer, after 1000 epochs of learning procedure, by using sigmoid function as transfer function

The predictive quality is better, but the errors are high compared to the perfect line of prediction (Fig. 1.9). The synaptic weights are high and constant for W_∞ variable.

Using one hidden layer, we improved the quality of prediction (Fig. 1.10): practically all observations are aligned on the perfect line (coordinate 1 : 1). For more details, see Lek et al. (1995).

1.3 Kohonen Self-Organizing Mapping (SOM)

1.3.1 Algorithm

The Kohonen SOM falls into the category of unsupervised competitive learning (Fig. 1.11) methodology, in which the relevant multivariate algorithms seek clusters in the data (Everitt 1993). Conventionally, at least in ecology, the reduction of the multivariate data is usually carried out using principal components analysis or hierarchical clustering analysis (Jongman et al. 1995). Unsupervised learning allows the investigator to group objects together on the basis of their perceived closeness in n-dimensional hyperspace (where n is the number of variables or observations made on each object).

Formally, a Kohonen network consists of two types of units: an input layer and an output layer. The array of input units operates simply as a flow-through layer for the input vectors and has no further significance. In the output layer, SOM often consists of a two-dimensional network of neurons arranged on a square (or other geometrical form) grid laid out in a lattice. A hexagonal lattice is preferred, because it does not favour horizontal or vertical directions. Each neuron is connected to its nearest neighbours on the grid (Fig. 1.12). The neurons store a set of weights (weight vector), an n-dimensional vector if input data are n-dimensional.

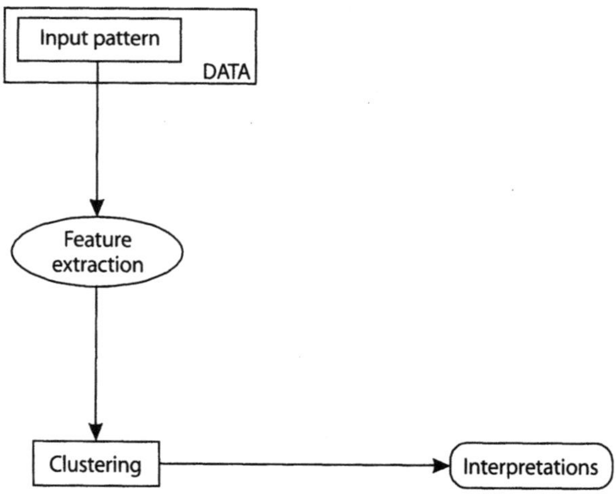

Fig. 1.11. Diagram showing how data are used to establish the model calibration in the unsupervised learning procedure. The goal of the unsupervised learning is to obtain a cluster or mapping in order that people can easily explain the data

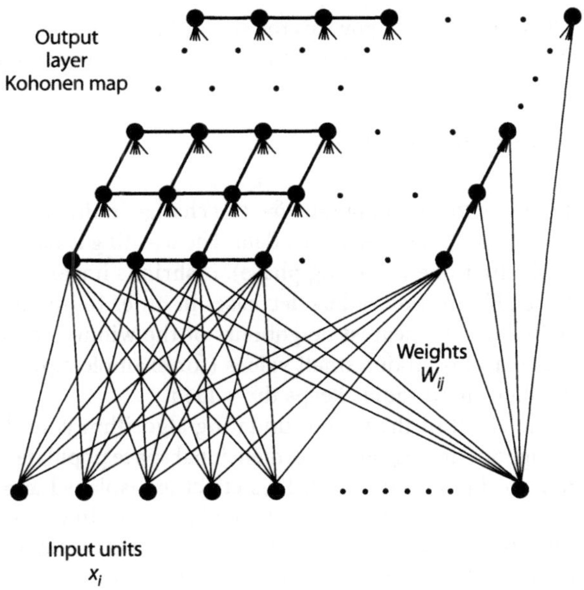

Fig. 1.12. A two dimensional Kohonen self-organizing feature map network

Since the introduction of the Kohonen neuronal network (Kohonen 1982), several training strategies have been proposed (see e.g. Hecht-Nielsen 1990; Freeman and Skapura 1992) which deal with different aspects of use of the Kohonen network. In this section, we will keep to the neuronal network proposed by Kohonen (1984). For an input x, each neuron j (weights: w^j) calculates its activation level, defined as:

$$\left\| w^j - x \right\| = \sqrt{\sum_{i=0}^{n} \left(w_i^j - x_i \right)^2} \tag{1.12}$$

Thus, this is simply the Euclidean distance between the points represented by the weight vector and the input in n-dimensional space. A node whose weight vector closely matches the input vector will have a small activation level, and a node whose weight vector is very different from the input vector will have a large activation level. The node in the network with the smallest activation level is deemed to be the winner for the current input vector.

During the training process the network is presented with each input pattern, and all the nodes calculate their activation levels as described above. The winning node and some of the node around it are then allowed to adjust their weight vectors to match the current input vector more closely. The nodes included in the set, which are allowed to adjust their weights, are said to belong to the neighbourhood of the winner. The size of the winner's neighbourhood is decreased linearly after each presentation of the complete training set (all available data being analysed), until it includes only the winner itself. The amount by which the nodes in the neighbourhood are allowed to adjust their weights is also reduced linearly through the training period.

The factor which governs the size of the weight variations is known as the learning rate. The adjustments to each item in the weight vector are made in accordance with:

$$\delta w_i = -\alpha\left(w_i - x_i\right) \tag{1.13}$$

where α is the learning rate, δw: the change in the weight. This is carried out for $i = 1$ to $i = n$, the dimension of the data. The learning is decomposed into two phases. During the first one (ordering phase), α shrinks linearly from 1 to the final value 0 and the neighbourhood radius decreases in order to initially contain the whole map and finally only the nearest neighbours of the winner. During the second phase, tuning takes place: α attains small values (for example 0.02) during a long period and the neighbourhood radius keeps the value 1.

The effect of the weight updating algorithm is to distribute the neurons evenly throughout the region of n-dimensional space populated by the training set (Kohonen 1984; Hecht-Nielsen 1990). This effect is displayed and shows the distribution of a square network over an evenly populated two-dimensional square input space, and a more complex input space. The neuron with the weight vector closest to a given input pattern will win for that pattern and for any other input patterns that it is closest to. Input patterns which allow the same node to win are then judged to be in the same cell, and when a map of their relationships is drawn, a line encloses them. By training with networks of increasing size, a map with several levels of groups or contours can be drawn. These contours, however, may sometimes cross, which appears to be due to a failure of the SOM to converge to an even distribution of the neuron over the input space (Erwin et al. 1992). Construction of these maps allows close examination of the relationships between the items in the training set.

1.3.2 Missing Data

In ecology, a difficult problem arises from missing data. For some data items, certain components of the data vectors are unknown. SOM accepts the fact that data may be missing and two approaches can be used. Firstly, data items with too many missing components are discarded during the learning process and are then mapped on the organizing map (Kaski and Kohonen 1996). Secondly, if only few components of a data vector are missing, a convenient solution consists in only using available components in Eqs. 1.12 and 1.13 (Samad and Harp 1992).

1.3.3 Outliers

Due to measurement errors or to values really different from the rest, outliers are often present in ecological data. Classical methods of clustering are sensitive to the presence of outliers in the data. And generally, it is useful to detect outliers before computing clusters. With SOM, the process is quite different. Each outlier takes its place in one unit of the map, and only the weights of that neuron and its nearest neighbours are affected. There is no effect on the other neurons. Moreover, the observation of scattered data in an area of the map should suggest the presence of an outlier.

1.3.4 Use of Different Metrics

Measuring ecological likeness often leads to the use of various similarity or distance coefficients. For example, the presence/absence measure or the computation of genetic data calls for the choice of distances other than the classical Euclidean distance. Describing community structure (for plants or animals) often leads the ecologist to use various distance measures between sample units (Ludwig and Reynolds 1988). These problems can be solved with SOM but it is necessary to pay attention. Not only should the activation level of each neuron (Eq. 1.12) be computed with the appropriate distance, but also a compatible metrics have to be used in the adjustment of the weights (Eq. 1.13) (Kohonen 1995).

1.3.5 Aspects of Use

The iris data published by Fisher (1936) have been widely used for examples in discriminant and cluster analysis. Sepal length, sepal width, petal length and petal width were measured in millimetres on 50 flower specimens from each of three species, *Iris setosa*, *I. virsicolor*, and *I. virginica*. The graphical representation of the two first PCA axes (Fig. 1.13) shows complete separation of the first class (*I. setosa*), and the two other classes are very close to each other. Using discriminant analysis, we obtain 98% of good

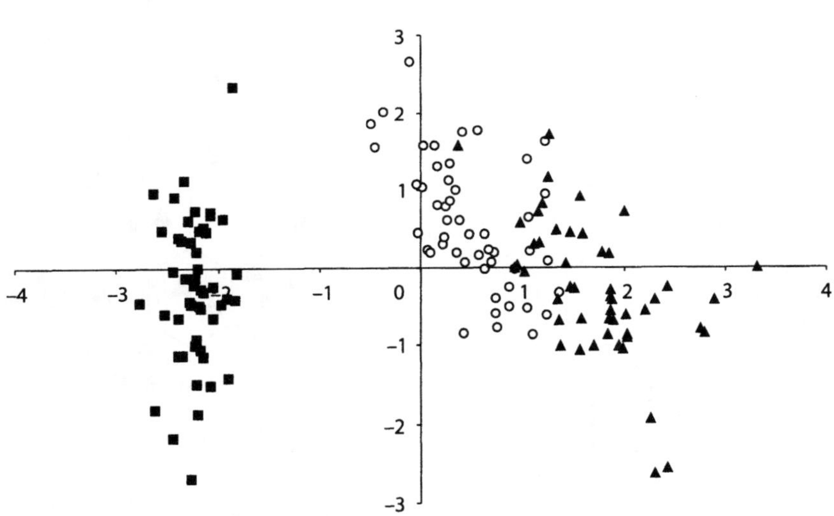

Fig. 1.13. Iris data projected linearly onto the two-dimensional subspace obtained with PCA. Variety: *I. setosa*, ■; *I. versicolor*, ○; *I. virginica*, ▲

classification (i.e. 3 misclassified observations, 2 *I. virsicolor* and 1 *I. virginia* specimens).

This data set was used to illustrate the SOM method. A Kohonen map with 8×10 neurons in a hexagonal lattice is trained with the 150 observations presented randomly and iteratively. The components of the data items were scaled to mean 0 and variance 1. The different varieties of iris, *I. setosa, I. versicola, I. virginica,* are not used during the learning (unsupervised learning). 2 000 iterations are made during the ordering phase then 40 000 iterations during the tuning phase. At the end, individuals are set in the appropriate unit of the SOM (Fig. 1.14). SOM allows a first clustering: individuals are present in only 61 hexagons. It is worth noting that only two hexagons contain irises of different species and *I. setosa* are present in the left lower part of the map, *I. versicolor* in the middle part and *I. virginica* in the right lower part.

Then, it is necessary to represent the relative distances between their neighbouring units. A scale of shades of grey is used (Iivarinen et al. 1994). Light shades indicate small distances and dark shades, large distances between two neighbouring hexagons. In that way, a "cluster landscape" is formed and clusters can be seen better (Fig. 1.15). Three plains appear (light areas) separated by hills or mountains (dark areas): *I. setosa* individuals residing mainly in the left lower plain, *I. versicolor* in the right upper plain and some *I. virginica* in a little plain area in the middle of the right side. The mountainous area from the upper left to the lower right part of the map mainly groups *I. versicolor* and *I. virginica.*

Another interesting representation with SOM is the distribution of each variable on the map (Fig. 1.16). SOM is coloured for each component of weight vectors, namely

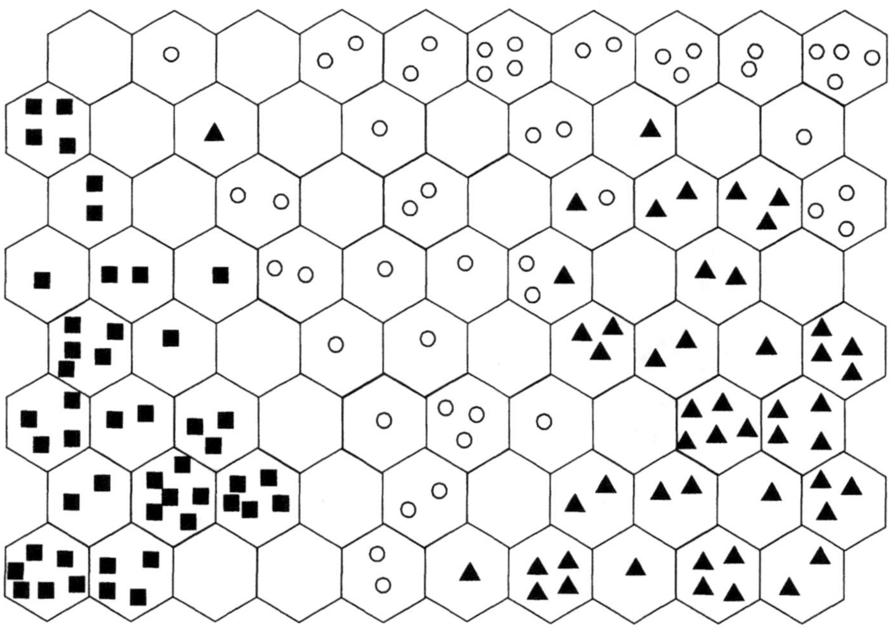

Fig. 1.14. Iris data mapped on the organizing map. Variety: *I. setosa*, ■; *I. versicolor*, ○; *I. virginica*, ▲

Fig. 1.15. Self-organizing map for iris data with shades of grey indicating the degree of clustering

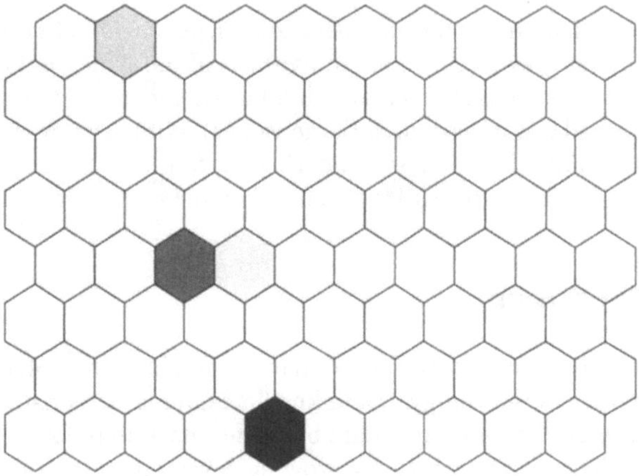

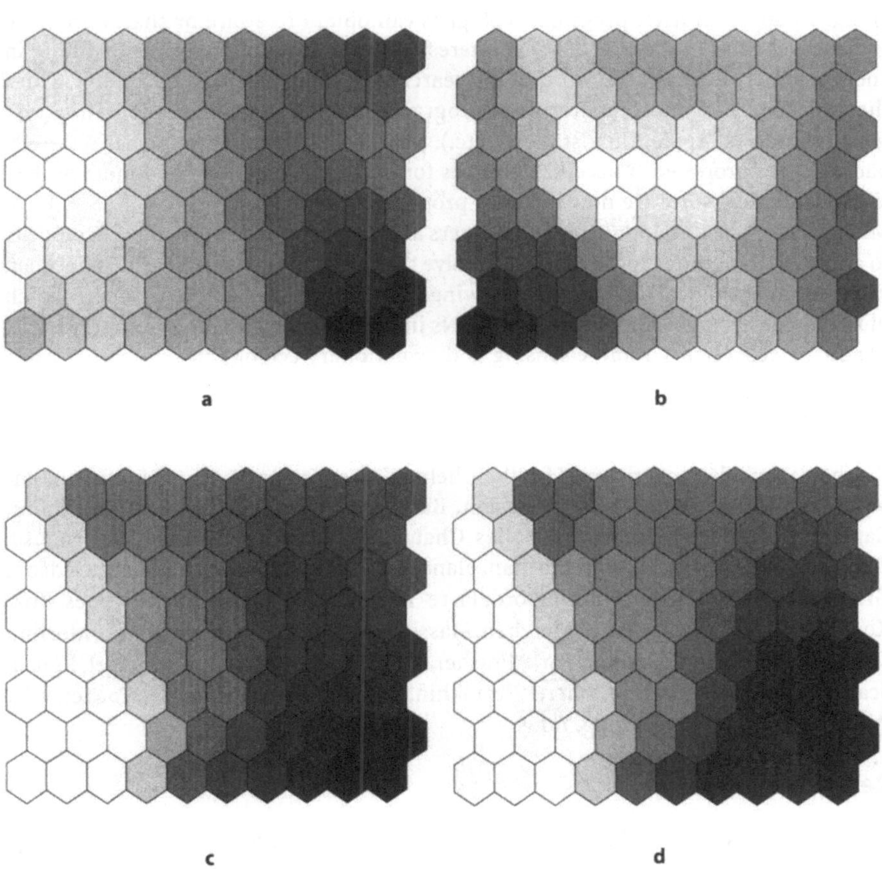

Fig. 1.16. Representation of the components of the weight vectors for each neuron; **a** sepal length; **b** sepal width; **c** petal length; **d** petal width. In each map, white colour indicates the smallest value and black colour the largest ones

sepal length, sepal width, petal length, petal width. In each display, two hexagons with similar grey level contain individuals of the same kind for this variable. For example, with sepal length, *I. setosa* individuals are in bright area (small values) and *I. virginica* individuals in dark area (large values). With sepal width, bright areas correspond to *I. versicola* individuals and dark areas to *I. setosa*. Then, well known characteristics of the different species are visualised on the map which can be useful for interpretation.

1.4
Conclusion

In this introductory part, we have highlighted the potential use of artificial neuronal networks in ecology. Using known examples (iris data or Q/B ratios, in annexe) we showed that one can obtain better results with the ANN. During the last two decades of the current century, the growing development of computer-aided analysis, which is easily accessible to all researchers, has facilitated the use of ANNs in ecological modelling. To apply an ANN program, ecologists can obtain freeware or shareware from various web sites in the world. Users interested can find these programs by filling in "neuronal network" as a keyword in the search procedure of the web explorer. Thus, they can obtain many computer ANN programs functioning with all operating systems (Windows, Apple, Unix stations, etc.). Moreover, increasingly specialized ANN packages are proposed at acceptable prices for personal computers and most professional statistical software now includes proposes ANN procedures (e.g. SAS, S-Plus, Matlab, etc.). The development of computers and ANN software must allow ecologists to apply ANN methods more easily to resolve the complexity of relationships between variables in ecological data. In the following chapters, readers will find papers which illustrate the ecological application of ANNs in several fields, ranging from terrestrial to aquatic ecosystems, remote sensing and evolutionary ecology.

Acknowledgements

We thank the following referees for their help in accepting to review the different papers: Ambroise, Arino, Auger, Aussem, Balls, Beacham, Belaud, Bourret, Canu, Capblancq, Careaux, Cereghino, Charles, Chau, Chon, Cisneros, Comrie, Culverhouse, Dimopoulos, Dreyfus, Dubois, Ehrman, Flanagan, Foody, Fuhs, Gan, Gascuel, Gedeon, Grandjean, Halls, Hanser, Huntingford, Jarre-Teichman, Kapetsky, Kaski, Kimes, Kok, Komatsu, Kropp, Lewis, Manel, Marzban, Masson, Megrey, Melssen, Morimoto, Morlini, Murase, Oberdorff, O'Connor, Park, Prasher, Puig, Sanchez Manes, Recknagel, Roush, Scardi, Schultz, Shamseldin, Starret, Teriokhin, Thibault, Thiria, Tomasel, Touzet, Vila, Wang, Werner, William, Wong, Yang.

References

Ackley DH, Hinton GE, Sejnowski TJ (1985) A learning algorithm for Boltzmann machines. Cognitive science 9:147–169

Albiol J, Campmajo C, Casas C, Poch M (1995) Biomass estimation in plant cell cultures: A neuronal network approach. Biotechnology Progress 11:88–92

Baran P, Lek S, Delacoste M, Belaud A (1996) Stochastic models that predict trout population densities or biomass on macrohabitat scale. Hydrobiologia 337:1–9

Bishop MC (1995) Neuronal networks for pattern recognition. Clarendon Press, Oxford
Bradshaw JA Carden KJ, Riordan D (1991) Ecological applications using a Novel expert system shell. Computer Applications in the Biosciences 7:79–83
Brey T, Jarre-Teichmann A, Borlich O (1996) Artificial neuronal network versus multiple linear regression: Predicting P/B ratios from empirical data. Mar Ecol Prog Ser 140:251–256
Chu WC, Bose NK (1998) speech signal prediction using feedforward neuronal-network. Electronics Letters 34:999–1001
Colasanti RL (1991) Discussions of the possible use of neuronal network algorithms in ecological modelling. Binary 3:13–15
Cosatto E, Graf HP (1995) A neuronal-network accelerator for image-analysis. IEEE Micro 15:32–38
Cybenko G (1989) Approximation by superpositions of a sigmoidal function. Mathematics of Control, Signals and Systems 2:304–314
d'Angelo DJ, Howard LM, Meyer JL, Gregory SV, Ashkenas LR (1995) Ecological use for genetic algorithms: Predicting fish distributions in complex physical habitats. Can J Fish Aquat Sc 52:1893–1908
Dekruger D, Hunt BR (1994) image-processing and neuronal networks for recognition of cartographic area features. Pattern Recognition 27:461–483
Edwards M, Morse DR (1995) The potential for computer-aided identification in biodiversity research. Trends in Ecology and Evolution 10:153–158
Efron B (1983) Estimating the error rate of a prediction rule: Improvement on cross-validation. J Am Statistical Association 78(382):316–330
Efron B, Tibshirani RJ (1995) Cross-validation and the bootstrap: Estimating the error rate of the prediction rule. Rep Tech Univ Toronto
Erwin E, Obermayer K, Schulten K (1992) Self-organizing maps: Ordering, convergence properties and energy functions. Biol Cyb 67(1):47–55
Everitt BS (1993) Cluster analysis. Edward Arnold, London
Faraggi D, Simon R (1995) A neuronal network model for survival data. Statistics in Medicine 14:73–82
Fisher RA (1936) The use of multiple measurements in axonomic problems. Annals of Eugenics 7:179–188
Freeman JA, Skapura DM (1992) Neuronal networks: Algorithms, applications and programming techniques. Addison-Wesley Publishing Company, Reading
Friedman JH (1997) On bias, variance, 0/1-loss and the curse-of-dimensionality. Data Mining and Knowledge Discovery 1:55–77
Gallant SI. (1993) Neuronal network learning and expert systems. The MIT Press, Cambridge
Geman S, Bienenstock E, Doursat R (1992) Neuronal networks and the bias/variance dilemma. Neuronal Computation 4:1–58
Giske J, Huse G, Fiksen O (1998) Modelling spatial dynamics of fish. Rev. Fish. Biol. Fish. 8:57–91
Golikov SY, Sakuramoto K, Kitahara T, Harada Y (1995) Length-frequency analysis using the genetic algorithms. Fisheries Science 61:37–42
Guégan JF, Lek S, Oberdorff T (1998) Energy availability and habitat heterogeneity predict global riverine fish diversity. Nature 391:382–384
Hecht-Nielsen R (1990) Neurocomputing. Addison-Wesley, Massachusetts
Hopfield JJ (1982) Neuronal networks and physical systems with emergent collective computational abilities. Proc Natl Acad Sci USA 79:2554–2558
Hornick K, Stinchcombe M, White H (1989) Multilayer feedforward networks are universal approximators. Neuronal Networks 2:359–366
Iivarinen J, Kohonen T, Kankas J, Kaski S (1994) Visualizing the clusters on the self-organizing map. In: Carlsson C, Järvi T, Reponen T (eds) Proceedings of the Conference on Artificial Intelligence Research in Finland. Finnish Artificial Intelligence Society, Helsinki, pp 122–126
Jongman RHG, ter Braak CJF, van Tongeren OFR (1995) Data analysis in community and landscape ecology. Cambridge University Press, Cambridge
Kaski S, Kohonen T (1996) Exploratory data analysis by the self-organzing map: Structures of welfare and poverty in the world. In: Refenes A-PN, Abu-Mostafa Y, Moody J, Weigend A (eds) Neuronal Networks in Financial Engineering. Proceedings of the Third International Conference on Neuronal Networks in the Capital Markets, World Scientific, Singapore, pp 498–507
Kastens TL, Featherstone AM (1996) Feedforward backpropagation neuronal networks in prediction of farmer risk preference. American Journal of Agricultural Economics 78:400–415
Kohavi R (1995) A study of cross-validation and bootstrap for estimation and model selection. Proc. of the 14th Int. Joint Conf. on Artificial Intelligence, Morgan Kaufmann Publishers Inc., Bari
Kohavi R, Wolpert DH (1996) Bias plus variance decomposition for zero-one loss functions. In: Saitta L (ed) Machine learning. Proceedings of the Thirteenth International Conference, Morgan Kaufmann Publishers Inc., Bari, pp 275–283
Kohonen T (1982) Self-organized formation of topologically correct feature maps. Biol Cybern 43:59–69
Kohonen T (1984) Self-organization and associative memory. Springer-Verlag, Berlin

Kohonen T (1995) Self-organizing maps. Springer-Verlag Berlin
Kung SY, Taur JS (1995) Decision-based neuronal networks with signal image classification applications. IEEE Transactions on Neuronal Networks 6:170–181
Kvasnicka V (1990) An application of neuronal networks in chemistry. Chemical Papers 44(6):775–792
Lek S, Belaud A, Lauga J, Dimopoulos I, Moreau J (1995) An improved estimation of the animal food of fish populations using neuronal networks. Marine and Freshwater Research, 46(8):1229–1236
Lek S, Belaud A, Baran P, Dimopoulos I, Delacoste M (1996a) Role of some environmental variables in trout abundance models using neuronal networks. Aquatic Living Resources 9:23–29
Lek S, Delacoste M, Baran P, Dimopoulos I, Lauga J, Aulagnier S (1996b) Application of neuronal networks to modelling nonlinear relationships in ecology. Ecol Model 90:39–52
Lerner B, Levinstein M, Rosenberg B, Guterman H, Dinstein I, Romem Y (1994) Feature selection and chromosomes classification using a multilayer perceptron neuronal network. IEEE International conference on Neuronal Networks, Orlando (Florida), pp 3540–3545
Lo JY, Baker JA, Kornguth PJ, Floyd CE (1995) Application of artificial neuronal networks to interpretation of mammograms on the basis of the radiologists impression and optimized image features. Radiology 197:242
Ludwig J A, Reynolds JF (1988) Statistical ecology: A primer on methods and computing. John Wiley & Sons, New York
Mastrorillo S, Dauba F, Oberdorff T, Guégan JF, Lek S (1998) Predicting local fish species richness in the Garonne river basin. C R Acad Sci Paris, Life Sciences 321:423–428
Palomares ML (1991) La consommation de nourriture chez les poissons: étude comparative, mise au point d'un modèle prédictif et application à l'étude des relations trophiques. Ph D thesis, Institut National Polytechnique de Toulouse, Toulouse
Palomares ML, Pauly D (1989) A multiple regression model for predicting the food consumption of marine fish populations. Australian Journal of Marine and Freshwater Research 40:259–73
Parker DB (1982) Learning logic. Invention report S81–64, File 1, Office of Technology Licensing, Stanford University
Rahim MG, Goodyear CC, Kleijn WB, Schroeter J, Sondhi MM (1993) On the use of neuronal networks in articulatory speech synthesis. Journal of the Acoustical Society of America 93:1109–1121
Recknagel F, Petzoldt T, Jaeke O, Krusche F (1994) Hybrid expert-system delaqua – a toolkit for water-quality control of lakes and reservoirs. Ecol Model 71:17–36
Recknagel F, French M, Harkonen P, Yabunaka KI (1997) Artificial neuronal network approach for modelling and prediction of algal blooms. Ecol Model 96:11–28
Ripley BD (1996) Pattern recognition and neuronal networks. Cambridge University Press, Cambridge
Rosenblatt F (1958) The perceptron: A probabilistic model for information storage and organization in the brain. Psychological Review 65:386–408
Rumelhart DE, Hinton GE, Williams RJ (1986) Learning representations by back-propagating errors. Nature 323:533–536
Samad T, Harp SA (1992) Self-organizing with partial data. Network: Computation in Neuronal Systems 3:205–212
Scardi M (1996) Artificial neuronal networks as empirical models for estimating phytoplankton production. Mar Ecol Prog Ser 139:289–299
Seginer I, Boulard T, Bailey BJ (1994) Neuronal network models of the greenhouse climate. Journal of Agricultural Engineering Research 59:203–216
Smith M (1994) Neuronal networks for statistical modelling. Van Nostrand Reinhold, New York
Smits JRM, Breedveld LW, Derksen MWJ, Katerman G, Balfoort HW, Snoek J, Hofstraat JW (1992) Pattern classification with artificial neuronal networks: Classification of algae, based upon flow cytometer data. Analytica Chimica Acta 258:11–25
Utans J, Moody JE (1991) Selecting neuronal network architectures via the prediction risk: application to corporate bond rating prediction. Proceedings of the First International Conference on Artificial Intelligence Applications on Wall Street, IEEE Computer Society Press, Los Alamitos
Webos P (1974) Beyond regression: New tools for prediction and analysis in the behavioral sciences. Thesis, Harvard University
Widrow B, Hoff ME (1960) Adaptive switching circuits. IRE Wescon conference record 4:96–104
Wythoff BJ, Levine SP, Tomellini SA (1990) Spectral peak verification and recognition using a multilayered neuronal network. Analytical Chemistry 62(24):2702–2709

Appendix

Table A1.1. 108 records of Q/B ratio data (for variable symbols, see text)

W	*T*	*A*	*d*	*p*	*H*	*P*	*D*	Q/B
63	20	2.32	0.334	0.271	0	0	0	8.23
362	25	1.41	0.35	0.267	0	0	0	8.1
1216	18	1.89	0.326	0.263	1	0	0	31.4
28	15	1.31	0.161	0.486	0	0	0	9.13
7500	13	2.4	0.261	0.34	0	0	0	6.49
10541	13	2.4	0.261	0.34	0	0	0	4.08
100	10	2.06	0.25	0.324	0	1	0	7.41
230	15.5	2.21	0.217	0.417	0	0	0	8.63
605	9	2.21	0.217	0.417	0	0	0	5.22
1824	16	2.21	0.217	0.417	0	0	0	23.9
1206	16	2.21	0.217	0.417	0	0	0	25.1
13312	15	1.5	0.182	0.429	0	0	0	2.03
10925	15	1.5	0.182	0.429	0	0	0	6.43
6049	7	1.5	0.182	0.429	0	0	0	2.95
8810	7	1.5	0.182	0.429	0	0	0	1.61
3288	7	1.5	0.182	0.429	0	0	0	0.581
2	25	1.03	0.216	0.409	0	0	0	33.9
1	25	0.78	0.183	0.425	0	0	0	24.3
2	25	0.75	0.211	0.524	0	0	0	12.1
8	25	1.02	0.238	0.326	0	0	0	37.6
10	25	0.83	0.188	0.444	0	0	0	16.2
11	25	0.81	0.235	0.354	0	0	0	12
13	25	1	0.238	0.28	0	0	0	26.9
4	25	0.93	0.187	0.4	0	0	0	19.9
250	25.8	2.52	0.232	0.437	0	0	0	5.93
32	15.4	2.13	0.443	0.3	0	0	0	2.5
32000	24.5	2.37	0.489	0.231	0	1	0	1.61
615	31.7	1.94	0.378	0.364	0	1	0	4.24
808	30.8	1.51	0.221	0.524	0	1	0	5.58
883	30.8	1.98	0.289	0.436	0	1	0	6.26
60	14	1.41	0.273	0.39	0	0	0	15.9
316	9	1.49	0.303	0.357	1	0	0	13.6
766	14	1.49	0.303	0.357	0	1	0	2.68
1769	12.4	1.49	0.303	0.357	1	0	0	14.5
710	12.4	1.34	0.256	0.456	0	0	0	12.7
6650	25.5	1.44	0.247	0.352	0	0	0	22.1
21887	25	1.26	0.161	0.353	0	0	0	1.32

Table A1.1. *Continued*

W	*T*	*A*	*d*	*p*	*H*	*P*	*D*	Q/B
6074	22.5	1.26	0.161	0.353	0	0	0	1.73
1688	22.5	1.01	0.178	0.595	0	0	0	1.33
12356	10	0.77	0.181	0.25	0	0	0	2.59
15714	12	0.77	0.181	0.25	0	0	0	2.26
2	10	1.69	0.3	0.166	0	0	0	3.85
3776	13	1.78	0.361	0.331	0	0	0	1.1
16595	25	1.09	0.365	0.357	0	0	0	4.26
1006	26	2.4	0.214	0.306	0	0	0	4.31
3067	15	1.76	0.25	0.4	0	0	0	10.2
3067	15	1.76	0.25	0.4	0	1	0	1.52
47000	19	0.69	0.275	0.395	0	0	0	4.02
12338	28	1.54	0.314	0.308	0	0	0	4.02
1880	28	1.07	0.312	0.302	0	0	0	2.77
17940	28	0.92	0.516	0.341	0	0	0	2.34
702	27	1.49	0.307	0.333	0	0	0	15.4
2296	27	1.69	0.307	0.388	0	0	0	6.22
3290	27	1.69	0.304	0.388	0	0	0	10.1
380	10	1.64	0.285	0.368	0	0	0	2.79
173	9	1.94	0.285	0.274	0	0	0	5.99
897	15.4	1.94	0.285	0.274	0	0	0	6.25
154	9	1.94	0.285	0.274	0	0	0	4.57
336	16.5	1.94	0.285	0.274	0	0	0	5.06
3036	27	2.21	0.292	0.123	0	0	0	10.6
147000	25	1.21	0.206	0.185	0	0	0	8.47
13000	20	1.28	0.412	0.309	0	0	0	5.26
3229	27	1.68	0.387	0.338	0	0	0	6.64
7400	24	1.19	0.369	0.292	0	0	0	2.34
9617	24	1.97	0.415	0.232	0	0	0	4.67
4000	16	1.97	0.415	0.232	0	0	0	1.61
3555	15	1.97	0.415	0.232	0	0	0	4.74
1093	27	0.91	0.267	0.267	0	0	0	13.9
16	30	0.76	0.361	0.331	1	0	0	17.5
153	20.5	1.32	0.381	0.367	1	0	0	29.6
479	25	1.2	0.454	0.367	0	1	0	7.5
1144	22.5	1.17	0.413	0.4	0	0	1	2.7
242	27	1.17	0.413	0.4	0	1	0	30.3
348	27	1.17	0.413	0.4	0	1	0	31.6

Table A1.1. *Continued*

W	*T*	*A*	*d*	*p*	*H*	*P*	*D*	Q/B
1193	27	1.17	0.413	0.4	0	1	0	2.24
996	27	1.17	0.413	0.4	0	0	1	75.5
271	26	1.28	0.457	0.337	0	1	0	28
2495	26	1.28	0.457	0.337	0	1	0	3.56
95	26.5	1.28	0.457	0.337	0	1	0	65.1
5700	28.5	1.28	0.457	0.337	0	1	0	3.3
361	28.5	1.28	0.457	0.337	0	1	0	24.8
2036	24.5	1.28	0.457	0.337	1	0	0	49.9
1517	30	1.28	0.457	0.337	1	0	0	12.8
2056	28.5	1.28	0.457	0.337	0	1	0	2.21
545	28.5	1.28	0.457	0.337	0	1	0	15.6
5700	20.5	1.28	0.457	0.337	1	0	0	17.2
431	26	1.28	0.457	0.337	1	0	0	61.8
95	27	1.28	0.457	0.337	0	1	0	42.8
101	32	1.28	0.457	0.337	0	1	0	15.3
145	32	1.28	0.457	0.337	0	1	0	28.4
145	27	1.28	0.457	0.337	0	1	0	54
2495	27	1.28	0.457	0.337	0	1	0	4.81
2495	32	1.28	0.457	0.337	0	1	0	4.15
1396	25.8	1.56	0.489	0.41	1	0	0	15.7
215	27	1.21	0.451	0.366	0	0	1	35.1
360	26	1.48	0.479	0.35	0	1	0	9.28
1265	26	1.48	0.479	0.35	0	1	0	4.46
429	27.5	1.65	0.458	0.344	0	0	1	113
5877	15	2.55	0.232	0.417	0	0	1	4.74
787	23	2.55	0.232	0.417	0	0	1	12.3
215	27	2.81	0.392	0.157	1	0	0	61.7
234	27	1.92	0.353	0.171	1	0	0	42
81920	24	5.8	0.26	0.088	0	0	0	11.6
622000	15	6.7	0.296	0.13	0	0	0	3.94
756	12	0.66	0.448	0.233	0	0	0	3.69
149	12.1	0.66	0.448	0.233	0	0	0	7.04
910	12	1.01	0.511	0.183	0	0	0	3.43
3430	12	1.01	0.511	0.183	0	0	0	2.12

Part II

Artificial Neuronal Networks in Landscape Ecology and Remote Sensing

Chapter 2

Predicting Ecologically Important Vegetation Variables from Remotely Sensed Optical/Radar Data Using Neuronal Networks

D.S. Kimes · R.F. Nelson · S.T. Fifer

2.1 Introduction

The large data sets engendered during the EOS era will enhance the temporal, spatial, and spectral coverage of the earth (Asrar and Greenstone 1995; Wharton and Myers 1997). The satellite digital data sets and ancillary data products will require the development of efficient algorithms that can incorporate and functionally utilize disparate data types. Numerous vegetation variables, e.g. leaf area, height, canopy roughness, land cover, stomatal resistance, latent and sensible heat flux, radiative properties, and many others, are required for global and regional studies of ecosystem processes, biosphere/atmosphere interactions, and carbon dynamics (Asrar and Dozier 1994; Hall et al. 1995). The success of efforts to extract vegetation variables such as these from remotely sensed data and available ancillary data will determine the degree and scope of vegetation-related science performed using EOS data.

In remote sensing missions of vegetation canopies, the problem is to accurately extract vegetation variables from remotely sensed data. These variables are, for the most part, continuous (e.g. biomass, leaf area index, fraction of vegetation cover, vegetation height, vegetation age, spectral albedo, absorbed photosynthetic active radiation, photosynthetic efficiency, etc.), and estimates may be made using remotely sensed data (e.g. nadir and directional optical wavelengths, multifrequency radar backscatter) and any other readily available ancillary data (e.g. topography, sun angle, ground data, etc.). Inferring continuous variables implies that a functional relationship must be made between the predicted variable(s), the remotely sensed data, and ancillary data. This is opposed to classification studies where the goal is to produce discrete categories of vegetation types as reviewed by Atkinson and Tatnall (1997).

A significant portion of the remote sensing community is active in developing techniques to accurately extract continuous vegetation properties. It is clear from the literature that significant problems exist with the "traditional techniques" being used. These are very topical and truly difficult problems that are being encountered in the remote sensing community. Neuronal networks can provide solutions to many of these problems. The intent of this paper is to raise the awareness of the ecological community to the advantages of using neuronal network techniques in this area of research. The advantages and power of neuronal networks for extracting continuous vegetation variables using optical/radar data and ancillary data are discussed and compared to traditional techniques. Several specific examples of research in this area are discussed.

2.2 Traditional Extraction Techniques

Several common approaches exist to extract continuous vegetation variables. These are classified as linear, nonlinear and physically-based models and are discussed in detail by Kimes et al. (1998). A brief summary of these models are discussed along with the respective advantages and problems of each. Classification studies where the goal is to produce discrete categories of vegetation types are reviewed by Atkinson and Tatnall (1997).

Ideally, there exists a functional relationship between the independent variables (e.g. remotely sensed signals) and the estimated variables (e.g. biomass, leaf area index, etc.). However even if a physical relationship exists, often it is not known. Consequently, one is often forced to make simplifying assumptions that allow one to develop a predictive equation in the form of a general linear model. Many physical biological processes are nonlinear. Therefore a general linear model often performs poorly in predicting vegetation variables because the relations between scattered radiation above vegetation canopies and vegetation variables may be nonlinear (e.g. Jakubauskas 1996).

More complicated linear models involve transformations on the independent and/or dependent variables. Transformations allow one to reduce a more complex model to a linear form. Many transformations used in the literature are some kind of vegetation index. For example, in the optical region Myneni et al. (1995) reported that there are more than 12 vegetation indices and they have been correlated with vegetation amount, fraction of absorbed photosynthetically active radiation, unstressed vegetation conductance and photosynthetic capacity, and seasonal atmospheric carbon dioxide variations. Indices have been developed to enhance the spectral contribution from green vegetation while minimizing those from soil background, sun angle, sensor view angle, senesced vegetation, and the atmosphere as reviewed by Kimes et al. (1998). Although these models can be related to a crude physical principle, it does not give the scientist any deep insight into the physical system. It is often difficult to decide what transformations to make, if any. Generally, the choice is made based on the results of previous studies in similar study areas and on trial and error.

The linear models above are linear in the coefficients and can be solved using least squares. In some studies, one has knowledge that a nonlinear form (nonlinear in the coefficients) is the more realistic and potentially more accurate model. Specifically, these models are intrinsically nonlinear in that it is impossible to convert them into a linear form. Numerous numerical iterative techniques exist to solve these nonlinear models. When using a nonlinear analysis, it is implied that the researcher knows the proper nonlinear form to implement. Generally, only simple nonlinear forms can be envisioned by the researcher. A few examples are described by Kimes et al. (1998).

Ideally, in the scientific community, one would like to develop accurate, physically-based models for the physical system being studied. This model serves as a hypothesis for our current understanding of the physical system and as a basis for extracting desired vegetation variables from other readily known/measured variables. These physically-based models are forced to address the entire radiative transfer problem which includes a large number of variables. In remote sensing applications many of these variables are not of interest. Physically-based models range in complexity from simple nonlinear models to complex radiative transfer models in realistic three-dimensional vegetation canopies. The optical and radar models are reviewed by Kimes et al. (1998).

To actually use physically-based models for extracting vegetation variables, the models must be inverted. In most cases these models are complex nonlinear systems which must be solved using numerical methods. The traditional approach employs numerical optimization techniques that, once initialised, search for the optimum parameter set that minimizes the error. There are difficulties in using these techniques. A stable and optimum inversion is not guaranteed and the technique can be computationally intensive when using complex radiative transfer models of vegetation. In addition, physical vegetation models have many parameters other than the variable(s) that is (are) being estimated. If one is to invert the model using numerical optimization techniques, many parameters have to be known and/or estimated using other methods. Initial conditions must be set to deduce the desired variable(s). Many studies have all or some of these problems.

Efforts to invert optical vegetation models are summarized by Privette et al. (1994, 1996), Pinty et al. (1990), Ross and Marshak (1989), and Goel (1987). An example of an effort to invert a radar model is described by Polatin et al. (1994). Several approaches have been adopted to overcome the difficulties in inverting a model with many parameters. Goel (1987) noted that for a model to be successfully inverted, the number of measurements must be greater than or equal to the number of canopy parameters that need to be determined. Furthermore, to invert nonlinear relationships, there should be many more measurements than unknown parameters to facilitate a numerical solution.

Several approaches have been taken to achieve these parameters/measurements criteria. Often physical constraints are imposed. For example, the number of physical parameters to describe the geometric and scattering properties of vegetation components are limited (e.g. Kuusk 1994; Pinty et al. 1990; Jacquemoud 1993; Moghaddam 1994; Saatchi and Moghaddam 1994). In addition, the radiative transfer functions are often simplified (e.g. Pinty and Verstraete 1991; Prevot and Schmugge 1994; Govaerts and Verstraete 1994). Most inversion strategies must consider some combination of multiangle data and multispectral data. In addition, some optical studies have used multiple sun angles. For inversions to be accurate, near optimal numbers of reflectance samples, spectral regions, signal anisotropy, and model sensitivity are required (Myneni et al. 1995). The amount of directional/multispectral data required to obtain accurate inversions is often appreciable and can not always be collected. Often, assumptions must be made about unknown variables (e.g. leaf-angle distribution, leaf size, plant spacing, reflectance and transmittance distributions of vegetation components, etc.). Generally, only one-dimensional models have been successfully inverted against measured data (Govaerts and Verstraete 1994).

There is a trade off between model accuracy and the number of model parameters considered. The most accurate and robust models generally have the most canopy parameters and are least appropriate for direct inversion. The models with few parameters are easier to invert but are also the most inaccurate models.

Because direct inversion of models is computationally intensive, it generally is not applied on a pixel by pixel basis over large regions. Generally, inversions in the literature are carried out on only a few selected canopies rather than the entire range of vegetation variations that would exist in real applications.

2.3 Neuronal Networks

In many areas of research an appropriate and accurate, physically-based model for the purpose of extracting continuous vegetation variables does not exist. Consequently, one is forced to adopt a linear or simple nonlinear form that must be explicitly designed by a researcher. Coefficients are then fitted by traditional regression or simple numerical routines. If the researcher has not correctly envisioned all of the complex functional relationships between the input and output data, this approach will not work well. What is needed is a structure which adaptively develops its own basis functions and their corresponding coefficients from data.

Neuronal networks have the ability to learn patterns or relationships given training data, and to generalize or extract results from the data (Anderson and Rosenfeld 1988; Wasserman 1989; Zornetzer et al. 1990). The approximation capability of neuronal networks is based on connectionism (Fu 1994). After training, the network is a machine that approximately maps inputs to the desired output(s). Kimes et al. (1998) discuss the structure of neuronal networks, important approximation properties, training algorithms and properties, and pruning strategies of network structures.

2.4 Uses of Neuronal Networks and Remote Sensing Data

Neuronal networks have several attributes which facilitate extraction of vegetation variables from remotely sensed data. The advantages of neuronal networks as compared to traditional techniques are discussed and example studies are presented. These studies, together with those cited by Kimes et al. (1998) and Atkinson and Tatnall (1997), provide a comprehensive review of this area of research. Neuronal network approaches have been shown to be equal or superior to conventional techniques, especially when strong nonlinear components exist in the system being studied.

2.4.1 Neuronal Networks as Initial Models

In many areas of research, physically-based radiative scattering models do not exist or are not accurate. In cases where models are lacking, neuronal networks can be used as the initial model. If accuracy is the only concern, then a neuronal network may be entirely adequate and desirable. A neuronal network can model the system on the basis of a set of encoded input/output examples of the system. The network maps inputs to the desired output by learning the mathematical function underlying the system. With this method, input and output variables can be related without any knowledge or assumptions about the underlying mathematical representation. Several examples follow.

Kimes et al. (1996) used an MLP (multilayer perceptron) network as an initial model to extract forest age in a Pacific Northwest forest using Thematic Mapper and topographic data. Understanding the changes of forest fragmentation through time are important for assessing alterations in ecosystem processes (forest productivity, species diversity, nutrient cycling, carbon flux, hydrology, spread of pests, etc.) and wildlife habitat and populations. The development of physically-based radiative scattering models that incorporate forest growth and topography, and that can be used to extract forest variables, is in its infancy. Consequently, accurate models that are invertible in this context are lacking.

The study area was the H.J. Andrews Experimental Forest on the Blue River Ranger District of the Willamette National Forest in western Oregon. Timber has been harvested from this forest for the past 45 years and the cutting and replanting history has been recorded. The study area was extracted from a georeferenced TM scene acquired on July 7, 1991. A coincident digital terrain model (DTM) derived from digital topographic elevation data was also acquired. Using this DTM and an image processing software package, slope and aspect images were generated over the study area. Sites were chosen to cover the entire range of forest stand age and slope and aspect. The oldest recorded clear-cut stands were logged in 1950. A number of sites were chosen as primary forest which had no recorded history of cutting. Various feed-forward neuronal networks trained with back propagation were tested to predict forest age from TM data and topographic data.

The results demonstrated that neuronal networks can be used as an initial model for inferring forest age. The best network was a $6 \rightarrow 5 \rightarrow 1$ structure with inputs of TM bands 3, 4, 5, elevation, slope and aspect. The *RMSE* (root mean squared errors) values of the predicted forest age were on the order of 5 years (Fig. 2.1). TM bands 1, 2, 6, and 7 did not significantly add information to the network for learning forest age. Furthermore, the results suggest that topographic information (elevation, slope and aspect) can be effectively utilized by a neuronal network approach. The results of the network approach were significantly better than corresponding linear systems. As discussed in Section 2.2, many transformations (ratios, indices etc.) of optical and radar wavelengths are used to infer vegetation variables of interest. The goal of these studies is to find the transformation that produces the maximum degree of accuracy when applied to a particular class of remote sensing problems. Researchers often use simple transformations (ratios, indices etc.) because they are fast and easy to apply and they are well known in the literature. However, they provide little if any physical insights that can be used effectively to increase the accuracy of inference. Consequently, we propose that an adaptive learning technique such as neuronal networks would be superior to these simple transformations in many applications. For example, Sader et al. (1989) found that the *NDVI* (TM4 – TM3)/(TM4 + TM3) was not significantly correlated with forest regeneration age classes. Neuronal networks have the potential to learn more accurate relationships because they are not confined to the fixed relationships represented by the above simple transformations. The neuronal network approach is free to learn complex relationships that could not be envisioned by researchers.

Kimes et al. (1999) employed neuronal nets in conjunction with SPOT multispectral data to discriminate secondary from primary forest in Rondonia, Brazil. Their work demonstrated that neuronal networks consistently outperformed linear discriminant functions with respect to forest classification. Neuronal nets differentiated primary forest, non-forest, and secondary forest at an overall accuracy of 91.0% using 2 SPOT bands, and at 95.2% using one SPOT band and two texture channels. The corresponding linear discriminant overall accuracies were 88.9% (3 SPOT bands) and 92.6% (3 SPOT bands and 5 texture channels). Neuronal nets also estimated secondary forest age more accurately than linear, parametric functions. Using 2 spectral and 2 tex-

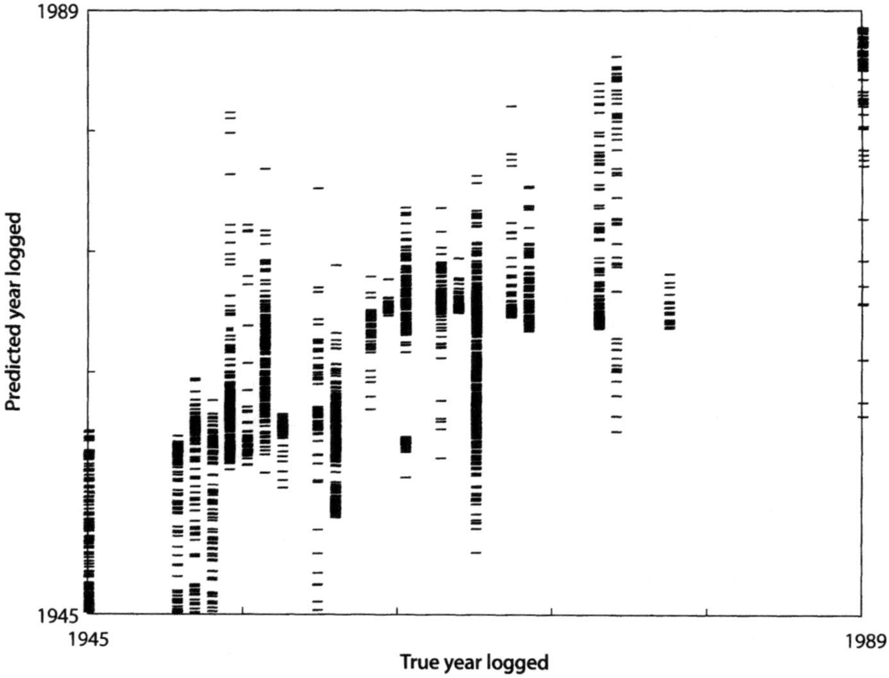

Fig. 2.1. Predicted year logged versus the true year logged for the testing data. The network structure was the 6 → 5 → 1 (#inputs, #hidden nodes, #outputs). The inputs were TM bands 3, 4, 5, elevation, slope and aspect. The number of pixels/points shown are 3555 for the testing data. The *RMSE* and R^2 values for the testing data were 5.6 and 0.69, respectively

ture channels, a 4 → 17 → 1 neuronal network predicted secondary forest age over a 9 year range with an *RMSE* of 2.0 years and $R^2_{(\text{actual vs. predicted})}$ of 0.38. The corresponding multiple linear regression employing 3 SPOT bands and 4 texture channels had an *RMSE* of 2.1 years and an R^2 of 0.31. Though neither technique could be said to accurately estimate secondary forest age, neuronal networks consistently outperform parametric, linear discriminant and regression procedures using fewer spectral and textural bands.

Additional work by Nelson et al. (2000) near the same area in Rondonia using Thematic Mapper multispectral data verifies these findings. For instance, a linear discriminant function using four spectral/textural measures differentiated primary forest, non-forest, and secondary forest with an overall accuracy of 96.6%. Using 3 channels (3 → 3 → 3), the comparable neuronal net yielded an overall accuracy of 97.2%. The TM spectral and textural data were also used to estimate secondary forest age. The multiple linear predictive regression utilized 6 spectral-texture channels and yielded an *RMSE* of 1.62 years and an $R^2_{(\text{actual vs. predicted})}$ of 0.35. The comparable neuronal net, using 4 channels, had an *RMSE* of 1.59 years and an R^2 value of 0.37. The differences between the linear and neuronal net results are small, but in this and the Kimes et al. (1999) studies, they are consistent. In general, using fewer bands, utilizing automatic variable selection procedures, and utilizing automatic weighting procedures, neuronal net results are, in general, comparable to or better than linear discriminant and linear regression results.

Ultimately, the scientific community needs to develop physically-based radiative scattering models for the above areas of research. These models need to be accurate and invertible for the desired variables. In research areas where these activities are immature, the neuronal network approach can provide an accurate initial model for predicting vegetation variables.

2.4.2 Neuronal Networks as Baseline Control

A network can be used as a baseline control while developing adequate physically-based models (Fu 1994). Where adequate field and ground truth data sets exist, a neuronal network can be trained and tested on these data sets. These networks attempt to find the optimum functional relationships that exist between the input variables and the output variables of interest. The networks can be trained in the forward direction on the field data (e.g. vegetation canopy variables are the inputs and radiative scattering is the output).

Improvements to the physically-based model are indicated if it cannot surpass the accuracy of a neuronal network. Specifically, model accuracies less than neuronal network accuracies indicate that the physical processes embedded in the model must be improved (i.e. made more realistic). In this manner, neuronal networks provide a performance standard for evaluating current and future physically-based models (Fu 1994).

2.4.3 Neuronal Networks for Inverting Physically-Based Models

In Section 2.2, the difficulties in inverting physically-based models were discussed. In summary, the following difficulties can occur when using numerical optimization techniques to invert models. These techniques can be time-consuming and generally can not be applied on a pixel by pixel basis for large regions. From a practical standpoint, often it is difficult to collect the measurements (multiple view angles and wavelengths) needed for an accurate inversion. Often models must be simplified before a stable and accurate inversion can be developed. The models are simplified by decreasing the number of parameters and/or simplifying the radiative transfer function. Simplified models tend to be more inaccurate than the full models. Neuronal network approaches provide potential solutions to all or some of these problems.

Significant simplification of physical models are made so that direct inversion using numerical techniques can be successfully applied. The disadvantage of this approach is that underlying relationships may be deleted that may be useful in extracting the variables of interest. In contrast, the neuronal network approach can be applied to the most sophisticated model without reducing the number of parameters or simplifying the physical processes. The models that have many parameters and include all physical processes tend to be the most accurate and robust models. Thus, the neuronal network approach applied to these models may, potentially, find more optimal relationships between the desired input and output variables. This approach provides a sound bench mark in terms of accuracy for extracting various variables. If direct inversion techniques of simplified models do not equal the accuracy obtained using the neuronal network approach on the full model, then important underlying relationships are being ignored in the direct inversion approach.

A neuronal network approach can be used to accurately and efficiently invert physically-based models. The approach is as follows. The physically-based model describes the mathematical relationships between all the vegetation and radiative parameters. The model is used to simulate a wide array of vegetation canopies (the range of all canopies that would be encountered in the application space) in the forward direction – that is, the vegetation canopy variables are the input and the radiative scattering above the canopy is calculated. Using the model, a wide range of canopies and their associated directional reflectances or backscatter values can be calculated. Using these model-based data, training and testing data sets can be constructed and presented to various neuronal networks. These data sets consist of pairs of data containing the desired network inputs (e.g. optical and/or radar) and the true outputs (e.g. vegetation variables of interest). Embedded in these data are mathematical relationships between the inputs and the outputs. In theory the neuronal network approximates the optimal underlying mathematical relationships to map the inputs to the output. If only weak mathematical relationships exist between the input and output values then the network results will be poor. Thus, using this approach a neuronal network can be used to invert a model. This inversion scheme can be applied using input data that can be practically obtained in remote sensing missions. Many studies have successfully used this approach.

Kimes et al. (1997) used a neuronal network approach to invert a combined forest growth model and a radar backscatter model. The forest growth model captures the natural variations of forest stands (e.g. growth, regeneration, death, multiple species, and competition for light). This model was used to produce vegetation structure data typical of northern temperate forests in Maine. Forest parameters such as woody biomass, tree density, tree height and tree age are important for describing the function and productivity of forest ecosystems. These data supplied inputs to the radar backscatter model which simulated the polarimetric radar backscatter (C, L, P, X bands) above the mixed conifer/hardwood forests. Using these simulated data, various neuronal networks were trained with inputs of different backscatter bands and output variables of total biomass, total number of trees, mean tree height, and mean tree age. Techniques utilized included transformation of input variables, variable selection with a genetic algorithm, and a cascade network and are described in detail by Kimes et al. (1997).

The accuracies (*RMSE* and R^2 values) for inferring various variables from radar backscatter were total biomass (1.6 kg m^{-2}, 0.94), number of trees (48 ha^{-1}, 0.94), tree height (0.47 m, 0.88), and tree age (24.0 yrs., 0.83). For example, Fig. 2.2 shows the true above ground biomass (kg m^{-2}) versus the predicted above ground biomass for a neuronal network with a structure of $5 \rightarrow 15 \rightarrow 1$ using frequencies C_{HH}, C_{VV}, L_{HH}, and P_{HH} (2 transformations were used for P_{HH}). The *RMSE* and R^2 values were 1.6 kg m m^{-2} and 0.94, respectively. These accuracies are considered good considering the complexity of the combined model and the fact that only simulated radar backscatter data were used without any other knowledge of the forest. Several networks were shown to be

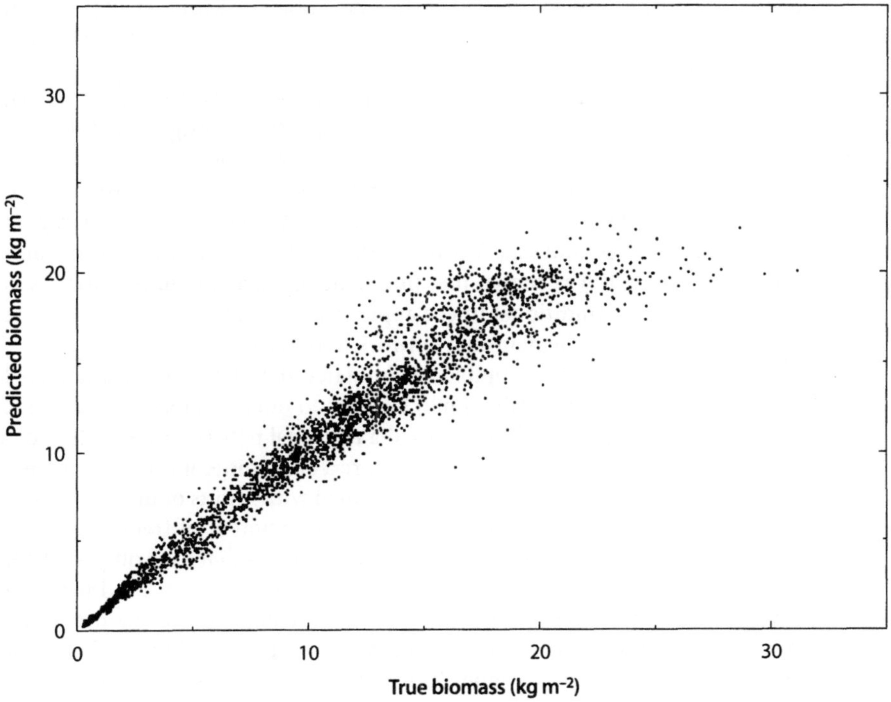

Fig. 2.2. True above ground biomass (kg m^{-2}) versus the predicted above ground biomass. The network structure was 5 → 15 → 1 (#inputs, #hidden nodes, #outputs). The inputs were frequencies C_{HH}, C_{VV}, L_{HH}, and P_{HH}. The *RMSE* and R^2 values were 1.6 kg m^{-2} and 0.94, respectively

relatively insensitive to the addition of random noise to radar backscatter. The accuracy of these networks were superior to traditional index techniques developed by Ranson et al. (1997).

2.4.4 Neuronal Networks for Defining Relevant Variables

Networks can be used as a variable selection tool to determine a set of variables that are relevant to the desired variable(s) to be inferred. If the mapping of a network is not accurate, then perhaps some input variable(s) is (are) missing. Also an input variable is relevant to the problem only if it significantly increases the network's performance. Alternately, if there is an unacceptably large number of input variables, several types of algorithms can be used to find desirable subsets of input variables. Genetic algorithms (Koza 1993) may be used to select an optimal subset of input variables. In this type of application, the genetic algorithm searches for a subset of input variables that behave synergistically to produce the highest network accuracy. The algorithm starts with a small subset of inputs of limited size and adds input variables according to network performance. This evolutionary process is detailed by Koza (1993) and a specific application relevant to this paper is described by Kimes et al. (1997). In these ways, networks can be used to identify input variables which best predict the variable(s) of interest. Several examples follow.

The network analysis in the forest age study discussed previously (Kimes et al. 1996), defined a set of variables that were relevant to modelling efforts designed to infer forest age. Specifically it was discovered that the best inputs were TM bands 3, 4, 5, elevation, slope and aspect. TM bands 1, 2, 6, and 7 did not significantly add information to the network for learning forest age. Furthermore, the study suggests that topographic information (elevation, slope and aspect) can be effectively utilized by a neuronal network approach. However, it was shown that this same topographic information was not useful when used in a traditional linear approach and had *RMSE* values on the order of 35–40% higher than the neuronal network approach.

Neuronal networks can also be applied to simulated data from physically-based models to define a set of variables which may be used to infer variable(s) of interest. As discussed previously, Kimes et al. (1997) used a neuronal network approach to develop accurate algorithms for inverting a complex forest backscatter model. Using these simulated data, various neuronal networks were trained with inputs of different backscatter bands and output variables of total biomass, total number of trees, mean tree height, and mean tree age. The authors found that the networks that used only AIRSAR bands (C, L, P) had a high degree of accuracy. The inclusion of the X band with the AIRSAR bands did not seem to significantly increase the accuracy of the networks. The networks that used only the C and L bands still had a relatively high degree of accuracy for all forest variables (R^2 values from 0.75 to 0.91). The significance of this fact is that there is no current instrument or planned instrument that is collecting or will collect P band data. However, there are planned instruments collecting C and L band data. Modest accuracies (R^2 values from 0.65 to 0.84) were obtained with networks that used only the L band, and poor accuracies (R^2 values from 0.36 to 0.46) were obtained with networks that used only the C band.

2.4.5 Neuronal Networks as Adaptable Systems

Neuronal networks are readily adaptable. They can easily incorporate new ancillary information that would be difficult or impossible to use with conventional techniques. For example, Kimes et al. (1996) included topographic data (slope, aspect, elevation) as ancillary information to infer forest age from TM data. They found that by introducing this ancillary information the network accuracy improved significantly (from 8.0 to 5.1 yrs. *RMSE*). It was not known how to incorporate topographic information effectively using traditional techniques. However, neuronal networks are ideally suited to learning new relationships between ancillary information, other input variables and the desired output variable. New input variables can be introduced to the network and tested with ease. This is especially useful when a researcher expects a new variable to add information to the problem of interest but does not have any knowledge of the functional form to use in introducing the new variable using traditional techniques.

Traditional numerical methods have difficulty in inverting multiple disconnected models. For example, Ranson et al. (1997) used a forest growth model to simulate growth and development of northern mixed coniferous hardwood forests. Output from this model were used as input to the canopy backscatter model that calculated radar backscatter coefficients for simulated forest stands. Classic numerical inversion of such a disconnected system (multiple models) is difficult when the functional connection between the different models is not explicitly defined. In these situations, researchers often adopt simple linear or nonlinear forms. For example, Ranson et al. (1997) choose to develop a simple index relationship to infer forest biomass from the radar backscatter coefficients. Neuronal networks are ideally suited for such problems. Networks find the best nonlinear function based on the network's complexity without the constraint of linearity or prespecified nonlinearity used in traditional techniques. No explicit functional relationships between the disconnected models is required. Kimes et al. (1997) found that networks were significantly more accurate than traditional techniques for inverting the disconnected models of Ranson et al. (1997). Specifically, using the above neuronal network approach, the above ground biomass (kg m^{-2}) was extracted with *RMSE* and R^2 accuracies of 1.6 kg m^{-2} and 0.94, respectively as opposed to 2.6 kg m^{-2} and 0.85, respectively, using the traditional index method of Ranson et al. (1997).

2.5 Disadvantages of Using Neuronal Networks with Remote Sensing Data

The main disadvantage of using neuronal network procedures with remotely sensed data is that they have not been generalized to handle any arbitrary subset of directional view data. Modern satellite-borne sensors (e.g. MODIS, MISR, POLDER, SeaWiFS) allow for rich spectral and angular sampling of the radiation field reflected by vegetation canopies. Airborne and satellite sensors that sample directional and spectral data create new demands on the retrieval techniques for vegetation variables of interest. It is highly desirable to develop methods for inferring vegetation variables that are able to handle any arbitrary subset of directional viewing. This capability is desirable because cloud cover, missing data, and the earth's curvature cause the subset of usable view angles from pixel to pixel to vary for global and regional data sets. There are many neuronal network structures and techniques that may accomplish this generalization; however, a single neuronal network structure designed to handle any arbitrary subset of directional view data has not yet been demonstrated. This represents a well defined and important area of future research.

In the literature, neuronal networks have been successfully developed using a fixed directional combination. For example, Abuelgasim et al. (1997) used a MLP network and directional data to invert the model of Li and Strahler (1992). This geometric optical model has been successful in predicting the bidirectional reflectance of canopies as a function of the geometry and spatial distribution of trees/shrubs, the component signatures of the canopy elements, and the illumination geometry. The model was used to generate input and output data from a conifer forest, savannah, and shrubland. The inputs to the network were 18 directional reflectances consisting of 9 views from the principal plane of the sun and 9 views from across the principal plane of the sun. In addition, 3 component signatures and solar illumination angle were inputs to the network. The output was the density of the canopy, crown shape of the trees/shrubs, and canopy height. The R squared values between the predicted canopy variables and the true canopy variables were 0.85, 0.75 and 0.75, respectively, for density, crown shape and height.

Using an MLP network, Chuah (1993) inverted a Monte Carlo radar backscatter model to infer leaf moisture content and radius and thickness of a circular leaf. The vegetation layer was modelled as a half-space of randomly oriented and randomly distributed disks. The Monte Carlo model handles multiple scattering between the scatterers (Chuah and Tan 1989). This model was used to simulate the radar backscatter coefficients given the vegetation variables. They trained two networks to invert the model for leaf moisture content. One network used a single frequency (1 GHz) with three view angles and three polarizations (HH, VV, HV) for a total of 9 inputs. This network had an accuracy of leaf moisture content to within ±2%. With the introduction of random noise to the inputs of ±0.5 dB and ±1 dB, the error in estimated leaf moisture content was less than ±5% and ±8%, respectively. The second network used multifrequencies (1–8 GHz with three view angles and three polarizations) and gave similar accuracies. Finally, a network was trained to estimate three variables simultaneously: leaf moisture content, and the radius and thickness of the circular disks using the multifrequency data. The accuracy of this network for inferring plant moisture content and radius of the disks was approximately 8%, while the accuracy of inferring the thickness of the disks was within 10%.

Many other optical and radar studies using fixed directional data and neuronal networks are reviewed by Kimes et al. (1998). However, a single neuronal network structure designed to handle any arbitrary subset of data needs to be designed and tested for many practical remote sensing missions.

2.6 Conclusions and Implications

Neuronal networks have attributes which facilitate extraction of vegetation variables. Neuronal networks have significant advantages as compared to traditional techniques when applied to both measurement and modelling studies.

In many areas of research physically-based radiative scattering models do not exist or are not accurate. In cases where accurate models are lacking, neuronal networks can be used as the initial model. A neuronal network can model the system on the basis of a set of encoded input/output examples of the systems.

Neuronal networks can provide a baseline against which the performance of physically-based models can be compared. The networks can be trained on field data. Improvements to the physically-based model are indicated if it cannot surpass the accuracy of a neuronal network.

A neuronal network approach can be used to accurately and efficiently invert physically-based models. The neuronal network approach can be applied to the most sophisticated model without reducing the number of parameters or simplifying the physical processes. The models that have many parameters and include all physical processes tend to be the most accurate and robust models. Thus, the application of neuronal networks to invert these models has the potential of finding more optimal relationships between the desired input and output variables.

Networks can be used as a variable selection tool to define a set of variables which accurately predict variable(s) of interest. If the mapping of a network is not accurate, then perhaps some input variable(s) is(are) missing. Also an input variable is relevant to the problem only if it significantly increases the network's performance. Thus, networks can be used to identify relevant variables in complex nonlinear systems.

Neuronal networks are readily adaptable. They can easily incorporate new information that would be difficult or impossible to use with conventional techniques. Neuronal networks are ideally suited to learning new relationships between ancillary information, other input variables and the desired output variable. New input variables can be introduced to the network and tested easily. This is especially useful when a researcher expects a new variable to add information to the problem of interest but does not have any knowledge of the functional form to use in introducing the new variable using traditional techniques.

References

Abuelgasim AA, Gopal S, Strahler AH (1997) Forward and inverse modeling of canopy directional reflectance using a neuronal network. Int J Rem Sen 19:453–471
Anderson JA, Rosenfeld E (eds) (1988) Neurocomputing: Foundations of research. MIT Press, Cambridge
Asrar G, Dozier J (1994) EOS-Science strategy for the earth observing system. AIP Press, Woodbury
Asrar G, Greenstone R (eds) (1995) MTPE EOS reference handbook. EOS Project Science Office, NASA/GSFC, Greenbelt
Atkinson PM, Tatnall ARL (1997) Neuronal networks in remote sensing. Int J Rem Sen 18:699–709
Chuah HT (1993) An artificial neuronal network for inversion of vegetation parameters from radar backscatter coefficients. Journal of Electromagnetic Waves and Appl 7:1075–1092
Chuah HT, Tan HS (1989) A Monte Carlo method for radar backscatter from a half-space random medium. IEEE Trans on Geoscience and Remote Sens 27:86–93
Fu L (1994) Neuronal networks in computer intelligence. McGraw-Hill, New York
Goel N (1987) Models of vegetation canopy reflectance and their use in estimation of biophysical parameters from reflectance data. Remote Sens Rev 3:1–212
Govaerts Y, Verstraete MM (1994) Evaluation of the capability of brdf models to retrieve structural information on the observed target as described by a three-dimensional ray tracing code. Society of Photo-Optical Instrumentation Engineers (SPIE) 2314:9–20
Hall FG, Townshend JR, Engman ET (1995) Status of remote sensing algorithms for estimation of land surface state parameters. Remote Sens Environ 51:138–156
Jacquemoud S (1993) Inversion of the PROSPECT + SAIL canopy reflectance model from AVIRIS equivalent spectra – theoretical study. Remote Sens Environ 44:281–292
Jakubauskas ME (1996) Thematic mapper characterization of lodgepole pine seral stages in Yellowstone National Park, USA. Remote Sens Environ 56:118–132
Kimes DS, Holben BN, Nickeson JE, McKee A (1996) Extracting forest age in a Pacific northwest forest from thematic mapper and topographic data. Remote Sens Environ 56:133–140
Kimes DS, Ranson KJ, Sun G (1997) Inversion of a forest backscatter model using neuronal networks. Int J Rem Sen 18:2181–2199
Kimes DS, Nelson RF, Manry MT, Fung AK (1998) Attributes of neuronal networks for extraction continuous vegetation variables from optical and radar measurements. Int J Rem Sen 19:2639–2663
Kimes DS, Nelson RF, Skole DL, Salas WA (1999) Mapping secondary tropical forest and forest age from SPOT HRV data. Int J Rem Sen 20:3625–3640
Koza J (1993) Genetic programming. MIT Press, Cambridge
Kuusk A (1994) Multispectral canopy reflectance model. Remote Sens Environ 50:75–82
Li X, Strahler AH (1992) Geometrical optical bidirectional reflectance modelling of the discrete crown vegetation canopy. Effect of crown shape and mutual shadowing. IEEE Trans on Geosciences and Remote Sens 30:276–292
Moghaddam M (1994) Retrieval of forest canopy parameters for OTTER. In: Society of Photo-Optical Instrumentation Engineers (SPIE) (ed) Multispectral and Microwave Sensing of Forestry, Hydrology, and Natural Resources, 26–30 Sept. Rome
Myneni RB, Hall FG, Sellars PJ, Marshak AL (1995) The interpretation of spectral vegetation indices. IEEE Trans on Geoscience and Remote Sens 33:481–486
Nelson RF, Kimes DS, Salas WA, Routhier M (2000) Secondary forest age and tropical forest biomass estimation using TM. BioScience (in press)

Pinty B, Verstraete MM (1991) Extracting information on surface properties from bidirectional reflectance measurements. J Geophy Res 96:2865–2874
Pinty B, Verstraete MM, Dickinson RE (1990) A physical model of the bidirectional reflectance of vegetation canopies. 2. Inversion and Validation. J Geophy Res 95:11767–11775
Polatin PF, Sarabandi K, Ulaby FT (1994) An iterative inversion algorithm with application to the polarimetric radar response of vegetation canopies. IEEE Trans on Geoscience and Remote Sens 32:62–71
Prevot L, Schmugge T (1994) Combined use of theoretical and semi-empirical models of radar backscatter to estimate characteristics of canopies. Physical Measurements and Signatures in Remote Sensing. Sixth Symposium of International Society of Photogrammetry and Remote Sensing, 17–21 January 1994, Val d'Isere, pp 415–430
Privette JL, Myneni RB, Tucker CJ, Emery WJ (1994) Invertibility of a 1-d discrete ordinates canopy reflectance model. Remote Sens Environ 48:89–105
Privette JL, Myneni RB, Emery WJ (1996) Optimal sampling conditions for estimating grassland parameters via reflectance model inversions. IEEE Trans on Geoscience and Remote Sens 34:272–284
Ranson KJ, Sun G, Weishampel JF, Knox RG (1997) forest biomass from combined ecosystem and radar backscatter modeling. Remote Sens Environ 59:118–133
Ross JK, Marshak AL (1989) Influence of leaf orientation and the specular component of leaf reflectance on the canopy bidirectional reflectance. Remote Sens Environ 24:213–225
Saatchi SS, Moghaddam M (1994) Biomass distribution in Boreal Forest using SAR imagery. In: Society of Photo-Optical Instrumentation Engineers (SPIE) (ed) Multispectral and Microwave Sensing of Forestry, Hydrology, and Natural Resources, 26–30 Sept. Rome
Sader SA, Waide RB, Lawrence WT, Joyce AT (1989) Tropical forest biomass and successional age class relationships to a vegetation index derived from landsat TM data. Remote Sens Environ 28:143
Wasserman PD (1989) Neuronal computing: Theory and practice. Van Norstrand Reinhold Publishers, New York
Wharton SW, Myers MF (eds) (1997) MTPE EOS data products handbook. NASA/Goddard Space Flight Center, Code 902, Greenbelt, Maryland
Zornetzer SF, Davis JL, Lau C (1990) An introduction to neuronal and electronic networks. Academic Press, New York

Chapter 3

Soft Mapping of Coastal Vegetation from Remotely Sensed Imagery with a Feed-Forward Neuronal Network

G.M. Foody

3.1 Introduction

Data on the distribution of vegetation in space and time are required in a range of studies. Such data are, however, typically unavailable or are of poor quality (Williams 1994; DeFries and Townshend 1994). Often the only practicable means of acquiring data on vegetation distribution at appropriate spatial and temporal resolutions is through remote sensing (Townshend et al. 1991; Skole 1994). The considerable potential of remote sensing for mapping and monitoring vegetation has, however, frequently not been fully realized. Of the many reasons for this, one major limitation has been the reliance on conventional supervised image classification approaches as the tool for mapping.

Supervised image classifications have essentially three stages. The first is the training stage in which areas of known class membership in the image are identified and characterized statistically. These training statistics describe the appearance of each class in the imagery and, together with the selected classification decision rule, are used in the second, class allocation, stage to allocate each image pixel to the class with which it has the greatest similarity. For example, one of the most widely used approaches in thematic mapping from remotely sensed imagery is the maximum likelihood classification in which each pixel is allocated to the class with which it has the highest likelihood of membership. The third stage of the classification is the testing stage in which the accuracy of the classification is assessed to indicate its quality. The final output of the conventional supervised image classification is a classified image in which each pixel has been allocated to the class with which it has the highest degree of membership. This type of classification is often referred to as being 'hard,' as a pixel may be associated with only a single class and will have such an allocation forced upon it. Often such an allocation is undesirable. The hard classification approach is, for instance, only appropriate under a set of stringent conditions, which include the requirement for the classes to be discrete and mutually exclusive. For mapping natural and seminatural vegetation, this is often not the case, as the classes are continuous and intergrade (Trodd et al. 1989; Foody et al. 1992; Kent et al. 1997). Hard classifications are, however, often used to map such vegetation. Inevitably the adoption of such an approach will result in the delineation of sharp boundaries between sites that, particularly near the boundary, differ only marginally and perhaps insignificantly resulting in a oversimplified representation of reality (Campbell and Mortenson 1989; Foody et al. 1992; Sheppard et al. 1995). The hard classifications are, therefore, particularly inappropriate for the representation of continuous vegetation classes, especially in the inter-class transitional areas where classes coexist. This is unfortunate as the transitional areas are often a focus of biogeographical interest (Kent et al. 1997). Moreover, many of the hard classifications used in remote sensing are based on conventional statistical techniques which may have a range of often untenable assumptions or requirements. The maximum likelihood classification, for example, assumes that the data have a unimodal and normal distribution and that the training set is sufficiently large for a representative statistical description of the classes, which is often not the case. Consequently, the conventional hard classification approaches are sometimes unsuitable for vegetation mapping.

Soft or fuzzy classifications represent an attractive alternative to the conventional hard classifications for thematic mapping. With these approaches, each image pixel is allowed multiple and partial class membership and therefore can represent the full range of class membership from pure stands of a particular class to complex mixtures (Wang 1990; Foody 1996). The output of such an analysis is typically an image or set of images displaying the spatial variation in the grade of membership to the selected classes. There are many methods for the production of such soft classifications (Foody 1996), and these have been applied at scales ranging from the local (Foody 1996) to the global (DeFries et al. 1995). Of the techniques used to derive a soft classification, neuronal networks are particularly attractive, given their independence of restrictive assumptions that constrain the applicability of some other methods. Neuronal networks, for instance, are free from the distribution assumptions which apply to softened maximum likelihood classification. This paper aims to illustrate the use of a neuronal network for the derivation of soft classifications of vegetation at a coastal test site where the land cover exists as a mosaic of discrete and continuous classes.

3.2 Test Site and Data

The vegetation of Whiteford Burrows in south Wales was the focus of this study. This narrow strip of land contains a wide range of vegetation types, many of which are continuous and intergrade gradually (e.g. dune communities) with some clumps of relatively discrete classes (e.g. woodland), and lies adjacent to saltmarsh communities. A Daedalus 1268 airborne thematic mapper (ATM) sensor was used to acquire multispectral imagery of this site (Fig. 3.1). Within one year of the ATM image acquisition, a map of the vegetation at the site, based on a standard (hard) classification scheme, was produced from field-based survey and this map was used as the ground data in support of the analyses of the ATM imagery.

From the ground data, 11 land cover classes were selected for mapping. These ranged from relatively discrete classes such as woodland, which were distributed in small clumps over the site, to continuous classes such as those associated with dunes and slacks. Moreover, some classes defined differed only slightly in composition (Table 3.1). Sites of the 11 classes were identified in the ATM imagery and a total of 960 pixels drawn from pure regions of each class extracted. The sample size varied significantly between classes, ranging from 6 pixels for the slack woodland and bare soil classes, which only occupied small areas that were often difficult to identify reliably, to 557 pixels of water. For each of these pixels, the image tone (DN) in all eleven wavebands of the ATM, spanning the visible to thermal infrared spectral region (Wilson 1997), was extracted.

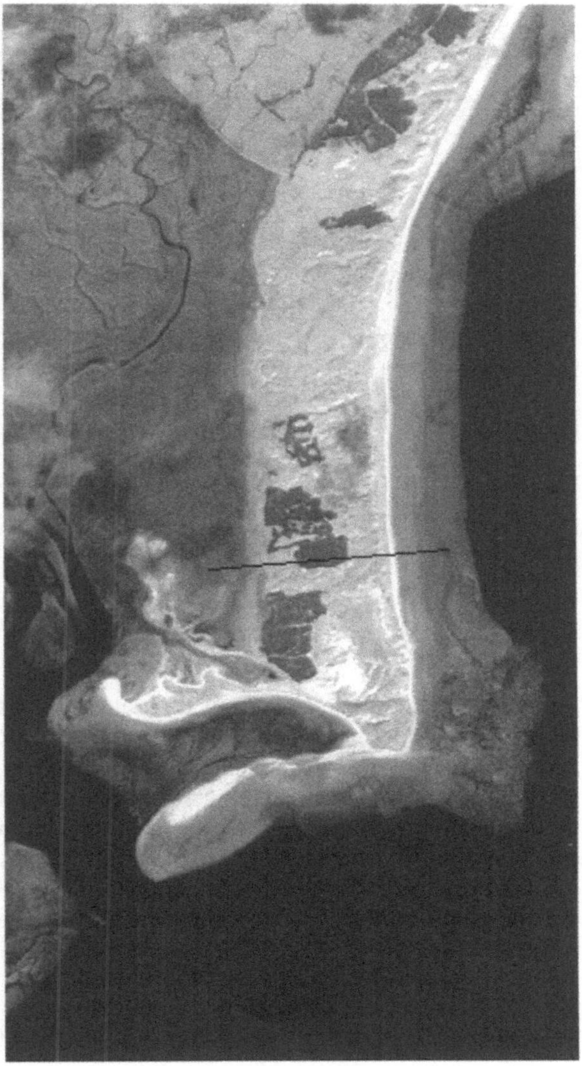

Fig. 3.1. ATM image (band 9) of the test site with the transect highlighted. Note that the areas of woodland (dark tone) contain small gaps which, while visible in the imagery, were not included in the ground data map

As in other studies, there was a high degree of intercorrelation between the data acquired in the ATM wavebands. To reduce redundancy in the data and the size of the data set, a feature selection was undertaken prior to the classifications (Campbell 1996). The feature selection, based on an analysis of the pairwise correlations of the data acquired in the ATM wavebands, was used to identify the most strongly intercorrelated wavebands. On the basis of these analyses, the data in a set of wavebands that were strongly correlated with data in the other wavebands were removed from the analyses. Seven ATM wavebands were selected for the analyses: these were wavebands 3 (0.52–0.60 μm), 4 (0.60–0.63 μm), 6 (0.69–0.75 μm), 7 (0.76–0.90 μm), 9 (1.55–1.75 μm), 10 (2.08–2.35 μm) and 11 (8.5–13.0 μm). The data acquired in the wavebands removed

Table 3.1. The classes mapped and number of training samples extracted from the ATM data

Class	Description and main species	Number of samples
Woodland	*Pinus nigra/sylvestris* woodland. Evidence of thinning and removal.	74
Slack woodland	*Anus glutinosa*, developed adjacent to salt marsh	6
Foredune	Transitional area with mixture of *Elymus farctus, Festuca rubra, Agrostis stolonifera, Limonium binervosum, Honkenya peploides* and *Cakile maritima* and *Atriplex postrata*	15
Mobile dune	*Ammophila arenaria*	22
Mobile dune	*Ammophila arenaria* with *Festuca rubra* understory	7
Semi-fixed dune	*Ammophila arenaria, Festuca rubra, Ononis repens* and *Tortula ruralis*	37
Dune slack	*Salix repens, Campylium stellatum, Calliergon cuspidatum, Carex nigra, Bryum pseudotriquetrum,* and *Aneura pinguis*	108
Dune slack	*Salix repens, Calliergon cuspidatum, Carex flacca* and *Pulicaria dysenterica*	14
Salt marsh		114
Bare soil		6
Water		557

from the analyses were very strongly correlated with the data in the seven wavebands selected ($r > 0.92$). However, to ensure that this reduction in the data volume did not adversely affect the class separability, classifications with a discriminant analysis were performed on the training data using the selected seven wavebands and using all eleven wavebands. Although not applied to an independent testing set, both analyses revealed a very high and constant level of separability, with 99.5% of the training pixels allocated correctly. No further preprocessing of the imagery was undertaken as the analyses did not require radiometric calibration and there was no significant geometrical distortion.

3.3 Methods

The data acquired in the seven selected ATM wavebands for the 960 pixels sampled from the image were used to train the neuronal network. This was a basic feed-forward neuronal network, of the type widely used for supervised image classification. This type of network can be envisaged as comprising a set of simple processing units arranged in layers, with each unit in a layer connected by a weighted channel to every unit in adjacent layers, and combined, these elements transform the remotely sensed input data into a class allocation (Fig. 3.2). The precise architecture of a neuronal network for supervised classification is determined by a range of factors which relate, in part, to the nature of the remotely sensed data and desired classification. There is usually, for instance, an input unit for every discriminating variable (e.g. the selected spectral wavebands of the remotely sensed imagery) and an output unit associated with each

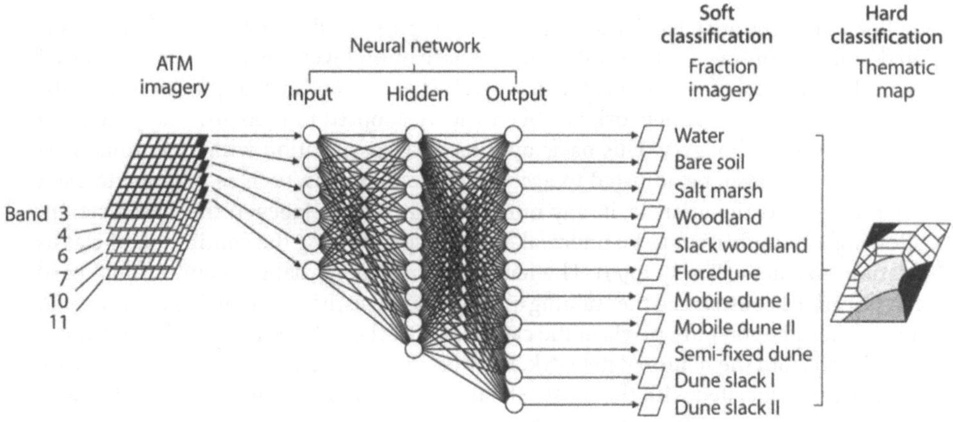

Fig. 3.2. Land cover classification with a neuronal network. The network shown is that used in the research to derive both soft and hard classifications

land cover class to be mapped. The number of hidden units and layers is typically defined subjectively on the basis of a series of trial runs.

The training stage of a neuronal network classification is based on the iterative application of a learning algorithm to the network in which the weights connecting the units are initially set at random. The aim of the training stage of the classification is to use a learning algorithm, such as back propagation, to adjust these weights until the network is able to correctly characterize the class membership properties of the training data set. On each iteration of the learning algorithm, the error in the network's output can be calculated, as the desired output is known for the training data set. This error is then fed backward through the network to the input layer with the weights connecting units changed in relation to the calculated error. The weight change on the *n*th iteration is generally achieved by,

$$\Delta w_{ji}(n) = -\eta \delta o_{ij} + \alpha \Delta w_{ji}(n-1) \tag{3.1}$$

where w_{ji} is the weight connecting the unit j with the unit i in the previous layer, o_i the output from unit i, δ is a computed error, the derivation of which varies for hidden and output units, η is the learning rate and α a parameter which, with the weight change from the previous iteration, determines the momentum which facilitates network learning (Schalkoff 1992). After completion of each iteration, the training data are then re-entered and the process repeated until the error value is minimized. The overall output error computed over all training patterns is

$$\underset{j}{E} = 0.5\sum(t_j - o_j)^2 \tag{3.2}$$

where t_j is the target output, known for training data, and o_j is the network output. When the overall output error has minimized or declined to a subjectively determined acceptable level, training ceases and the network may be used for the classification of previously unseen cases of unknown class membership.

For a conventional 'hard' classification, each pixel is allocated to the class associated with the output unit with the highest activation level (i.e. largest o_j, measured typically on a scale from 0 to 1) for that pixel. This conventional approach to classification with a neuronal network has been widely adopted in thematic mapping from remotely sensed imagery. This basic approach to classification with a neuronal network may, however, be adjusted to accommodate fuzziness, such as that due to class mixing in transitional areas, in any number of the three stages of the classification, allowing a classification to be undertaken at any point along the continuum of classification fuzziness (Foody 1999). The nature of the ground data (a conventional hard classification) constrained the training and testing stages, however, such that they could only be based upon pure pixels of the classes. A soft class allocation may, however, still be derived from the neuronal network. For this, instead of simply allocating each case to the class associated with the most activated output unit, the magnitude of the activation level of each output unit may be used as a measure of the strength of membership to the class associated with the unit and mapped. With the activation level of every output unit derived for a pixel, it is possible to illustrate the manner in which membership is partitioned between the classes and so indicate the class composition of the pixel. A pure pixel, representing an area of a single class, would ideally display a very high output unit activation level for the unit associated with the actual class of membership, with negligible output unit activation levels to all other classes. However, with a mixed pixel, perhaps in the inter-class transitional area, the relative magnitudes of the output unit activation levels over all classes would reflect the class composition. Thus the activation level of each network output unit may be treated as a measure of the strength of membership to the associated class and used as a surrogate for the fractional cover of that class to form a soft classification (Foody 1996). The spatial distribution of the class may be represented by fraction images in which image tone is positively related to the class coverage as reflected in the magnitude of the activation levels (e.g. Foody et al. 1997). Unlike some other approaches used for the derivation of fraction images, the neuronal network is not constrained to provide an output in which the activation levels sum to unity over all classes. The derived activation levels may, however, be rescaled to ensure that they do sum to unity. This rescaling enables the values to be treated as indicating the proportional coverage of a class from which entropy may be calculated to help describe the partitioning of the total membership between the various classes. The entropy, H, may be derived for a pixel from,

$$H = -\Sigma p(x) \log_2 p(x) \tag{3.3}$$

where $p(x)$ is the proportional coverage of class x; the choice of logarithm base is arbitrary but commonly base 2 is used (Klir and Folger 1988). Entropy is minimized when the pixel is associated with a single class and maximized when membership is partitioned evenly between all of the classes. Entropy based measures have been used in the evaluation of image classifications, both as a measure of classification accuracy and as an indicator of the confidence that may be associated with a class allocation (Maselli et al. 1994; Foody 1996).

The use of the output unit activation levels to indicate the class composition of pixels is similar in nature to the softening of conventional classifications, such as the maximum likelihood classification (Foody et al. 1992; Foody 1996). It is, however, achieved without making the often untenable assumptions that underlie many conventional classifications. The soft output derived from the neuronal network should also provide a realistic representation of the spatial distribution of both continuous and discrete classes.

A series of trial runs was undertaken to define an appropriate network architecture and set of learning algorithm parameters for the analyses. The neuronal network used comprised 7 input units (one to present the DN for each of the selected wavebands of ATM imagery to the network), a single hidden layer containing 9 units and 11 output units (one associated with each class to be mapped). This network was trained for up to 10 000 iterations with a stochastic back propagation algorithm with the parameters for the momentum and learning rate set at 0.9 and 0.1 respectively. Training was interrupted at various stages and the network used to classify the ATM imagery before continuing the training process. In this way, outputs derived from networks trained with a different number of iterations of the learning algorithm were derived. In each, the output unit activation levels derived were used to form a soft classification, in which the magnitude of the activation level of the output unit associated with a class was used as a surrogate for the fractional coverage of that class in the pixel. These soft classification outputs could also be hardened, by giving each pixel the label of the class associated with the most activated output unit, to yield standard hard classifications. In evaluating the accuracy of the classifications, attention focused particularly on a transect of 137 previously unseen pixels in length spanning a range of land cover classes that could be confidently located in the ground data (Fig. 3.1). Since the ground data set was a conventional hard classification derived from a field based survey, the accuracy of the soft classifications for the pixels along this transect could not be rigorously evaluated as the ground data did not adequately represent the continuous classes. However, to gain a quantitative assessment of classification accuracy, an independent sample of pixels drawn from the 'core areas' of the classes (i.e. away from boundaries/transitional areas), which could be considered to be pure, was extracted and used to form a conventional confusion matrix for accuracy assessment (Campbell 1996).

3.4 Results and Discussion

As an initial step the conventional hard classification of the site was produced from the neuronal network trained for 10 000 iterations of the learning algorithm. Using a sample of 100 pure pixels not used in training the network, the classification was found from the confusion matrix derived to have a very high accuracy, with 95% of these testing pixels allocated to the correct class. Such results may appear highly satisfactory, but the hard classification derived may not be conveying all the information on land cover contained in the imagery, and the accuracy assessment could be misleading.

Although the accuracy statement indicated a very accurate classification, it is, for instance, possible that accuracy has been exaggerated. Many factors can influence the calculated or apparent accuracy of a classification. Leaving aside important issues such as the sampling design used to acquire the test set, a major problem with the data used here is that, like the pixels used in training the neuronal network, the testing pixels were extracted from pure or exemplar sites of the classes. They are, therefore, only representative of a small part of the test site. Consequently, the accuracy statement derived is only really applicable to regions composed of homogeneous cover of a single class, and excludes, for example, areas where classes intergrade. The real accuracy of the classification may, therefore, be lower than the stated value, with error concentrated around the boundaries of discrete classes and the transitional areas between continuous classes where mixing is most prevalent. The hard classification of the data also provided an inappropriate representation of some classes, particularly the continuous classes.

The output unit activation levels used to derive the conventional hard class allocation were also used to form a soft classification. In this soft classification the magnitude of the activation level of an output unit reflected the fractional coverage of the class associated with the unit. In the absence of an appropriate ground data set, the value of the soft classification output can be illustrated with reference to some of the classes encountered along the transect. For a relatively discrete class such as woodland, the output unit activation levels for the class were generally either very high, near 1.0, or very low, near 0.0, indicating complete coverage or absence of the class, in agreement with the situation observed on the ground (Fig. 3.3a). For continuous classes, however, a different pattern was observed. In the 'core areas' of the classes, the activation levels were very high or very low, as for the discrete classes. However, less extreme activation levels were frequently observed in the inter-class transitional areas, reflecting the intergradation of the classes (Fig. 3.3b).

Further information on the variation of class membership along the transect is presented in Fig. 3.3c. This shows the total activation level for each pixel, summed over all classes, and the entropy statistic as a guide to how membership was partitioned between the classes. In general, there was a close correspondence between the variations in the total activation level and entropy along the transect. For most pixels, the total activation level was near 1.0 with a low, near 0.0, entropy, indicating a relatively hard class allocation. Since the magnitude of the activation level was not constrained to sum to 1.0, these outputs may indicate membership of the class associated with the most activated output unit with a relatively high degree of confidence. There were, however, a number of zones along the transect that deviated from the general trend. Four of these are particularly noteworthy. First, there was a zone characterized by having a total activation level >1.0 and a high entropy value. This corresponded to the location of a transition in the water body. This transition is visible in the imagery (Fig. 3.1) and likely to arise due to a change in water depth and turbidity. Since the total activation level exceeded 1.0, the maximum that can be derived from an output unit, the total membership must be partitioned between two or more classes which explains the relatively high entropy observed in this zone. Inspection of the output unit activation levels revealed that the pixels in this zone had a very high level of activation (>0.93), and so membership of the actual class, water, together with a smaller degree of membership to the foredune communities. As before, the magnitude of the activation level to a class relative to the total activation level derived with respect to all classes may be used to indicate the confidence in a class allocation, which here indicated some uncertainty, but a high degree of membership to water. Second, there were zones where the total activation level was ~1.0 but the entropy >0. These results indicate mixing of the communities, which is evident in Figs. 3a and 3b. Third, there

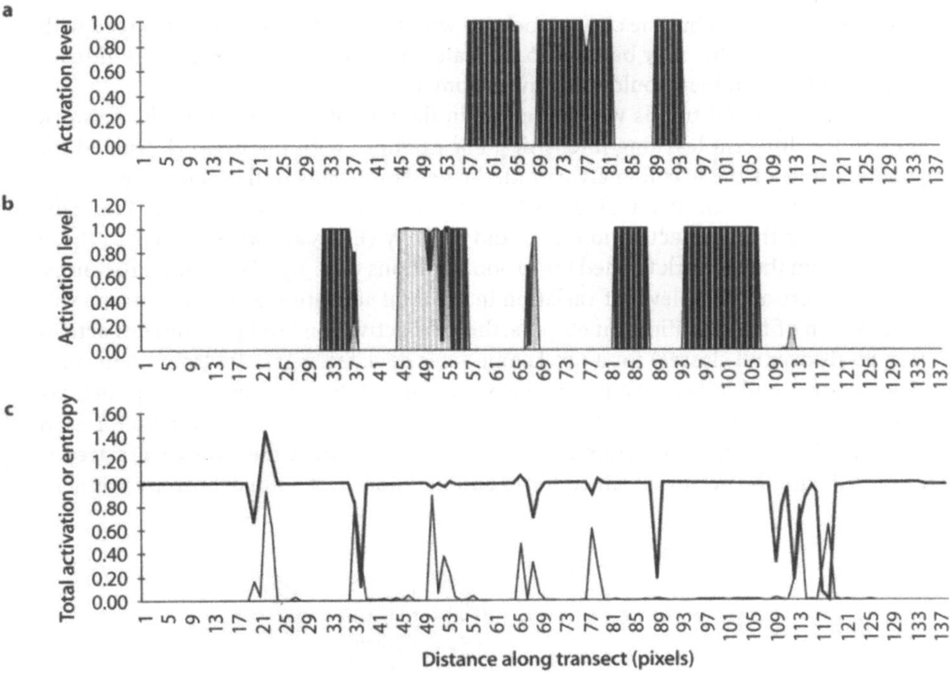

Fig. 3.3. Variations in class membership along the transect from the neuronal network trained for 10 000 iterations of the learning algorithm; **a** bar chart showing the activation level to the woodland class, a discrete class; **b** compound bar chart showing the activation level to the dune slack (*light tone*) and semifixed dune (*dark tone*) classes, continuous classes; **c** total activation level (*thick line*) and entropy (*thin line*)

were zones where the total activation level was <1.0 but the entropy low, near 0.0. This may indicate that the pixels in this zone are essentially associated with a single class but with a relatively low degree of confidence. Fourth, there were zones where the total activation level was <1.0 and entropy >0.0. Since all the activation levels were low and the entropy was calculated from rescaled activation levels, these outputs are difficult to interpret. However, some appear to correspond to observable changes in the vegetation cover. For example, the middle section of the transect crosses through a region of woodland, and within the region depicted as woodland in the ground data, the membership to the woodland class is either very high or low (Fig. 3.3a). This could indicate either that the regions with low membership to woodland have been misclassified or that the classification is revealing the vegetation mosaic differently to the map. It is, for example, apparent that the woodland contains small gaps. These gaps in the forest canopy are not depicted on the vegetation map used as ground data, presumably due to their size in relation to the minimum mapping unit and for the purposes of generalization, but are visible in the imagery (Fig. 3.1). This may indicate that the ground data are insufficiently detailed or not accurate enough, as noted in other studies (e.g. Bauer et al. 1994; Bowers and Rowan 1996). The zone near the end of the transect also had total activation levels <1.0 and entropy >0.0. Here the total activation level may be low but the class associated with the most activated unit agreed with the ground data. This may be used to indicate a relatively low degree of confidence with the allocation that would be derived from a hard classification.

The same general trends were observed in the outputs derived from the network trained for different learning intensities. For example, with the network trained for 300 iterations of the learning algorithm, the general trends in the magnitude of the activation levels associated with discrete (Fig. 3.4a) and continuous classes (Fig. 3.4b) as well as for the total activation level and entropy (Fig. 3.4c) were similar to those derived from the network trained for 10 000 iterations (Fig. 3.3). The main differences, however, were a higher level of variation in the total activation level and entropy values. In each of the classification outputs, the total activation level (i.e. the sum activation levels over all classes) for a pixel sometimes deviated markedly from 1.0 in transitional areas and indicated a possible means of locating and characterizing such areas. When the network's outputs were rescaled to sum over all classes to 1.0 for each pixel and the variations in entropy along the transect evaluated, the same general trends observed with the network trained for 10 000 iterations were noted. Entropy was gen-

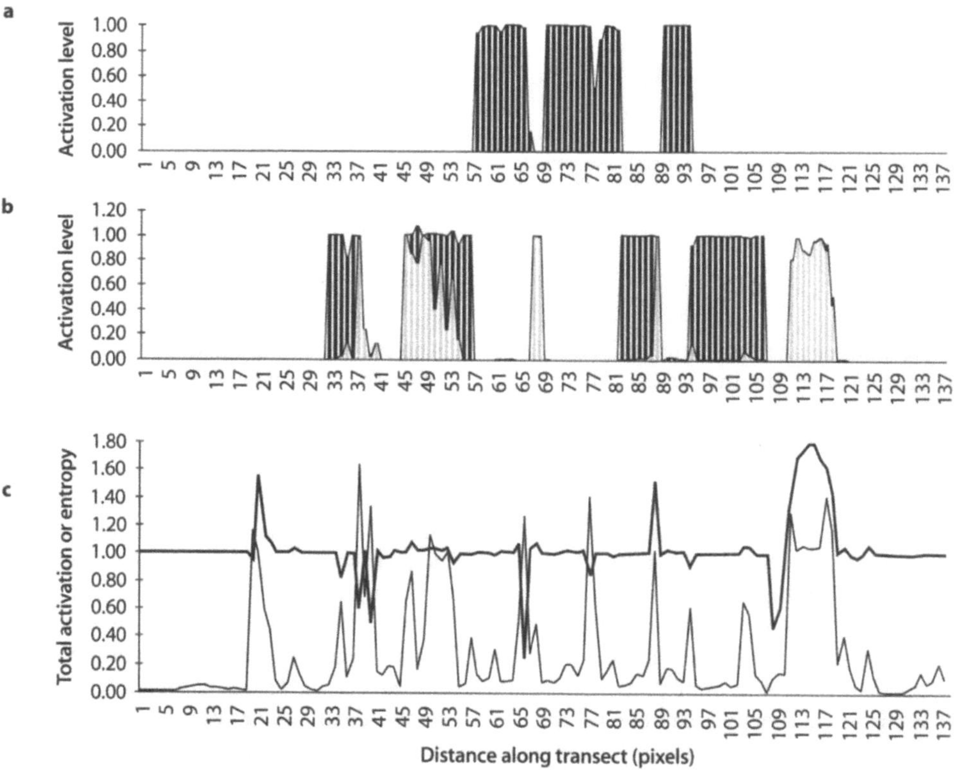

Fig. 3.4. Variations in class membership along the transect from the neuronal network trained for 300 iterations of the learning algorithm; **a** bar chart showing the activation level to the woodland class, a discrete class; **b** compound bar chart showing the activation level to the dune slack (*light tone*) and semifixed dune (*dark tone*) classes, continuous classes; **c** total activation level (*thick line*) and entropy (*thin line*)

Table 3.2. Summary of the entropy statistics for pixels along the transect derived from the neuronal network trained over a range of iterations of the learning algorithm

Training Iterations	Error	Entropy Mean	Maximum
300	0.0311	0.296	1.647
500	0.0241	0.257	1.534
1 000	0.0207	0.196	1.262
2 000	0.0157	0.135	1.510
5 000	0.0011	0.080	1.159
10 000	0.0004	0.064	0.946

erally low in the 'core areas' of the classes (i.e. the pixel was associated mostly with a single class), but higher in the transitional areas (i.e. membership was divided among more than one class). This may also provide a basis for identifying and characterizing transitional areas. It was, however, apparent that the distinctiveness of the transitional areas varied with training intensity. The entropy values calculated for the pixels along the transect, for instance, declined with training intensity over the range investigated (Table 3.2). As training intensity increased, therefore, the class allocation generally became crisper with membership increasingly associated with a single class, often leading to a decrease in both the total activation level and entropy. The selection of an appropriate training intensity for a classification must, therefore, be geared to suit the needs of the particular investigation (e.g. the desired level of fuzziness) and, as with the avoidance of the overtraining problem, cannot be based on the training error (Table 3.2). The results do, however, highlight that a neuronal network may be used to derive a soft classification that can represent both discrete and continuous vegetation.

3.5 Summary and Conclusions

Hard supervised image classification techniques are sometimes inappropriate for the mapping of continuous vegetation classes from remotely sensed imagery. Furthermore, the hard classification techniques commonly used generally make untenable assumptions about the vegetation to be mapped as well as of both the remotely sensed and ground data sets. The conventional approach to neuronal network classification may be softened, allowing the derivation of a soft classification that models appropriately the distribution of both discrete and continuous vegetation classes without making unrealistic assumptions about the data. In the soft classification, transitional areas may be identified and characterized conveying considerably more information on the vegetation at the site than a conventional hard classification.

Acknowledgements

I am grateful to the University of Wales Swansea for access to the data sets and the referees for their comments on the paper. The ATM data were acquired as part of the 1990 NERC airborne remote sensing campaign and the analyses undertaken with the NCS NeuronalDesk package.

References

Bauer ME, Burk TE, Ek AR, Coppin PR, Lime SD, Walsh TA, Walters DK (1994) Satellite inventory of Minnesota forest resources. Photogrammetric Engineering and Remote Sensing 60:287–298

Bowers TL, Rowan LC (1996) Remote mineralogic and lithologic mapping of the Ice River Alkaline Complex, British Columbia, Canada, using AVIRIS data. Photogrammetric Engineering and Remote Sensing 62:1379–1385

Campbell JB (1996) Introduction to remote sensing, second edition. Taylor and Francis, London

Campbell WG, Mortenson DC (1989) Ensuring the quality of geographic information system data: A practical application of quality control. Photogrammetric Engineering and Remote Sensing 55:1613–1618

DeFries RS, Townshend JRG (1994) Global land cover: Comparison of ground based data sets to classifications with AVHRR data. In: Foody GM, Curran PJ (eds) Environmental remote sensing from regional to global scales. Wiley, Chichester, pp 84–110

DeFries RS, Field CB, Fung I, Justice CO, Los S, Matson PA, Matthews E, Mooney HA, Potter CS, Prentice K, Sellers PJ, Townshend JRG, Tucker CJ, Ustin SL, Vitousek PM (1995) Mapping the land-surface for global atmosphere-biosphere models – toward continuous distributions of vegetations functional-properties. Journal of Geophysical Research-Atmospheres 100:20867–20882

Foody GM (1996) Approaches for the production and evaluation of fuzzy land cover classifications from remotely sensed data. Int J Rem Sen 17:1317–1340

Foody GM (1999) The continuum of classification fuzziness in thematic mapping. Photogrammetric engineering and remote sensing 65:443–451

Foody GM, Campbell NA, Trodd NM, Wood TF (1992) Derivation and applications of probabilistic measures of class membership from the maximum likelihood classification. Photogrammetric Engineering and Remote Sensing 58:1335–1341

Foody GM, Lucas RM, Curran PJ, Honzak M (1997) Mapping tropical forest fractional cover from coarse spatial resolution remote sensing imagery. Plant Ecology 131:143–154

Kent M, Gill WJ, Weaver RE, Armitage RP (1997) Landscape and plant community boundaries in biogeography. Progress in Physical Geography 23:315–353

Klir GJ, Folger TA (1988) Fuzzy sets, uncertainty and information. Prentice-Hall International, London

Maselli F, Conese C, Petkov L (1994) Use of probability entropy for the estimation and graphical representation of the accuracy of maximum likelihood classifications. ISPRS Journal of Photogrammetry and Remote Sensing 49:13–20

Schalkoff R (1992) Pattern recognition: Statistical, structural and neuronal approaches. Wiley, New York

Sheppard CRC, Matheson K, Bythell JC, Murphy P, Myers CB, Blake B (1995) Habitat mapping in the Caribbean for management and conservation: Use and assessment of aerial photography. Aquatic Conservation: Marine and Freshwater Ecosystems 5:277–298

Skole DL (1994) Data on global land-cover changes: Acquisition, assessment and analysis. In: Meyer WB, Turner II BL (eds) Changes in land use and land cover: A global perspective. Cambridge University Press, Cambridge, pp 437–471

Townshend J, Justice C, Li W, Gurney C, McManus J (1991) Global land cover classification by remote sensing: Present capabilities and future possibilities. Remote Sensing of Environment 35:243–255

Trodd NM, Foody GM, Wood TF (1989) Maximum likelihood and maximum information: Mapping heathland with the aid of probabilities derived from remotely sensed data. In: Remote Sensing Society (ed) Remote sensing for operational applications. Nottingham, pp 421–426

Wang F (1990) Fuzzy supervised classification of remote sensing images. IEEE Transactions on Geoscience and Remote Sensing 28:194–201

Williams M (1994) Forest and tree cover. In: Meyer WB, Turner II BL (eds) Changes in land use and land cover: A global perspective. Cambridge University Press, Cambridge, pp 97–124

Wilson AK (1997) An integrated data system for airborne remote sensing. Int J Rem Sen 18:1889–1901

Ultrafast Estimation of Neotropical Forest *DBH* Distributions from Ground Based Photographs Using a Neuronal Network

M.A. Dubois · L. Cournac · J. Chave · B. Riera

4.1 Introduction

There is no ecosystem model able to describe tropical forests in a comprehensive way, and the complexity of the problem is even higher than climatic prediction. Tropical forests have comparable spatial and temporal dynamics; moreover they exhibit another dimension of complexity, namely the high number of dynamic variables (Oldeman 1990): instead of a few fluids as in climatology, theoretical ecologists have to consider a high diversity of animal and vegetal species, with relevant and long-ranged interactions (Charles-Dominique 1995a). A drastic simplification is thus necessary if one wants to develop a manageable model, and for this it is essential to find sets of easily measured synthetic macroscopic variables.

A rich information for forests is embodied in the distribution functions of tree diameters at breast height or *DBH* (e.g. Rollet 1974; Cusset 1980), which also constitute reference data for forestry. But *DBH* inventory by tree counting in dense tropical forests represents a considerable effort and is almost impossible to achieve routinely on large areas. We therefore investigated the possibility of getting part of this information, but in a much faster way.

In this paper, we propose to characterize vegetation transects with a sampling of standardized photographs (fixed hour, azimuth, film, focal length, aperture). These photos are digitised, then analysed with a neuronal network, in order to obtain a *DBH* histogram. The network is trained on a set of photographs taken on 1 ha plots where *DBH* distributions have been obtained independently by traditional methods.

Neuronal networks have been chosen here for their ability to approximate a broad range of relationships, without a priori knowledge of the function and without need for linearity (Fu 1994). Indeed, we do not attempt a geometrical reconstruction of the forest from image processing, but we rather conduct a spectral analysis of the photograph, which is considered as fuzzily and nonlinearily related to the subjacent geometry. The various components of the photograph (stem presence, light diffusion, interferences, etc.) are synthesized within a statistical vector which constitutes the raw data given to the network. By analogy, the method can be compared to the "guess estimate" of an experienced forester. In this paper, we show the first results of implementing this procedure using the Nouragues research station database.

4.2
Methods

4.2.1
Study Site

The Nouragues station, 4°5' N, 52°42' W, in French Guyana, was established in 1986 (Charles Dominique 1995b). The annual rainfall is about 3 000 mm, and the average daily temperature is 25 °C, with a night minimum at 21 °C and a maximum at 34 °C during the day. The vegetation is characteristic of a dense humid evergreen tropical forest. Two different geological zones are separated by a fault which is now occupied by the Nouragues river: north of the river (on the inselberg side) granitic and crystalline rocks of Caribbean series give a sandy clay soil; to the south, metamorphic rocks of the "Paramaca" series (green rocks) bear a clay soil. This leads to several vegetation types: a rocky savannah on the top of the inselberg, a transition forest in the inselberg periphery, a high forest, then a disturbed zone of smaller forest with bamboos ("cambrouzes") or a gradient where the density of lianas increases so that they become leading types in some places (low elevation points of "pinotières" swamps and riparian forests).

4.2.2
Tree Inventory

Onto this site nearly 32 km of trails were open. Part of them forms a 1 000 m × 700 m grid in which each plot is 100 m by 100 m. Trees within the plots were mapped and measured on an overall surface of 100 hectares. Data include trees over 30 cm *DBH* for the whole installation and over 10 cm *DBH* on 22 ha. The histogram of diameter classes (10 cm intervals) was constructed for each ha (Riéra 1995). For this work, the histograms were fitted by a single exponential. This was the most simple approximation of the distributions of our data set, but it exhibits rather satisfactory properties: the mean sum of square errors per distribution is 45 (that means average errors of 2 to 3 trees per diameter class, which is really low), and the estimated basal area (sum of stems sections areas) per ha is well predicted: the linear regression between actual and fitted basal area has a slope of 0.94, with $R^2 = 0.65$. The two parameters of the exponential fit are normalized and constitute the values which are to be predicted from photograph analysis.

4.2.3
Photograph Sampling

In July 1995 a series of ground-based photographs was taken at each trail crossing (format 24 × 36, focal length 35 mm, aperture 8, shutter speed between 1/15 and 1 s, camera on a tripod at 1.3 m height, pointing north toward horizon line, slide colour film 100 asa, time: solar noon ±1 h). A second series of pictures indexed with a '*' was taken 10 m before each intersection with north-south axes (Fig. 4.1). A leaf area index (LAI) measurement was performed simultaneously with each photograph using a Licor LAI2000 device. We decided then to investigate how the information contained in the photographs, which essentially intercept stems, could be used to estimate *DBH* distributions.

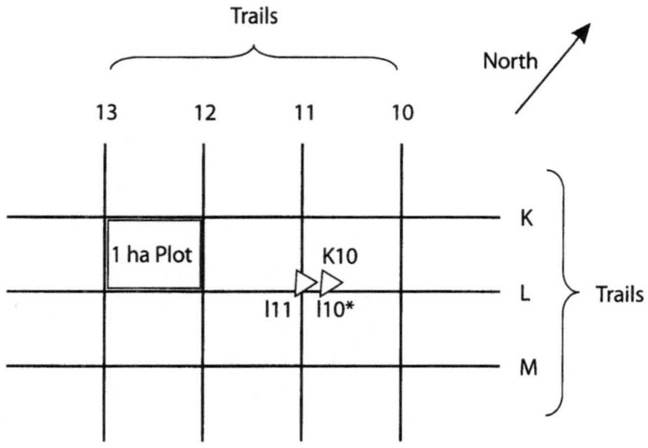

Fig. 4.1. Nomenclature of photographs and plots in the Nouragues study site. Plots are referenced by the names of the trails which cross at north corner. Photographs (*triangles*) are referenced by the name of the plot from which they were taken, not of the plot which is pointed at. The presence of a "*" means that the photo is used for the test database, whereas it is not

We will not attempt to reconstruct *DBH* distribution through a geometrical inference from the photographs. Indeed, such an attempt would be considerably difficult: inclination from the trees need to be taken into account (necessity of shape recognition); there is a problem of relative distance from the trees to interpret stem width (apparent width vs real width), that could be experimentally overridden from stereoscopic photographs, but it is not realistic in tropical forest conditions. On the other side, the apparent stem widths on the pictures could be analysed without direct reference to actual widths: a probability distribution of *DBH* results in a probability distribution of apparent stem widths, but this is not an unambiguous relationship. Moreover the trees visible on the photo are only a part of the angular sector covered by the field of view (shading of trees of the back plane by trees of the first plane): in this respect, getting the precise information of apparent stem widths would provide a partial and presumably not general view of the distribution of the trees, unless a huge number of pictures are taken. This can be summarized by the French proverb "l'arbre cache la forêt" (*the tree hides the forest)* which can also be metaphorically interpreted as: a large amount of sophisticated operations on a partial aspect of the signal hides its statistical generality.

Other kinds of information which are also represented on the photograph prove to be at least as important, as they reflect the ambient conditions which are conditioned by "hidden trees": abundance of leafs, diffusion of light between the trunks, etc. To synthesize these various aspects a statistical rather than geometrical image analysis should be preferable.

4.2.4 Image Processing

Pictures are digitised and stored on CD-Rom. The chosen format is 768×512 8-bit encoded Bitmap (Fig. 4.2). A 700×200 pixels window is imposed on the upper part of each picture: we extract from each image a 700×200 matrix of integers from 0 to 255, and a colour map encoding the RGB (red, green, blue) intensities on the [0, 255] interval. We obtain then three intensity matrices: red, green and blue. It is necessary

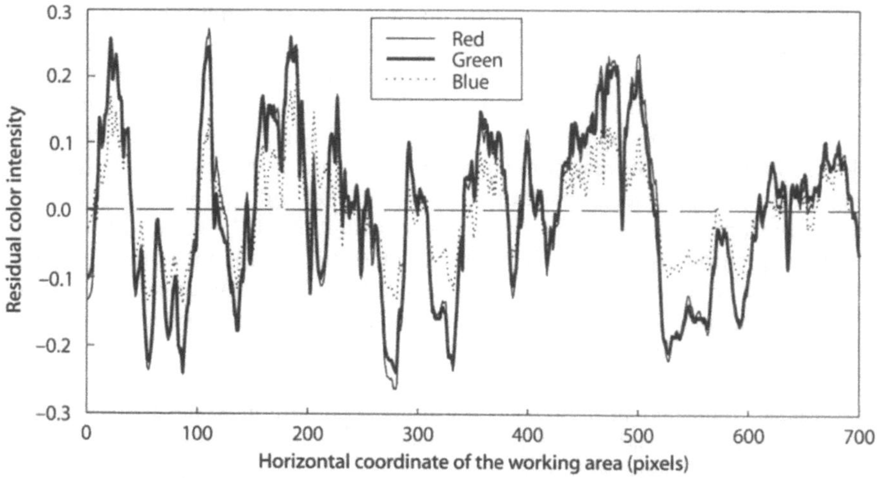

Fig. 4.2. Digitised images and signals (red, green, blue) extracted from the colour components of the picture. This signal is the residual column average of pixel component intensities, once average value and long-range trends have been removed by a 3rd order polynomial approximation

to extract from the original matrices a vector compact enough to be used as input of a neuronal network but keeping as much as possible of the initial information.

The vertical average of the green components is extracted from the digitised photographs and provides a 700-components signal. We subtract the average value (= average green intensity of pixels) and long-range horizontal trend (= large scale light intensity gradients across the photo) of the signal. For this a 3rd order polynomial fit proved to be sufficient. After subtraction of the polynomial approximation, the residual signal remarkably reflects the presence of the trees observed on the original picture (Fig. 4.2), negative peaks being roughly correlated to the presence of trunks and positive peaks to gaps. In forests with low stem diameters (liana forests, disturbed areas) the deviation from zero of the residual signal is small, whereas in high stand forests the presence of large trees creates broad peaks. Neither wavelet transforms nor Fourier transforms are adapted for analysis of these data due to the low number of components.

4.2.5 Extraction of the Input Vector

A simple way to synthesize the information contained in the residual signal consists in building a histogram of the values it takes. We found that extracting a 19-bin histogram was sufficient to unequivocally separate all the different photos we had. This is achieved by dividing the [−0.225, 0.225] interval which covers the range of observed residuals in 19 equally spaced intervals and counting for each residual signal the number of values which fall in each interval. This leads to a 19-component vector which was called VH (for Vector Histogram) which reflects the vertical characteristics of the image. Moreover, photos taken at 10 m (indexed "*", Fig. 4.1), in which the field of view is almost completely separated from that of the photos taken at trail crossings, mainly (but not systematically) produce histograms which are correlated with the latter: we have then a good candidate to synthetically describe tree distribution. Diagrams in Fig. 4.3 illustrate these correlations. The presence of "wings" on a histogram (high frequency of elevated values on the residual signal: on plots *L10* and *J10* for instance) corresponds to distributions with a high frequency of large-diameter trunks. On the other hand, plots with a low frequency of large trees result in peak or bell-shaped histograms (see plots *M18* or *O18*). Intermediate cases can be observed on plots (*N19*, *K14*). Globally, the left part of the histograms is mainly determined by large apparent stems, and the right part by large apparent gaps, the middle part resulting from the combination of small trees/gaps and from transition zones. However, some of the plots show unexpected shapes or discrepancies between the signals extracted from corner and "*" photographs (*L18*, *J12*). This especially occurs when the plot is strongly heterogeneous and produces then very different photographs even from a single 10 m shift. We decided to use the VH histograms as input values of the neuronal network.

4.2.6 Neuronal Network Design

Neuronal network design and training was achieved by using the SNNS software package (Stuttgart Neuronal Network Simulator v4.1, freely distributed by the Institute for Parallel and Distributed High Performance Systems -IPVR-, Stuttgart University, Germany, obtainable by anonymous ftp at ftp://ftp.informatik.uni-stuttgart.de/pub/SNNS, Zell et al. 1991).

The training database contains 72 data sets resulting from the photos taken at trail crossings. Each data set comprises the 19-component VH vector associated with the

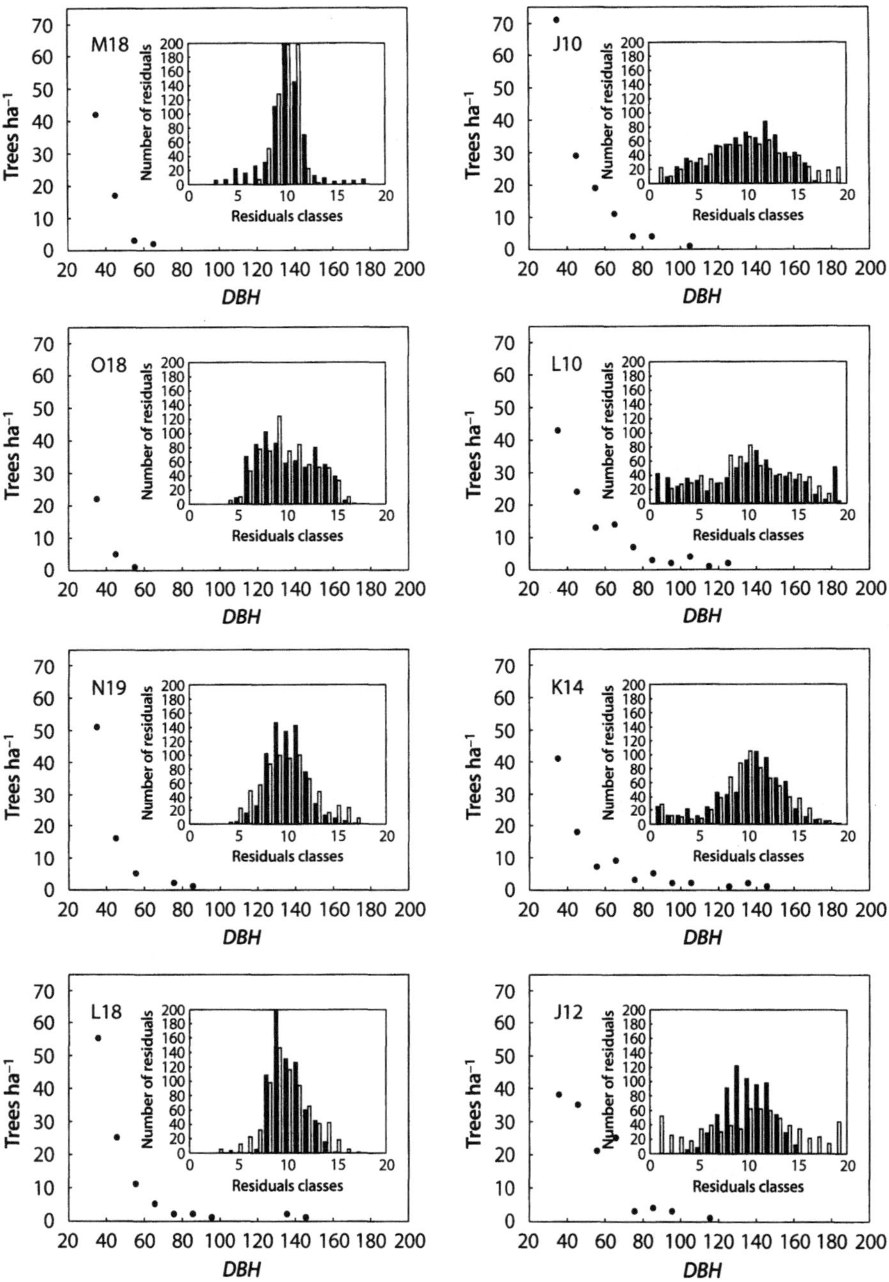

Fig. 4.3. Histograms of the 19-components VH vectors extracted from the photos of different plots in the Nouragues station and corresponding *DBH* classes distribution. Each graph summarizes the data collected for one plot. The name of the plot is given by the intersection of the trails at north corner, *filled circles* represent in situ tree measurements, the two VH histograms extracted from the pictures taken on each plot are inserted : *black bars* correspond to the photo taken at the south corner, *white bars* to the 10 m-shifted photos

photo, which is the input of the neuronal network, and the two coefficients of the exponential fit of the distribution of trees above 30 cm *DBH* in the photographed plot, which are to be approximated by the output. The database used to test the generalization capacity of the network consists in 72 VH vectors extracted from 10 m-shifted photographs, taken on the same area. During learning, evolution of the sum of square errors (*SSE*) on training set and test set are computed simultaneously.

The structure of the network was chosen as a 3-layer (sufficient for approximating any continuous function, Fu 1994) feed-forward neuronal network: a 19-node input layer, a hidden layer, and a 2-node output layer; the activation function used was the sigmoid. Several trials have been done with varying the hidden layer size from 2 to 30 nodes: the final design chosen was the one which showed the best performances for generalization (lowest *SSE* on test set during learning), that is 15 nodes. We adopted a rather slow learning rate: indeed, we found that learning was rather unstable until we decreased the learning rate to around 0.05, a value at which the *SSE* did not show chaotic variations during learning and consistently had the same evolution between independent learnings (i.e. independent initialisations of the network) performed on a given data set.

4.3 Results

First, after some epochs of learning, the network gives a hardly variable output which corresponds to an average distribution which is the same in training and test sets and is given by the mean values of the two output variables. Learning process is visualised in Figs. 4.4 and 4.5 where tree counting, exponential fits, and neuronal network estimates are superimposed at various learning times. As learning proceeds, the training set is more and more accurately described (Fig. 4.4), whereas test set description first globally improves (Fig. 4.5, 80 000 epochs), then is gradually separated between distributions which are better fitted as the learning goes on and distributions which are getting worse (Fig. 4.5, 180 000 epochs). Overlearning results in all distributions perfectly fitted on the training set (Fig. 4.4, 400 000 epochs), and some distributions well fitted on the test set, while many of them are not (Fig. 4.5, 400 000 epochs). Not surprisingly, well-correlated histograms between photos of the train set and of the test set resulted in convergent learning, whereas decorrelations between the histograms (more heterogeneous plots) resulted in divergence.

In our conditions, optimum learning (lowest *SSE* on test set) occurs at about 80 000 epochs with the following parameters: learning rate 0.05, hidden layer 15 nodes. At this stage, variability in the test set starts being well described, only extreme distributions being under or overestimated. The state of progression of learning is shown on Fig. 4.6, which compares the two parameters of the exponential decay to their neuronal network-estimated counterparts. Correlations between actual and NN-predicted parameters was significant in all cases ($R^2 = 0.57$ for N_{35}, $R^2 = 0.4$ for k, $P < 0.001$ for both in the training set, $R^2 = 0.15$ for N_{35}, $R^2 = 0.2$ for k, $P < 0.05$ for both in the test set). After this stage, overfitting occurs: points corresponding to estimated parameters in the training set stack onto the 1 : 1 line and R^2 consequently tends to 1 as the number of learning epochs increases, whereas in the test set an increasing proportion of outputs diverge (20% of outputs at 180 000 epochs, and 85% at 400 000 epochs for instance are less well predicted than at 80 000 epochs).

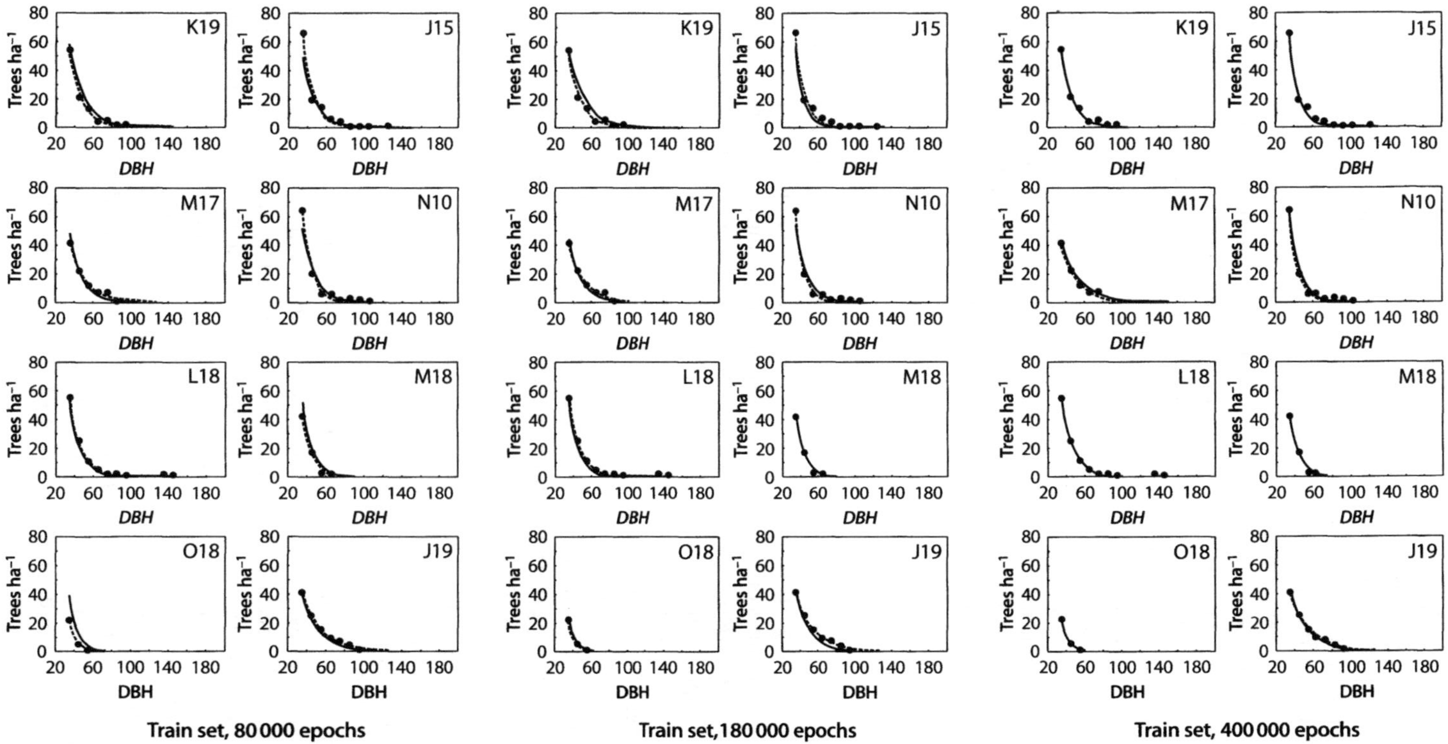

Fig. 4.4. Estimates of tree distribution given by the network on a representative subset of the training database (south corner photos), at two learning times. *Dotted lines* represent the exponential fit of *DBH* distributions with a classical least squares method, the parameters of which were given for learning of the network; *straight lines* represent the shape of the exponential distribution given by the parameters estimated by the neuronal network

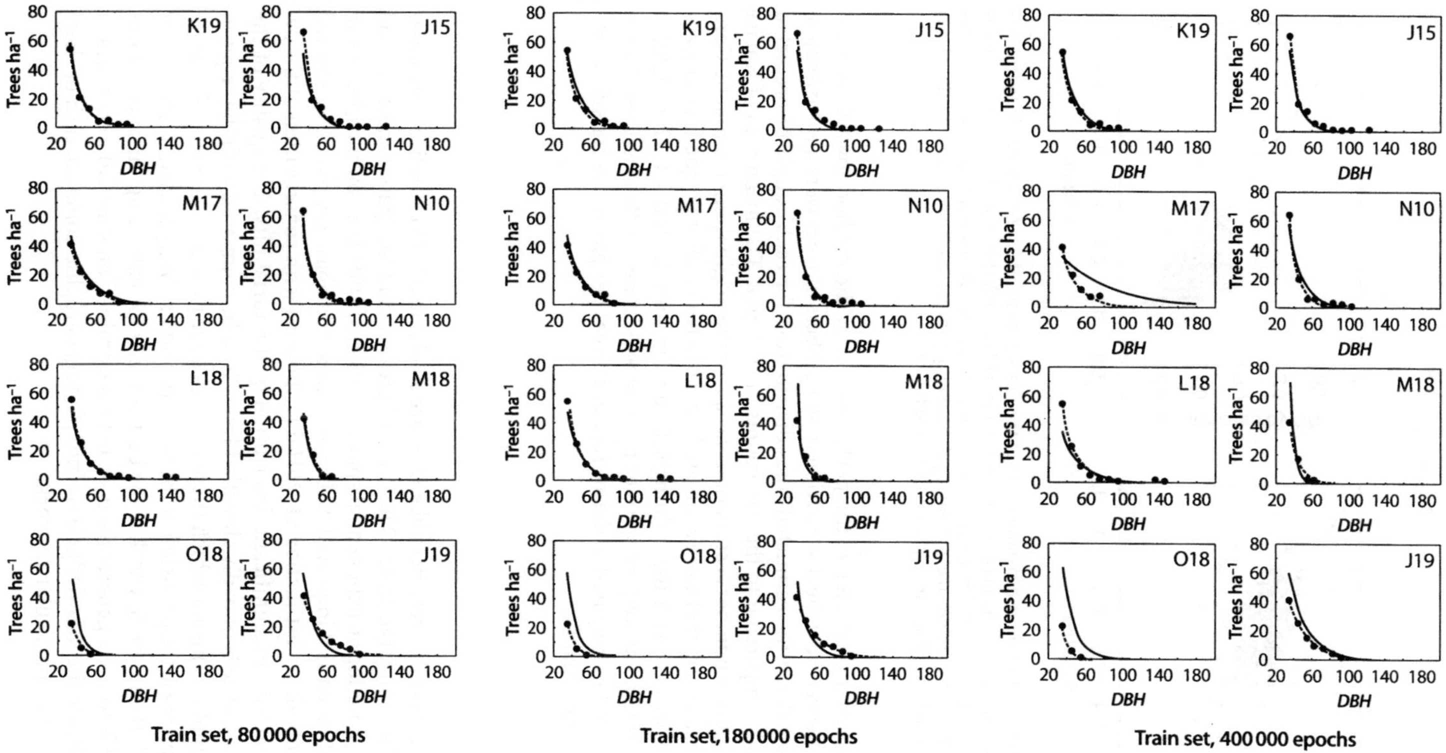

Fig. 4.5. Estimates of tree distribution given by the network on a representative subset of the test database (10 m-shifted photos), using the network weight values obtained from training, at three learning times. *Dotted lines* represent the exponential fit of *DBH* distributions with a classical least squares method; *straight lines* represent the shape of the exponential distribution given by the parameters estimated by the neuronal network

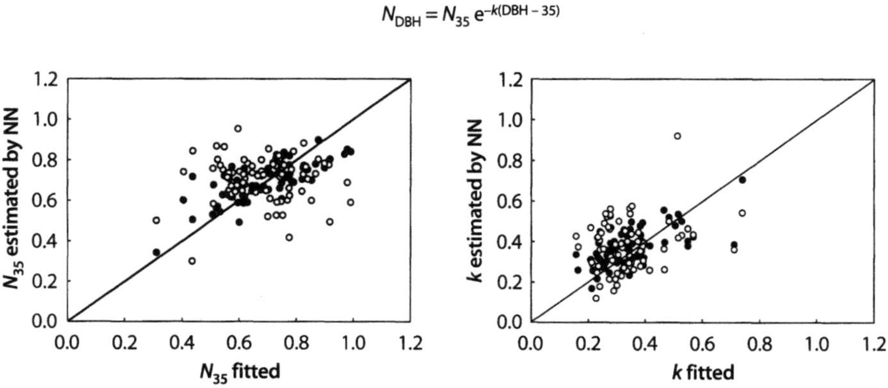

Fig. 4.6. Comparison between the two parameters of the single exponential least-squares fit to tree distributions and values given by the network on the train set (*filled circles*) and on the test set (*open circles*), at optimal learning time. N_{35}: number of trees in the 30–40 cm *DBH* class; k: time constant of exponential decay. N_{35} and k are normalized during the learning process to be comprised within a [0, 1] interval: these normalized values are represented on the figure

In order to better characterize the performances of the method toward its main objective, we checked how the distributions and numbers of trees were approximated. For this, we calculated the estimated trees distribution for each couple of predicted parameters and compared it to the tree counts in each plot. The mean sum of square errors (*MSSE*) of tree numbers estimation was 45 for the original exponential fit. Considering the parameters that were estimated by the network after 80 000 learning epochs, the *MSSE* was 173 with the training set, and 277 with the test set. This *MSSE* of 277 corresponds to an average of 30 trees per plot which are "badly classified" by the network. This is to be compared to the average of 13 trees per plot badly classified with the original exponential fit.

4.4 Discussion

Few data on structural parameters like *DBH* are available due to excessive difficulties in obtaining them on large scales. The method presented in this paper shows a good potential for estimating an exponential approximation of *DBH* distributions, but it still needs to be validated and improved. Using our limited database (72 plots), we showed that we could obtain an estimation of *DBH* distributions with conservation of the average, and a partial but significant description of the variability. The quality of the prediction (around 20% of badly classified trees) is to be compared with the requirements of the applications for which these estimates are provided. The use of a larger database will increase the precision of the network and its ability to characterize other forests. We will then have to check if the best strategy for extending the applications of the method is to use a general purpose network, trained on very different types of forests, or to divide the knowledge base into phytogeographic domains, each of these being related to one particular network.

The question of the sampling frequency is also crucial for optimizing network learning. In fairly undisturbed places like the Nouragues research station, the distributions of trees and the photograph sampling characteristics are rather homogeneous at the ha scale, thus a frequency of 1 photo ha^{-1} seems to be sufficient to predict *DBH* distributions using the NN method, except in some locations (plot *J12* for instance). But many forests exhibit heterogeneity at a smaller scale, due for instance to anthropogenic perturbations. In such cases, increasing sampling frequency and consequently adapting the learning procedures will be necessary, as more parameters will be necessary to correctly describe the distributions. In these cases, an exponential decay will no longer be a good approximation, and more complex functions should be used. The choice of the minimal but relevant functions (ideally functions in which the parameters are related to the regeneration dynamics or the predation pressure) will be an important task for the future developments of the method. Until now, higher sophistication attempts in image processing or in network design did not provide significantly better learning efficiency, but this feature could change by incorporating new data sources and new forest types. Improvement of the image processing could be obtained by higher pixel density or by considering the two dimensions in spectral analysis.

Until now, two main kinds of applications of neuronal networks in forest research can be found in the literature. First, they have been used in predictions-extrapolations of vegetation growth response to the environment when these interactions are complex and difficult to calibrate from deterministic equations. This use of neuronal networks as metamodels has been performed to simulate ecophysiological processes at the stand scale (Huntingford and Cox 1997), or forest dynamics on landscape scale, sometimes in conjunction with GIS (Gimblett and Ball 1995). The second major use of neuronal networks in forest research has been the interpretation of teledetection data (Pierce et al. 1994; Gopal and Woodcock 1996; Kimes et al. 1997). But one important point for the validation of remote sensing algorithms is the constitution of reliable ground databases and this also stands for neuronal network approaches (Kimes et al. 1998). The method that we propose to develop has the potential to produce significant amounts of ground data for remote sensing studies calibration.

4.5 Conclusion

The description of structural parameters in tropical forests remains an important objective for the characterization of these environments. Automatic sampling from ground-based photographs proves to be of real interest for studying these ecosystems on larger scales than with traditional counting. The processing of these photos could allow researchers to locate the various stages of the forests during sylvigenetic cycles and the different vegetation types. Indeed, the distribution of individuals or of species is a dynamic mechanism highly depending on the perturbations (Riéra 1995). The characterization of perturbations, through their "footprints" in diameter class histograms, will give new insights in the understanding of biodiversity. An important diversity of sylvigenetic stages and of plant communities in a given place favours the installation, development and persistence of a greater number of species than in homogeneous environments. The study of climate modifications is also possible through the perturbations that they have induced (e.g. drier periods associated or not to forest fires). These perturbations together with climate modifications also constrain the diversity. Acquisition of forest structural parameters on representative scales could help us to unravel these relationships

Acknowledgements

This work is part of the ECOFIT program (ECOsystèmes et paléoécosystèmes des Forêts InterTropicales, which is a joint program of the French institutes CNRS, ORSTOM, MNHN, CEA, CIRAD). It was funded by a grant from the "Programme SOFT" of the French Ministry of Environment.

References

Charles-Dominique P (1995a) Interactions plantes-animaux frugivores, conséquences sur la dissémination de graines et la régénération forestière. Revue d'Ecologie (Terre et Vie) 50(3):223–235
Charles-Dominique P (1995b) "Les Nouragues", une station de recherche pour l'étude de la forêt tropicale. In: Legay JM, Barbault R (eds) La révolution technologique en écologie. Masson, Paris, pp 183–199
Cusset G (1980) Sur les paramètres intervenant dans la croissance des arbres : relation hauteur/diamètre de l'axe primaire aérien. Candollea 35:231–255
Fu L (1994) Neuronal networks in computer science. McGraw-Hill, New York
Gimblett HR, Ball GL (1995) Neuronal network architectures for monitoring and simulating changes in forest resource management. AI Applications 9(2):103–123
Gopal S, Woodcock C (1996) Remote sensing of forest change using artificial neuronal networks. IEEE Transactions on Geoscience and Remote Sensing 34:398–404
Huntingford C, Cox PM (1997) Use of statistical and neuronal network techniques to detect how stomatal conductance responds to changes in the local environment. Ecol Model 97:217–246
Kimes DS, Ranson KJ, Sun G (1997) Inversion of a forest backscatter model using neuronal networks. Int J Rem Sen 18:2181–2199
Kimes DS, Nelson RF, Manry MT, Fung AK (1998) Attributes of neuronal networks for extracting continuous vegetation variables from optical and radar measurements. Int J Rem Sen 19:2639–2663
Oldeman RAA (1990) Forests: Elements of sylvology. Springer Verlag, Berlin, Heidelberg
Pierce LE, Sarabandi K, Ulaby FT (1994) Application of an artificial neuronal network in canopy scattering inversion. Int J Rem Sen 15:3263–3270
Riéra B (1995) Rôle des pertubations actuelles et passées dans la dynamique et la mosaïque forestière. Revue d'Ecologie (Terre et Vie) 50(3):209–222
Rollet B (1974) L'architecture des forêts denses humides sempervirentes de plaine. C.T.F.T. Nogent-sur Marne
Zell A, Mache N, Sommer T, Korb T (1991) Recent developments of the SNNS neuronal network simulator. Proc. Applications of Neuronal Networks 1469:708–719

Normalized Difference Vegetation Index Estimation in Grasslands of Patagonia by ANN Analysis of Satellite and Climatic Data

F.G. Tomasel · J.M. Paruelo

5.1 Introduction

The Normalized Difference Vegetation Index (*NDVI*), derived from the red and infrared bands of the AVHRR on-board sensor of NOAA satellites, shows a high correlation with biophysical rates of the target area, such as transpiration or primary productivity (Sellers et al. 1992). *NDVI* has been shown to be a linear estimator of the fraction of the photosynthetic active radiation (PAR) absorbed by the canopy (Potter et al. 1993; Ruimy et al. 1994). Monteith (1981) showed that the amount of PAR absorbed throughout the growing season is the major control of net primary production. *NDVI* data also allows the tracking of intra-annual changes in carbon gains (Lloyd 1990; Paruelo and Lauenroth 1995).

NDVI has also been shown to be strongly correlated to the Aboveground Net Primary Production (*ANPP*) in grassland and shrubland areas (Tucker et al. 1985; Box et al. 1989; Prince 1991a; Prince 1991b; Burke et al. 1991; Paruelo et al. 1997). *ANPP*, the rate of carbon accumulation in plants, is a key attribute of the ecosystem. It represents the amount of energy available to the upper trophic levels and integrates many important functional characteristics such as nutrient cycling, secondary production (McNaughton et al. 1989), and root biomass and soil organic carbon dynamics (Sala et al. 1997). The importance of *ANPP* is also related to applied reasons. For example, *ANPP* is the major control of forage availability for both domestic and wild herbivores in grasslands, savannahs and shrublands (Oesterheld et al. 1992; MacNaughton et al. 1993; Oesterheld et al. 1998). The understanding of the environmental controls of *ANPP* and the prediction of future values is, therefore, a crucial issue for both theoretical and applied ecologists. The development of predictive models of *ANPP* is clearly restricted by the availability of long-term data sets. The reason behind the lack of extensive databases is quite simple: estimation of *ANPP* is time-consuming, and therefore expensive (Lauenroth et al. 1986; Sala et al. 1988).

NDVI has been proved to be a reliable alternative in cases where long records for *ANPP* are unavailable (Paruelo et al. 1997). Several agencies have compiled and reprocessed original data to produce global databases of *NDVI* images at a spatial resolution of 8 × 8 km (i.e. James and Kalluri 1994; Tucker and Newcomb 1994). The NOAA/NASA EOS AVHRR Pathfinder data set include 36 images per year for the period 1981–1994. This database is specially suited to analysing the temporal dynamics of *ANPP*.

Southern Argentina is dominated by temperate, arid and semiarid steppes and semideserts (Soriano 1983; León et al. 1998). The design of sustainable systems for this area clearly depends on a better understanding of the structure and functioning of main ecosystems of the region (Soriano and Paruelo 1990). This area is characterized by scarce and variable precipitation, ranging from 700 mm toward the western edge of the region, to 150 mm in the centre of the area (Jobbagy et al. 1995). Most of the region is influenced by Pacific air masses (Prohaska 1976). The Pacific influence determines a clear concentration of precipitation during winter months. The area dominated by Pacific air masses corresponds to the Patagonian Phytogeographical Province (Paruelo et al. 1991). The north-eastern part of the region is also influenced by Atlantic air masses, which determine a more even distribution of precipitation (Paruelo et al. 1998). This area corresponds to the Monte Phytogeographical Province (León et al., 1998) and is covered by steppes dominated by evergreen shrubs of the genus *Larrea.*

Desertification has been a major concern for the scientific community, federal agencies and environmental groups for more than two decades (Soriano and Movia 1987). Sheep have grazed native vegetation since the beginning of the century (Soriano and Paruelo 1990). Grazing is blamed as the major determinant of vegetation degradation across the area (León and Aguiar 1985; Perelman et al. 1997). Aguiar et al. (1996) have showed, using simulation models, the impact of the structural changes associated to overgrazing on ecosystem functioning.

Jobbágy et al. (1999) have analysed long-term *NDVI* data for the Patagonia steppes using regression models. Even though regression models resulted in valuable tools to understand the system, they showed a low predictive power. A better knowledge of temporal dynamics of *NDVI* is advantageous in the management of natural resources. Predictive models of *NDVI* may also provide the basis for the development of "warning systems" for Patagonian rangelands. The objective of this paper is to investigate the temporal dynamics of the *NDVI* and its internal and external controls across northern Patagonia by using ANNs. We also explore the use of ANNs as predictive tools of the intra-annual dynamics of the *NDVI.*

5.2 Methodology

5.2.1 Artificial Neuronal Networks

Applications of ANNs to ecological and environmental problems have started early this decade, mainly through the use of feed-forward multilayer networks. Some examples are classification of remotely sensed data (Liu and Xiao 1991; Kanellopoulos et al. 1992; Foody et al. 1995), resource management (Gimblett and Ball 1995), ecosystems modelling (Lek et al. 1996; Recknagel et al. 1997; Paruelo and Tomasel 1997), weather forecasting (McCann 1992; Derr and Slutz 1994), prediction of daily solar radiation (Elizondo et al. 1994), and many others. In particular, there has been a clear interest in using ANNs for nonlinear prediction of time series. One of the most impressive results has been shown by Wan in the prediction of a chaotic time series through the use of a finite-duration impulse response (FIR) multilayer perceptron (Wan 1994). Although many of these ANNs are able to make very good predictions, training is in general based on availability of very long data sets. Unfortunately, this is not the common case in ecological modelling where, for example, population time series for terrestrial animals are usually composed of tens of samples (see, for example, Turchin and Taylor (1992) for a compilation of some of the longest data sets available on vertebrate and insects).

In this paper we use a feed-forward network, trained by a newly proposed learning technique based on Information Theory (IT). This direct learning approach has been shown to improve the performance of simple perceptrons, providing very good predictions based on a rather small quantity of known data (Diambra et al. 1995; Diambra and Plastino 1995). We will only outline the method here; the interested reader is encouraged to read the original references for an in-depth description of the procedures involved.

Following Diambra, Fernandez, and Plastino, let us consider a simple perceptron with N inputs I_i connected to a single output unit O whose state is determined according to $O = g(h)$, where $g(x)$ is the activation function, $h = W_j I_j$ is the weighted sum of the inputs I_i, and repeated dummy indices imply a summation over those indices. In the structures discussed herein, we have chosen $g(x) = \tanh(x)$. For each set of weights W the perceptron maps I on O. The perceptron is trained with a set of P examples, with input vectors I^μ and the corresponding outputs $O^\mu \equiv O(I^\mu)$. From here we can write

$$g^{-1}(O^\mu) = W_j I^\mu{}_j \tag{5.1}$$

where I^μ is an input patterns matrix and $g^{-1}(O^\mu)$ is a vector of components $g^{-1}(O^1), \ldots,$ $g^{-1}(O^P)$, given by the output patterns, which constitute our available information. The central idea in this approach is to use an Information Theory approach to determine the weights W on the basis of an incomplete information supply (rank $(I^\mu) < N$, in general). In order to determine weights consistent with Eq. 5.1, it is assumed that each set of weights W is realized with probability $P(W)$. In other words, a normalized probability distribution is introduced over the collection of possible sets W. The normalization condition is written as

$$\int P(W) dW = 1 \tag{5.2}$$

where $dW = dW_1, dW_2, \ldots, dW_N$. Expectation values $<W_i>$ are defined as

$$<W_i> = \int P(W) W_i dW \tag{5.3}$$

The differential entropy associated with the probability density function $P(W)$ is written as

$$S = -\int P(W) \ln(P(W) / P_0(W)) dW \tag{5.4}$$

where $P_0(W)$ is an appropriately chosen a priori distribution. The problem of determining the set of weights W is now transformed into a constrained optimization problem: we must now determine the form of the probability density function for the differential entropy of W to assume its largest value for the prescribed constraints of Eq. 5.1 and Eq. 5.2. The authors' central idea is to reinterpret Eq. 5.1 according to:

$$g^{-1}(O^\mu) = <W_j> I_j^\mu \tag{5.5}$$

where explicit account is taken of the fact that one is dealing with many sets of weights, each one being realized with a given probability, and borrowing from statistical mechanics the idea that measured data are to be reproduced by theoretical averages.

It can be shown that, after maximization of the differential entropy, the expectation vector $<W>$ can be expressed solely in terms of the training examples, and that it can be written as

$$<W> = g^{-1}(O^{\mu})I_{MP}[O^{\mu}] \tag{5.6}$$

where $I_{MP}(O^{\mu}) = (O^{\mu})^t[O^{\mu}(O^{\mu})^t]^{-1}$ is the Moore-Penrose pseudoinverse. The most probable configuration of weights, compatible with the constraints of Eq. 5.1, is thus given directly by the pseudoinverse matrix of O^{μ}, with no iterative processes associated with the training of the network. IT-trained networks have been successfully applied to the prediction of some classical chaotic time series, even when a small quantity of examples was made available for the training process (Diambra and Plastino 1995).

5.2.2
The Data Set

As we mentioned earlier in the introduction, *ANPP* has been shown to be strongly correlated with the *NDVI*. Therefore, an analysis of the *NDVI* cycles and their relationship with the climatic variables may translate into a better understanding of the environmental controls of *ANPP* and into a better predictive power. In this paper, we used *NDVI* from 10 sites covering a broad range of climatic conditions across northern Patagonia (Fig. 5.1, Table 5.1). These locations were selected based on the availability of precipitation data. We obtained the *NDVI* data from the Pathfinder AVHRR Land database (James and Kalluri 1994), from which data was available for a period of 11 years

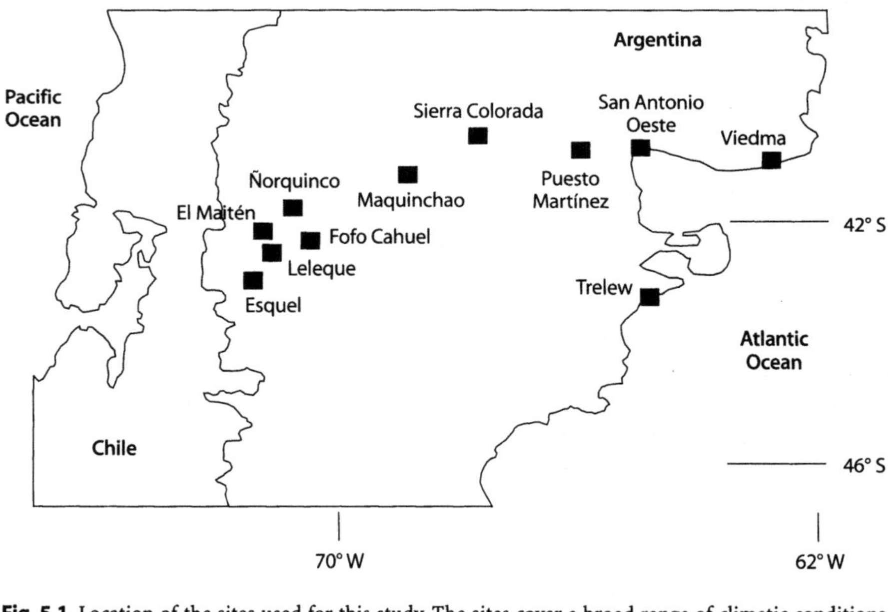

Fig. 5.1. Location of the sites used for this study. The sites cover a broad range of climatic conditions across northern Patagonia, with mean annual precipitation ranging from 130 mm in Fofo Cahuel to 420 mm in Leleque

Table 5.1. Characteristics of the precipitation regime for the sites selected for the present study. Mean annual precipitation ranges from 130 mm in the case of Fofo Cahuel to 420 mm in the case of Leleque

Site	Mean annual precipitation (mm)	Precipitation falling in summer (%)
Leleque	418	8
El Maitén	356	9
Viedma	328	21
Esquel	268	9
San Antonio Oeste	268	22
Trelew	214	20
Sierra Colorada	205	24
Ñorquinco	196	9
Maquinchao	173	18
Puesto Martínez	153	35
Fofo Cahuel	129	14

(1981–1991). For each year 36 images were available, each corresponding to a 10-day composite (Holben 1986). The spatial resolution of the images was 8 × 8 km., and every site was characterized by a single pixel (6 400 ha).

5.3
Results and Discussion

Our first approach to the problem was to analyse the predictive power of ANNs trained solely on k past values of the *NDVI* time series. In this case, the training data were of the form

$$I^i = \{NDVI(t_i), NDVI(t_i - T), \ldots, NDVI(t_i - kT)\} \text{ and}$$
$$O^i = NDVI(t_i + mT), i = 1 : P \qquad (5.7)$$

where P is the number of patterns used for training, T is the sampling period and m denotes a suitable number of time steps. So given k past values of *NDVI*, the ANN was asked to extrapolate the value of the *NDVI* m steps ahead.

On each site, an ANN was trained by using 8 years of data and tested on the remaining three. The dynamics of the *NDVI* time series was best captured when 36 past values (one year) of data were used as the input. Figure 5.2 shows the *NDVI* 9-step-ahead (three months) extrapolation for the case of sites Esquel, Leleque and Fofo Cahuel. Although the correlation between calculated and observed values of *NDVI* differed among sites, the results show that in general the agreement is very good. To evaluate the performance of our predictors, we calculate the mean square error,

$$MSE = E\{[NDVI_{\text{calc}}(t, kT) - NDVI_{\text{obs}}(t + mT)]^2\}$$

where E is the expected value operator. For convenience in the comparison among sites, we normalize this by the mean square deviation of the data, $\sigma^2 = E\{(NDVI - E[NDVI])^2\}$,

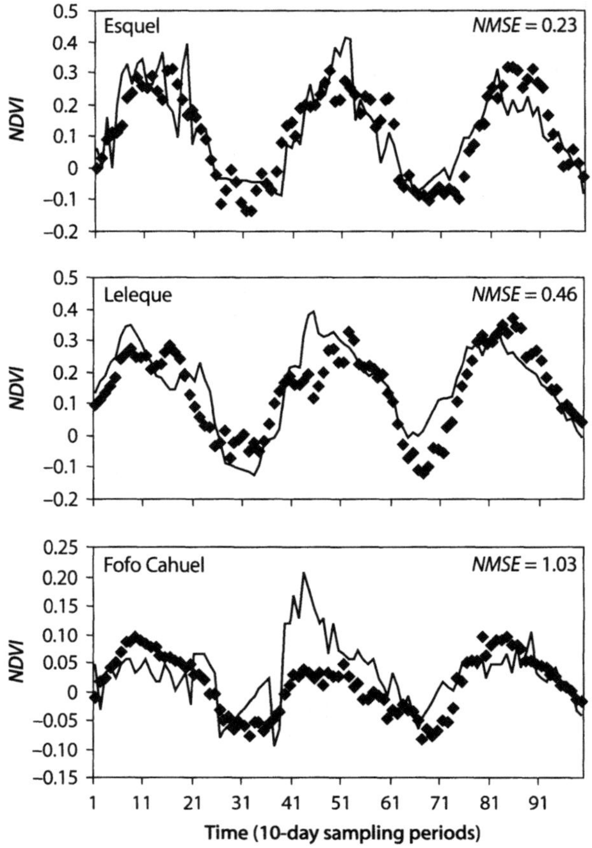

Fig. 5.2. Calculated (*square dots*) and observed (*solid line*) values of *NDVI* for the case of a 9-step-ahead extrapolation in sites Esquel, Leleque, and Fofo Cahuel. Values for the *NMSE* are indicated

forming the normalized mean square error $NMSE = MSE / \sigma^2$ (Farmer and Sidorowich 1987). In this way, smaller values of *NMSE* correspond to better predictions.

The *NMSE* values for the case of 9-step-ahead extrapolation ranged between 0.23 and 4.58 (Table 5.2). For sites with relatively small values of *NMSE*, extrapolation could be made up to 18 steps in advance (six months) without significant degradation of the forecasting error.

The results of calculating the *NDVI* from its internal dynamics highlight interesting aspects of the ecology of the different phytogeographical regions of Patagonia. Figure 5.3 shows that the *NMSE* strongly increases as the proportion of precipitation falling during summer increases. Sites located in the southwestern portion of the area analysed presented a better agreement between observed and calculated values than those located in the north-eastern area (Fig. 5.1). These two areas differ on the seasonal pattern of the precipitation. The southwestern area corresponds to the Patagonian Phytogeographical region. In this area, because of the strong influence of Pacific air masses, precipitation is mainly concentrated during winter. In contrast, the north-eastern portion of the region has a more evenly distributed precipitation regime. This area corresponds to the Monte phytogeographical region.

Table 5.2. Normalized mean square error for extrapolations based solely on *NDVI* data, and predictions based both on past values of *NDVI* and accumulated precipitation

Site	*NDVI* trained	*NDVI* + PPT trained
Leleque	0.46	0.47
El Maitén	0.45	0.46
Viedma	1.01	0.81
Esquel	0.23	0.24
San Antonio	4.58	1.39
Sierra Colorada	2.37	0.81
Ñorquinco	0.56	0.58
Maquinchao	0.94	0.94
Puesto Martínez	4.12	1.93
Fofo Cahuel	1.06	0.49

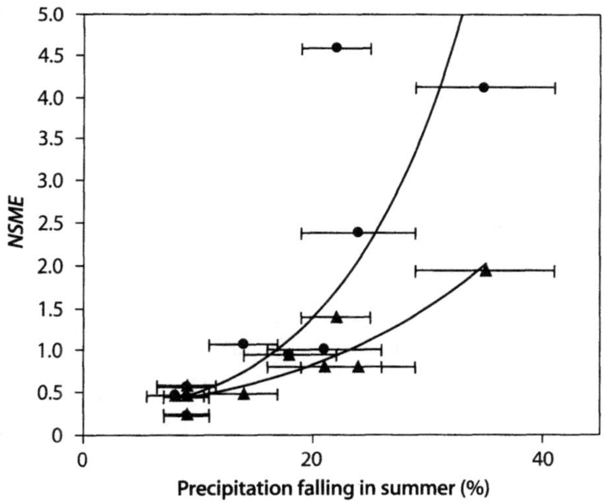

Fig. 5.3. Normalized mean square error as a function of percentage of precipitation falling in summer. *Circles* correspond to extrapolation based solely on past values of *NDVI* data, and *triangles* correspond to predictions based both on past *NDVI* data and accumulated precipitation *Solid lines* are exponential fits intended to show the general trend of the data

When ANNs are trained exclusively on past values of *NDVI* data, the *NMSE* provides a measure of the intrinsic predictability of the system. Areas showing a low *NMSE* would display a similar phenological pattern every year. Predictability is a very important attribute of the ecosystems. It would determine, for example, the kind of evolutive pressure that organisms will experience. Opportunistic strategies will be favoured in areas where the resources are not reliable in time or space. Predictability is also important for applied reasons: to define the stocking density on a given rangeland, the nutritional need of the flocks have to match as closely as possible the seasonal dynamics of forage availability. Given the same total production, the average stocking density will be higher in an environment where the timing of maximum and minimum forage availability is similar among years.

A winter concentration of precipitation seems to increase the predictability of the systems. Areas with winter precipitation in Patagonia showed a decoupling between the growing season and the wet season (Paruelo and Sala 1995). During winter, water is accumulated in the soil because transpiration losses are low. Soil water is then transferred from winter to spring. When temperature raises, this water becomes available to plants. In areas not extremely dry, the amount of water available at the beginning of the growing season is set by the holding capacity of the soil. Excess water is lost as deep drainage and/or runoff (Paruelo et al. 1998). Consequently, in areas with winter distribution of precipitation, a very stable component of the system (the soil) becomes the main control of water availability.

Predictive power increases when precipitation data are used along with *NDVI* past values as inputs. Precipitation was sampled in a 10-day period, corresponding to the sampling period of the *NDVI* data. Analysing the available data for precipitation, we observed that accumulated values are more relevant as inputs than ten-day values. When precipitation is accumulated by assigning to a given sampling period the precipitation of the past k periods, it can be seen that for accumulations of 9–10 periods (about three months) a very well defined structure appears which shows a temporal correlation with the *NDVI* series (Fig. 5.4). The cross-correlation function shows a peak for a lag of approximately 11–14 periods on the *NDVI* with respect to the accumulated precipitation.

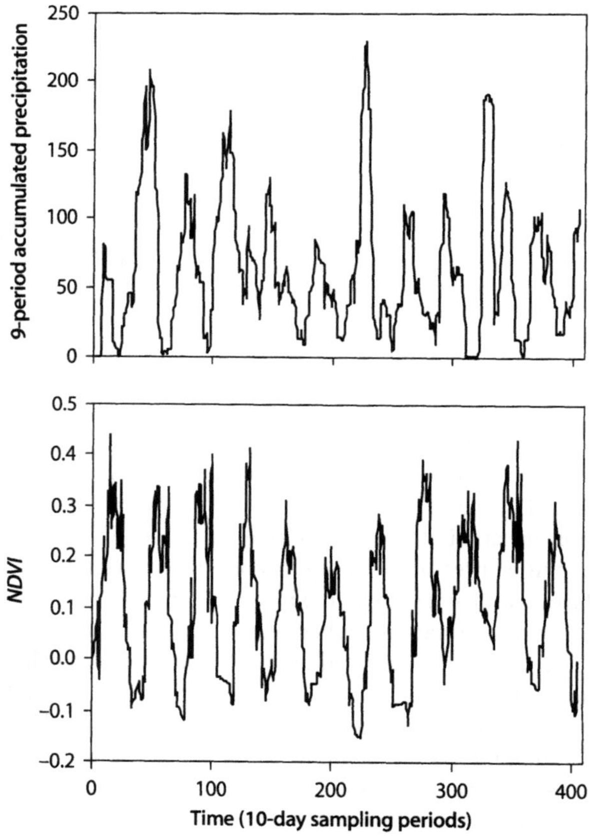

Fig. 5.4. a 9-period accumulated precipitation as a function of time for site Esquel. Values are obtained by assigning to a given sampling period the total precipitation of the past 9 periods; **b** *NDVI* data for site Esquel. Note the marked correspondence between the upper and lower parts of this figure, characterized by a lag of about 11 periods on the *NDVI* series respect to the accumulated precipitation

Augmenting the input vector with six past periods of accumulated precipitation (taken with a lag of 9 periods with respect to the predicted date) significantly improves the agreement between calculated and observed *NDVI* for most of the sites (Table 5.2, Fig. 5.4). Figure 5.5 shows a 9-step-ahead prediction for the case of sites Esquel, Leleque and Fofo Cahuel. In the case of site Fofo Cahuel, where the fraction of precipitation falling in summer is comparatively higher, the *NMSE* has been reduced by approximately 50%. Higher reductions in *NMSE* are observed in places with even lower summer precipitation, as it is the case of sites San Antonio, Sierra Colorada and Puesto Martínez (Table 5.2).

In summary, the study of the internal controls of the seasonal dynamics of the *NDVI* through ANN analysis of satellite and climatic data identified important differences between two phytogeographical areas (Patagonian and Monte steppes) and allowed for a satisfactory prediction of the *NDVI* values up to six months ahead.

ANN analysis is likely to become a valuable tool to be added to the standard toolbox of the researcher in ecological modelling. In particular, IT-trained ANNs appear as a promising approach for the analysis of time series in ecology. Preliminary results from a study we are presently undertaking also show promising results on the prediction of population time series through the use of IT-trained ANNs.

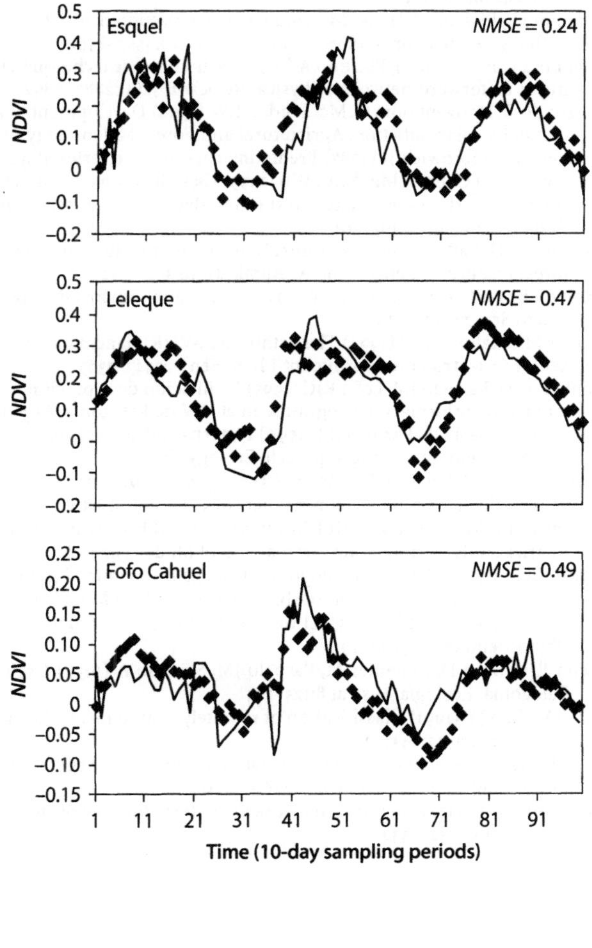

Fig. 5.5. 9-step-ahead prediction based on both past values of *NDVI* and accumulated precipitation for the case of sites Esquel, Leleque and Fofo Cahuel. Values for the *NSME* are indicated. Note that in Fofo Cahuel, where precipitation in summer is relatively higher, inclusion of precipitation data significantly improves the agreement between observed and predicted data respect to extrapolations based solely on *NDVI* data (see Fig. 5.2)

ANNs and satellite data also provide a very promising alternative for the prediction of *ANPP* and forage availability over extensive rangelands. Forecast of forage availability will provide to ranchers and natural resources managers a critical piece of information to devise sustainable systems in arid and semiarid lands.

Acknowledgements

F.G. Tomasel would like to acknowledge A. Plastino for directing his attention toward pseudoinverse training techniques. This work was funded by Fundación Antorchas (Argentina) and the Inter-American Institute for Global Change Studies (IAI – ISP III # 3 077).

References

Aguiar MR, Paruelo JM, Sala OE, Lauenroth WK (1996) Ecosystem consequences of plant functional types changes in a semiarid grassland. Journal of Vegetation Science 7:599–608

Box EO, Holben BN, Kalb V (1989) Accuracy of the AVHRR Vegetation Index as a predictor of biomass, primary productivity and net CO_2 flux. Vegetation 80:71–89

Burke IC, Kittel TGF, Lauenroth WK, Snook P, Yonker CM, Parton WJ (1991) Regional analysis of the central great plains.Bioscience 41:685–692

Derr VE, Slutz RJ (1994) Prediction of El Niño events in the Pacific by means of neuronal networks. AI Applications 8:51–63

Diambra L, Plastino A (1995) Maximum entropy, pseudoinverse techniques, and time series predictions with layered networks. Physical Review E 52(4):4557–4560

Diambra L, Fernandez J, Plastino A (1995) Pseudoinverse techniques, information theory, and the training of feeforward networks. Physical Review E 52(3):2887–2892

Elizondo D, Hoogenboom G, McClendon RW (1994) Development of a neuronal network model to predict daily solar radiation. Agricultural and Forest Meteorology 71:115–132

Farmer JD, Sidorowich JJ (1987) Predicting chaotic time series. Physical Review Letters 59:845–848.

Foody GM, McCulloch MB, Yates WB (1995) Classification of remotely sensed data by an artificial neuronal network: issues related to training data characteristics. Photogrammetric Engineering and Remote Sensing 61:391–401

Gimblett RH, Ball GL (1995) Neuronal Network architectures for monitoring and simulating changes in forest resource management. AI Applications 9:103–123

Holben B (1986) Characteristics of maximum value composite images from temporal AVHRR data. Int J Rem Sen 7:1417–1434

James ME, Kalluri SNV (1994) The pathfinder AVHRR land data set: an improved coarse resolution data set for terrestrial monitoring. Int J Rem Sen 15:3347–3363

Jobbágy EG, Paruelo JM, León RJC (1995) Estimacion de la precipitacion y de su variabilidad interanual a partir de informacion geografica en el NW de Patagonia, Argentina. Ecología Austral 5:47–53

Jobbágy EG, Sala OE, Paruelo JM (1999) Patterns and controls of primary production in the Patagonian steppe: A remote sensing approach. Ecology (in press)

Kanellopoulos I, Varfis A, Wilkinson GG, Megier J (1992) Land-cover discrimination in SPOT HRV imagery using an artificial neuronal network–a 20-class experiment. Int J Rem Sen 13:917–924

Lauenroth WK, Hunt HW, Swift DM, Singh JS (1986) Estimating aboveground net primary productivity in grasslands: A simulation approach. Ecol Model 33:297–314

Lek S, Delacoste M, Baran P, Dimopoulos I, Lauga J, Aulagnier S (1996) Application of neuronal networks to modeling nonlinear relationships in ecology. Ecol Model 90(1):39–52

León RJC, Aguiar MR (1985) El deterioro por uso pasturil en estepas herbaceas patagonicas. Phytocoenologia 13:181–196

León RJC, Bran D, Collantes M, Paruelo JM, Soriano A (1998) Grandes unidades de vegetación de la Patagonia. Ecología Austral 8:125–144

Liu ZK, Xiao JY (1991) Classification of remotely sensed image data using artificial neuronal networks. Int J Rem Sen 12:2433–2438

Lloyd D (1990) A phenological classification of terrestrial vegetation cover using shortwave vegetation index imagery. Int J Rem Sen 11:2269–2279

McCann DW (1992) A neuronal network short-term forecast of significant thunderstorms. Weather and Forecasting 7:525–534

McNaughton SJ, Oesterheld M, Frank DA, Williams KJ (1989) Ecosystem-level patterns of primary productivity and herbivory in terrestrial habitats. Nature 341:142–144
McNaughton SJ, Sala OE, Oesterheld M (1993) Comparative ecology of African and South American arid to subhumid ecosystems. In: Goldblatt P (ed) Biological relationships between Africa and South America. Yale University Press, New Haven, pp 548–567
Monteith JL (1981) Climatic variation and the growth of crops. Quarterly Journal of the Royal Meteorological Society 107:749–774
Oesterheld M, Sala OE, McNaughton SJ (1992) Effect of animal husbandry on herbivore-carrying capacity at a regional scale. Nature 356:234–236
Oesterheld M, DiBella CM, Kerdiles H (1998) Relation between NOAA-AVHRR satellite data and stocking rate in grasslands. Ecological Applications 8:207–212
Paruelo JM, Lauenroth WK (1995) Regional patterns of *NDVI* in North American shrublands and grasslands. Ecology 76:1888–1898
Paruelo JM, Sala OE (1995) Water losses in the Patagonian Steppe: A modeling approach. Ecology 76:510–520
Paruelo JM, Tomasel F (1997) Prediction of functional characteristics of ecosystems: A comparison of artificial neuronal networks and regression models. Ecol Model 98:173–186
Paruelo JM, Aguiar MR, Leon RJC, Golluscio RA, Batista WB (1991) The use of satellite imagery in quantitative phytogeography: A case study of Patagonia (Argentina). In: Crovello TJ, Nimis PL (eds) Quantitative approaches to phytogeography. Kluwer Academic Publishers, The Hague,pp 183–204
Paruelo JM, Epstein HE, Lauenroth WK, Burke IC (1997) ANPP estimates from *NDVI* for the central grassland region of the US. Ecology 78(3):953–958
Paruelo JM, Beltrán AB, Jobbagy EG, Sala OE, Golluscio RA (1998) The climate of patagonia: General patterns and controls on biotic processes. Ecología Austral 8:89–101
Perelman SB, León RJC, Bussacca JP (1997) Floristic changes related to grazing intensity in a Patagonian shrub steppe. Ecography 20:400–406
Potter CS, Randerson JT, Field CB, Matson PA, VitousekPM, Mooney HA, Klooster SA (1993) Terrestrial ecosystem production: A process model based on global satellite and surface data. Global Biogeochemical Cycles 7:811–841
Prince SD (1991a) Satellite remote sensing of primary production: comparison of results for Sahelian grasslands 1981–1988. Int J RemSen 12:1301–1311
Prince SD (1991b) A model of regional primary production for use with coarse resolution satellite data. Int J Rem Sen 12: 1313–1330
Prohaska F (1976) The climate of Argentina, Paraguay and Uruguay. In: Schwerdtfeger E (ed) Climate of Central and South America. Elsevier, Amsterdam, pp 57–69
Recknagel F, French M, Harkonen P, Yabunaka K (1997) Artificial neuronal network approach for modeling and prediction of algal blooms. Ecol Model 96(1–3):11–28
Ruimy A, Saugier B, Dedieu G (1994) Methodology for the estimation of terrestrial net primary production from remotely sensed data. J Geophy Res 99:5263–5283
Sala OE, Biondini ME, Lauenroth WK (1988) Bias in estimates of primary pruduction: An analytical solution. Ecol Model 44:43–55
Sala OE, Lauenroth WK, Golluscio RA (1997) Plant functional types in temperate arid regions. In: Smith TM, Shugart HH, Woodward FI (eds) Plant functional types. Cambridge University Press, Cambridge, pp 217–233
Sellers PJ, Berry JA, Collatz GJ, Field CB, Hall FG (1992) Canopy reflectance, photosynthesis, and transpiration. III. A reanalysis using improved leaf models and a new canopy integration scheme. Remote Sens Environ 42:187–216
Soriano A (1983) Deserts and semi-deserts of Patagonia. In: West NE (ed) Ecosystems of the world: Temperate deserts and semi-deserts. Elsevier, Amsterdam, pp 423–460
Soriano A, Movia CP (1987) Erosión y desertización en Patagonia. Interciencia 11:77–83
Soriano A, Paruelo JM (1990) El manejo de campos de pastoreo en Patagonia: Aplicación de principios ecológicos. Ciencia Hoy 2:44–53
Tucker CJ, Newcomb WW (1994) AVHRR data sets for determination of desert spatial extent. Int J Rem Sen 15:3547–3565
Tucker CJ, Vanpraet CL, Sharman MJ, van Ittersum G (1985) Satellite remote sensing of total herbaceous biomass production in the Senegalese Sahel: 1980–1984. Remote Sens Environ 17:233–249
Turchin P, Taylor AD (1992) Complex dynamics in ecological time series. Ecology 73(1):289–305
Wan EA (1994) Time series prediction by using a connectionist network with internal delay lines. In: Weigend AS, Gershenfeld NA (eds) Time series prediction: Forecasting the future and understanding the past. Addison-Wesley Pub. Co., Reading

Chapter 6

On the Probabilistic Interpretation of Area Based Fuzzy Land Cover Mixing Proportions

J. Manslow · M. Brown · M. Nixon

6.1 Introduction

Techniques traditionally used to extract land cover information from remotely sensed images have tended to produce crisp (or hard) classifications of image pixels. This has been criticised, however, since the resulting maps of ground cover consist of grids of pixels of homogeneous class membership, and are hence inherently dissimilar to the true ground cover which they intend to model (Fisher 1997; Foody 1997b; Cracknell 1998). Much effort has been made to increase the richness of such pixel based classifications by, for example, relating the probability of class membership of pixels in particular classes to the sub-pixel area occupied by those classes (Foody 1996a).

An alternative approach, which is discussed in this paper, is to represent the composition of pixels by the proportions of the sub-pixel area occupied by each cover class, a process sometimes referred to as fuzzy classification (Kent and Mardia 1988; Wang 1990). Such sub-pixel area proportion estimates are highly desirable, since not only do they more accurately represent true ground cover than either crisp classifications or probability estimates, but also that many applications have a specific interest in the area of land cover types.

Although sub-pixel area proportion estimation is conventionally performed using parametric and often linear models (Horwitz et al. 1971; Settle and Drake 1993), some recent studies have used advanced semiparametric nonlinear models such as neuronal networks (Foody 1997a). Despite the performance improvements resulting from the application of such advanced methods, there is some confusion in the literature about the relationship between crisp pixel classification and fuzzy pixel classification. This paper introduces a probabilistic interpretation of sub-pixel area proportions and uses it to show that the cross entropy function commonly used for crisp pixel classification may also be used for obtaining maximum likelihood fuzzy classifications. In addition, it is argued that the posterior probabilities of class membership used for crisp pixel classifications are not, in general, optimal fuzzy classifications.

Section 6.2.1 describes the way in which sub-pixel area proportions can be given a probabilistic interpretation. Some practical and theoretical implications of this interpretation are discussed in Section 6.2.2. In particular, notation for representing area proportions which has a strong analogy to that used for probabilities is introduced and used to describe the main properties of area proportions. The probabilistic interpretation is also used to show that maximum likelihood estimates of sub-pixel area proportions over a set of exemplars minimize the cross entropy error function over that set. A simple analysis of the relationship between posterior probabilities and optimal sub-pixel area proportion estimates is presented and used to show that they are not, in general, equal. The lower bound on the expected cross entropy error is described and related to the problem of spectral confusion and the Bayes' error rate of traditional classifiers. Section 6.3 presents the results of a series of experiments to compare the performance of networks trained using the traditional sum of squares error function and the cross entropy error function at estimating sub-pixel area proportions for a real world remotely sensed data set. Finally, the results of an experiment to assess the performance of a network trained to estimate posterior probabilities of class membership is assessed for the sub-pixel area proportion estimation problem.

6.2 Conceptual Classification

In order to estimate the proportion of a pixel's area occupied by a class it must be possible, in principle, to measure the area of the class given perfect information. Figure 6.1 shows a single pixel consisting of two cover types, grass and water. When the land area is remotely observed, a mixed pixel, shown at the bottom of the figure, is generated which has spectral contributions from both of the sub-pixel classes. If perfect information were available in the form of the true distribution of the two cover types within the pixel area, each point within the pixel could be uniquely classified as belonging to one of the cover types, and hence a sub-pixel map of true class membership could conceptually be constructed.

It has been argued that there are classes which, due to their tendency to continuously intergrade, prevent points within the sub-pixel map from being assigned unambiguously to a single class (Wood and Foody 1989). Clearly, if this is accepted, no boundary can be defined around such classes which makes the division between the set of points which are members of the class, and those which are not. It is thus not possible to measure the area of such classes given perfect information and hence their area must be treated as undefined.

6.2.1 The Probabilistic Interpretation of Sub-Pixel Area Proportions

If a point is chosen at random from a uniform distribution over the conceptual sub-pixel cover map described above, it will fall within a region occupied by one of the sub-pixel classes. In the limit of an infinite number of such points being chosen, the proportion of points falling within each class region will be equal to the proportion of the sub-pixel area the region occupies, and also equal to the probability of an individual point falling within each region. This suggests that there is a direct equivalence between these probabilities and the sub-pixel area proportions.

It is important to emphasise that this probabilistic model does not equate the proportion of the sub-pixel area occupied by a specific class with the posterior probability of class membership of the entire pixel in that class, as would be estimated by most classical classification algorithms. Although estimates of these probabilities have been used to model sub-pixel area proportions (see, for example, Foody 1996b; Masselli et al. 1996; Canters 1997; Gorte and Stein 1998), it can be shown that they are not, in general, optimal estimates. This issue is discussed in greater detail in the Section 6.2.2.3.

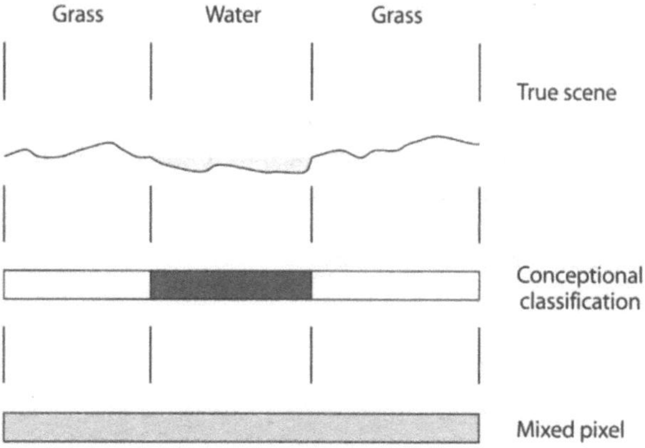

Fig. 6.1. The relationship between the real world land cover, the conceptual classification and the mixed pixel

6.2.2
Implications of the Probabilistic Interpretation

The following sections describe some of the implications of the probabilistic interpretation. Section 6.2.2.1 introduces notation for representing and manipulating sub-pixel area proportions which is analogous to the notation used to represent probabilities. This is then used to list the axioms governing sub-pixel area proportions that, once again, are analogous to their probabilistic equivalents. Section 6.2.2.2 shows how the probabilistic interpretation can be used to show that maximum likelihood sub-pixel area proportion estimates for a set of exemplar pixels minimize the cross entropy error over the set. Section 6.2.2.3 analyses the relationship between posterior probabilities of membership of pixels in classes and optimal sub-pixel area proportions and argues that the two quantities are, in general, not equal. Finally, Section 6.2.2.4 discusses the lower bound of the cross entropy error function and relates it to the phenomenon of spectral confusion.

6.2.2.1
Area Proportions: Notation and Properties

In order to describe the properties of area proportions, it is convenient to introduce a compact notation: if the area of a pixel P is represented by $\mu(P)$ and the area of the intersection of pixel P and class C_n by $\mu(C_n, P)$, then the proportion of P occupied by C_n will be denoted by $\mu(C_n|P)$. Here, the equivalence of area proportions and (conditional) probabilities is made explicit in the choice of notation. The proportion of P occupied by class C_n is found using

$$\mu(C_n|P) = \frac{\mu(C_n, P)}{\mu(P)} \tag{6.1}$$

From Eq. 6.1 the area proportion equivalent of Bayes' theorem may be derived. This can be used to convert quantities of the form 'the proportion of class C_n occupied by pixel P' to 'the proportion of pixel P occupied by class C_n' as follows:

$$\mu(C_n|P) = \frac{\mu(P|C_n)}{\mu(P)}\mu(C_n)$$

Clearly, the total area occupied by any object or class is found by summing the areas of its intersections with other classes. Thus, the total areas of a pixel P (where there are N classes that form a closed world partition) and of a class C_n (where '*allP*' is the set of all pixels) are given by

$$\mu(P) = \sum_{n=1}^{N} \mu(C_n, P)$$

$$\mu(C_n) = \sum_{allP} \mu(C_n, P)$$

when classes and pixels are nonintersecting. For two classes, C_n and C_m with m, $n \in [1, N]$ the area of their union may be computed from the sum of their individual areas minus the area of their intersection. More concisely,

$$\mu(C_n \cup C_m|P) = \mu(C_n|P) + \mu(C_m|P) - \mu(C_n, C_m|P)$$

A set of classes C_n: $1 \le n \le N$ is considered to be closed world upon the target domain D if

$$\mu\left(\bigcup_{n=1}^{N} C_n|P\right) = 1\,, \quad \forall P \in D$$

Such a set of classes may trivially be constructed by the addition of a class that contains any sub-pixel region that is not assigned to any other class. Unless otherwise stated, this condition will be assumed to be satisfied throughout this paper. Finally, area proportions lie in the closed interval [0, 1] as can be seen from Eq. 6.1. Thus,

$$\mu(C_n|P) \in [0, 1]\,, \ \forall n : 1 \le n \le N$$

All of these axioms are directly equivalent to those for manipulating probabilities (as given in Cox 1946; DeGroot 1989).

6.2.2.2
Maximum Likelihood Sub-Pixel Area Proportion Estimates and the Cross Entropy Function

Data driven models such as neuronal networks typically contain a number of parameters which may be found by minimizing some error function which quantifies the difference between the model's actual behaviour and its desired behaviour as given by a set of exemplars (Haykin 1994; Bishop 1995). Although the cross entropy function is the clear choice for pixel classification (and, in fact, classification in general, see Bishop 1995), it is less obvious as to which error function should be used to develop sub-pixel area proportion estimation models. This section uses the probabilistic interpretation to show that the cross entropy function is appropriate for sub-pixel area proportion estimation as well as pixel classification.

Consider the set of N classes, C_n: $1 \leq n \leq N$ where C_n is the label of the nth class. If this set is closed world and the classes are mutually exclusive, the probability distribution of class memberships obtained from random samplings of the conceptual sub-pixel map will be multinomial, as given below:

$$p(\underline{C}|x,y) = \prod_{n=1}^{N} p(C_n|x,y)^{C_n(x,y)}$$

where $\underline{C}$ is a vector indicating the class membership of the sub-pixel point (x, y). For example, if $(x, y) \in C_n$ then $(\underline{C}|x,y)$ has a one in the nth position and zeros in all others. $p(C_n|x, y)$ is the posterior probability of membership of the sub-pixel point (x, y) in class C_n. The probability of M such points having class membership $\underline{C}$, where $\underline{C}$ is now a matrix of M rows of vectors each indicating the class membership of one of the M points is given by:

$$p(\underline{C}|M) = \prod_{m=1}^{M} \prod_{n=1}^{N} p(C_n|x_m, y_m)^{C_n(x_m,y_m)}$$

A neuronal network would typically be trained to classify such a set of points by using the maximum likelihood procedure. That is, the network would model the distribution parameters $p(C_n|x_m, y_m)$ so as to maximize the probability that the distribution would reproduce the set of training patterns. Using the probabilistic interpretation, the distribution parameters are equal to the sub-pixel area proportions such that:

$$p(\underline{C}|M) = \prod_{m=1}^{M} \prod_{n=1}^{N} \hat{\mu}(C_n|P)^{C_n(x_m,y_m)} \tag{6.2}$$

where the $\mu(C_n|P)$ are the neuronal network estimates of the distribution parameters. In other words, given a set of M sub-pixel points of class membership $C_n(x_m, y_m)$, maximum likelihood sub-pixel area proportion estimates may be obtained finding the area proportions which maximize Eq. 6.2. It is, however, possible to go further than this by taking the product over M inside the power, and letting M become infinitely large:

$$p(\underline{C}|M) = \prod_{n=1}^{N} \hat{\mu}(C_n|P)^{M\mu(C_n|P)} \tag{6.3}$$

This makes it possible to simulate the effect of training a neuronal network on an infinitely large number of sub-pixel samples. To find the maximum likelihood sub-pixel area proportion estimates, it is convenient to minimize the negative logarithm of the likelihood given in Eq. 6.3 rather than maximize the likelihood itself. The negative log-likelihood is given by:

$$-\ln\left|p(\underline{C}|M)\right| = -\sum_{n=1}^{N} M\mu(C_n|P)\ln\hat{\mu}(C_n|P) \tag{6.4}$$

The multiplicative constant M is independent of the distribution parameters and hence does not change the set of parameters that maximize the likelihood. For this reason the M term may be ignored when maximizing Eq. 6.4. The problem of finding the sub-pixel area proportions that maximize the likelihood can therefore be summarized as finding the $\mu(C_n|P)$ which minimize:

$$E = -\sum_{n=1}^{N} \mu(C_n|P)\ln\hat{\mu}(C_n|P)$$

which is clearly the same problem as minimizing the cross entropy error between the true and estimated sub-pixel area proportions. Multiple exemplar pixels may be easily accommodated by accumulating the expected error over the set of exemplars. Note that although Foody (1995, 1996b) suggests that the cross entropy function may be suitable for use in sub-pixel area proportion estimation, the discussions focus on the interpretability of the resulting error measure and its relation to information theory and do not derive the cross entropy function for the sub-pixel area estimation problem from first principles as is done here.

The outputs of networks that estimate the parameters of a multinomial distribution, should, due to the closed world assumption, be constrained by the softmax function (Bishop 1995; Dunne and Campbell 1997). That is,

$$\hat{\mu}_m = \frac{\exp(f_m)}{\sum_{n=1}^{N}\exp(f_n)}$$

where the summation is over the N classes, and the f_m and f_n are the pre-softmax output neuron activations. This has the dual benefits of incorporating a priori knowledge about the normalization of the sub-pixel area proportions, and avoids the trivial minimum of the cross entropy function which occurs when:

$$\hat{\mu}_n = 1,\ \forall n \in [1, N]$$

6.2.2.3
The Relationship between Posterior Probabilities and Optimal Sub-Pixel Area Proportion Estimates

In Section 6.2.1 it was suggested that it was generally inappropriate to use the posterior probabilities of class membership of pixels as estimates of the proportions of sub-pixel areas occupied by classes. This section presents a more detailed discussion of this issue and provides a simple illustration of the nonoptimality of posterior probabilities as sub-pixel area proportion estimates.

$$E(\underline{s}) = -\sum_{n=1}^{N} (\ln \hat{\mu}_n) \int \mu_n p(\mu_n | \underline{s}) d\mu_n \qquad (6.5)$$

It can be shown that the vector of sub-pixel area proportion estimates μ which minimizes the expected cross entropy error over some distribution of true proportions $p(\underline{\mu}|\underline{s})$ for some spectral measurement $\underline{s}$ (as given in Eq. 6.5) is equal to the mean of the $p(\underline{\mu}|\underline{s})$ distribution (given in Eq. 6.6) when the μ vector is constrained to be of unit length. For the purpose of the discussion that follows, the optimal sub-pixel area proportion estimates are defined in this way, i.e. as those that minimize Eq. 6.5.

$$\hat{\underline{\mu}} = \int \underline{\mu} p(\underline{\mu}|\underline{s}) d\underline{\mu} \qquad (6.6)$$

When pixels are classified according to the proportions of sub-pixel cover, the classification rule can be described using the vector of conditional probabilities $p(\underline{C}|\underline{\mu})$ where each element in the vector is the conditional probability of one of the target classes. With this term, it is possible to decompose the posterior probability of class membership of pixels of spectral signature $\underline{s}$ to explicitly represent its construction from sub-pixel area proportions:

$$p(C_n|\underline{s}) = \int p(C_n|\underline{\mu}) p(\underline{\mu}|\underline{s}) d\underline{\mu} \qquad (6.7)$$

Equations 6.6 and 6.7 show that the optimal sub-pixel area proportion estimates and the posterior probabilities of class membership of pixels of spectral signature $\underline{s}$ are only guaranteed to be equal for arbitrary choices of $p(\underline{\mu}|\underline{s})$ if,

$$\underline{\mu} = p(\underline{C}|\underline{\mu}), \ \forall \underline{\mu} \in [0,1]^N \qquad (6.8)$$

If classification is based on sub-pixel area proportions, and is unambiguous given those proportions, $p(\underline{C}|\underline{\mu})$ will always have a one in the nth position where $1 \leq n \leq N$ and zeros in all others, and hence cannot satisfy Eq. 6.8. This shows that the posterior probability of the membership of pixels in classes where class membership can be determined unambiguously from sub-pixel area proportions cannot be guaranteed to equal the optimal sub-pixel area proportion estimates for arbitrary choices of $p(\underline{\mu}|\underline{s})$.

It is interesting to note that the equivalence does hold for special forms of $p(\underline{\mu}|\underline{s})$. One such form occurs when all pixels with spectral signature $\underline{s}$ are pure (consist of a single sub-pixel cover class). Under these circumstances, $p(\underline{\mu}|\underline{s})$ is zero except when $\underline{\mu}$ has a one in the nth position where $1 \leq n \leq N$, and zeros in all others. In general, the form of $p(\underline{\mu}|\underline{s})$ is determined by the properties of the target classes and their mixtures, and is unlikely to satisfy Eq. 6.8.

6.2.2.4
Expected Generalization Performance: The Upper Bound

When a model produces the optimal sub-pixel area proportion estimates given in Eq. 6.6, the expected cross entropy error of Eq. 6.5 is not zero, but is equal to the quantity given by Eq. 6.9. This is the lower limit of the cross entropy error function for sub-pixel area proportion estimators and is analogous to the Bayes' error rate of classifiers (which is discussed in Ripley 1996). It can be shown, for pixels of spectral signature $\underline{s}$, to be a monotonically increasing function of the variance of the conditional distribution of area proportions $p(\mu_n|\underline{s})$.

$$E_{\min}(\underline{s}) = -\sum_{n=1}^{N}\left(\ln \int \mu_n p(\mu_n|\underline{s})d\mu_n\right) \int \mu_n p(\mu_n|\underline{s})d\mu_n \tag{6.9}$$

This limit, which is dependent upon the choice and definition of the target classes and the set of measurements from which they are modelled, increases with the level of spectral confusion, and is particularly important, since it indicates the maximum performance that a sub-pixel area proportion estimation algorithm can achieve.

6.2.3
Summary

The practical implications of the results in the previous sections may be summarized as follows: in order to construct an empirical model for sub-pixel area proportion estimation, it is necessary to collect a set of exemplars consisting of the true sub-pixel area proportions of a closed world set of classes and their associated predictors (usually pixel spectra). Note that if the set of exemplars consists entirely of pure pixels, the model will estimate the posterior probabilities that pixels are pure rather than sub-pixel area proportions. An empirical model should be chosen with outputs which are constrained by the softmax function, and the model's parameters found by minimizing the cross entropy error function over the set of exemplars. The upper bound on the performance of a sub-pixel area proportion estimator is given by the minimum expected cross entropy error (given in Eq. 6.9) and is a monotonically increasing function of the variance of the conditional distribution of sub-pixel area proportions $p(\underline{\mu}|\underline{s})$ and hence also the level of spectral confusion.

6.3 Sub-Pixel Area Proportion Estimation on the FLIERS Project

The FLIERS (Fuzzy Land Information from Environmental Remote Sensing) project is a European Union funded research project which aims to develop novel techniques for land cover mapping using remotely sensed imagery. The main emphasis of this research is the investigation of the potential of advanced modelling algorithms such as neuronal networks to extract the relationship between pixel spectra and the associated sub-pixel cover. For this purpose, neuronal networks are considered particularly promising since they make relatively weak assumptions about the nature of the relationship and require only examples of pixel spectra and associated sub-pixel cover proportions from which the relationship may be inferred.

6.3.1 The Data

The data used in this paper was generated as part of the FLIERS project and covers the Stoughton area (near Leicester, UK), which consists mainly of large scale agriculture, and is shown in Fig. 6.2. Although the original data contained as many as 26 classes, many were conflated to reduce this to only the four classes of 'built areas' (class 1), 'grasses' (class 2), 'other' (class 3) and 'crops' (class 4) used in this paper. The properties of the data set are summarized in Table 6.1.

Of the four classes, 'built areas' (consisting of asphalt, concrete and other construction materials) were the rarest, having an average sub-pixel membership of only 0.029, and being completely absent from 20 244 of the 22 000 pixels available. 'Crops' and 'grasses' on the other hand were common, with average sub-pixel memberships of 0.58 and 0.28 respectively and a significant number of pure pixels were present. Finally,

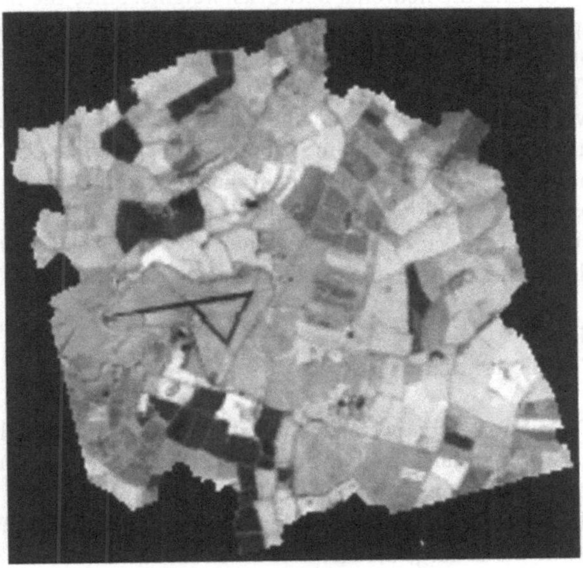

Fig. 6.2. The data set

Table 6.1. Data set summary

Class name	Mean pixel membership	Pure exemplars	Absent exemplars
Built areas	0.0286	66	20244
Grasses	0.2840	3631	11919
Other	0.0774	261	15496
Crops	0.5787	9482	7407

the 'other' class was present in all but 15 496 of the pixels with an average sub-pixel membership of 0.077. Since the 'other' class is a conflation of other relatively loosely related cover classes, it is expected to show significant spectral variation, making the sub-pixel area proportion estimation problem particularly difficult. Unfortunately, the inclusion of the class is necessary to ensure that the sub-pixel proportions always sum to unity – a prerequisite for the use of the cross entropy function.

6.3.2 The Neuronal Networks

In all experiments described in this paper, MLPs (multilayer perceptrons) were constructed using specially written C++ code and trained using the stochastic back propagation algorithm. All hidden neurons had logistic activation functions and all output neurons had softmax activation functions regardless of the error function used, so as to isolate the effect of changing the error function. The available data was divided into three sets: the training set, the test set and the validation set. The training set constituted 60% of all the data and was used with the back propagation algorithm to directly find the neuronal networks' weights. Since the neuronal network weights are specifically tailored to minimize the training set error, the error provides a negatively biased (overly optimistic) estimate of the true performance of the neuronal network. To overcome this, the networks' performances on a test set of 20% of the total data set were periodically assessed and used to estimate the point during training when the network offered maximum generalization performance. Once these 'optimal' networks had been identified, their generalization performances were re-assessed using the validation set which was also made up of 20% of the original data set. This final evaluation is necessary because the network weights are indirectly tailored specifically to the test set through the use of that set to decide when to stop training.

6.3.3 The Experiments

The first set of experiments was designed to compare the performance of neuronal networks trained using the cross entropy and sum of squares error functions on the area proportion estimation problem. For this purpose, it was necessary to preprocess the available data to remove all pure pixels, since it was found that the number of pure pixels was so large that the networks could achieve low error rates by learning to identify pure pixels even if sub-pixel area proportion estimates were poor for mixed pixels. Once the networks were retrained on mixed pixels only, their overall sub-pixel area proportion estimation performance improved in those pixels where significant mixing occurred. Note that this is only acceptable because of the specific interest in the performance of the models at estimating sub-pixel area proportions in mixed pixels.

The correlation between the validation set targets (the true sub-pixel cover proportions) and the proportions predicted by the neuronal networks was chosen as the measure of the networks' performances. This choice was motivated by the relatively widespread of the use of this measure in the remote sensing literature. Although many other measures could have been used, the appropriate performance measure is highly application dependent, and each network will always outperform the other on some of the measures. For this reason, the relative performance of the models was not evaluated on a large suite of alternative measures. The second set of experiments was designed to compare the relative performances of networks trained on pure pixels (which estimate the posterior probabilities that pixels are exemplars of pure pixels of each class) with networks trained only on mixed pixels. Once again, the performance of the models was assessed by measuring the correlation between the networks' predictions and the true sub-pixel area proportions.

6.3.4 Results

Initially, ten networks, five with each of the two error functions and two hidden neurons, were trained on data consisting of mixed pixels only. The typical correlation between the sub-pixel area proportions predicted by the networks and the true sub-pixel area proportions for an unseen validation set are shown in Fig. 6.3. Generally, the cross entropy trained networks offered higher performance on the built and other classes, slightly poorer performance on the grass class and essentially the same performance as the sum of squares trained networks on the crops class. The reason for this is that the built and other classes form very small sub-pixel areas in many pixels,

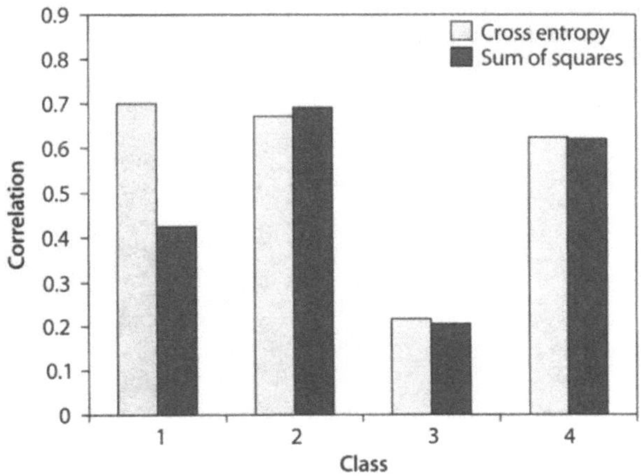

Fig. 6.3. Correlation between estimated and true proportions for 2 hidden neurons

resulting in a tendency for networks to predict small values for the sub-pixel membership of these classes in all pixels. The derivatives of the cross entropy function with respect to the model predictions are equal to $\mu(C|P) / \hat{\mu}(C|P)$ (where $\mu(C|P)$ is the true sub-pixel area proportion and $\hat{\mu}(C|P)$ is the neuronal network estimate) as compared to $\mu(C_n|P) - \hat{\mu}(C|P)$ for the sum of squares error. Thus, training patterns where the training target is large, but the model prediction is very small (which tend to be relatively common for the built and other classes) dominate learning in gradient based algorithms which use the cross entropy error function. More computational resources in the cross entropy trained networks are therefore dedicated to predicting the sub-pixel area proportions of the built and other classes than in the sum of squares trained networks, resulting in better performance on these classes. Since the computational resources of a network are fixed, this focusing of resources on the built and other classes would be expected to degrade performance on other classes, and this is indeed what is observed with the grass class.

A further ten networks (again, five with each of the error functions) with ten hidden neurons were trained to see how the increased flexibility of the networks changed the effect of the choice of error function. Once again, the correlation between the predicted and true sub-pixel area proportions were used to assess the models' performances and a typical set of results is shown in Fig. 6.4. The most obvious feature of this figure is that the apparent performance advantage of the cross entropy function trained networks over the sum of squares trained networks on the built and other classes has all but disappeared. It can be shown theoretically that infinitely flexible networks trained using the cross entropy and sum of squares functions produce the same models and would hence have identical performance (Bishop 1995). It would therefore be expected that increasing a network's complexity would reduce the performance advantage observed in using the cross entropy function for training. Overall, the networks consisting of ten hidden neurons tended to produce better performance on both the training data and the independent validation data than the networks with only two hidden neurons.

In the final experiment, ten neuronal networks (five with each error function) with ten hidden neurons each were trained on only pure pixels and applied to mixed pixels

Fig. 6.4. Correlation between estimated and true proportions for 10 hidden neurons

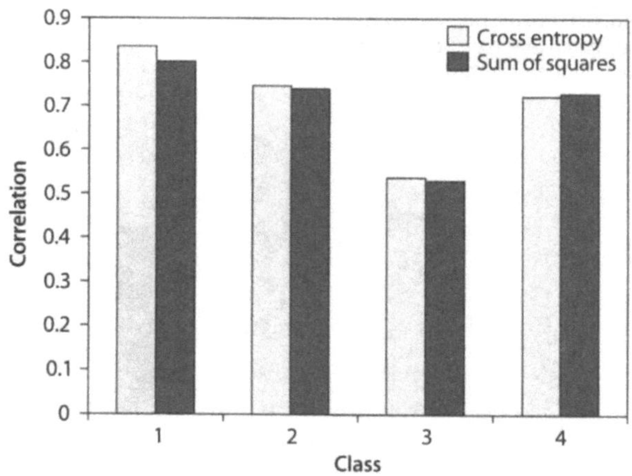

in an attempt to predict the sub-pixel area proportions. Typical correlation of the sub-pixel area proportion estimates produced by networks trained on only pure pixels with the true sub-pixel area proportions (shown in Fig. 6.5 for a cross entropy network with ten hidden neurons) were, as expected, consistently lower than those achieved by networks trained on mixed pixels, ranging from between 0.67 and 0.40 as compared with 0.82 and 0.52. Note that some positive correlation is expected since any network which achieved high correlation on the training set of pure pixels should also achieve high correlation on similar pixels in the independent validation set.

A scatter plot of the network predictions against true sub-pixel proportions (as shown in Fig. 6.6) reveals a phenomenon noted by other authors (Foody 1996a; Bastin 1997). Specifically, the predicted sub-pixel area proportions tend to cluster around the extremes of zero and one, lending the scatter plot almost a noisy sigmoidal shape. This phenomenon is caused by the relative positions of the clusters of pure pixels in the network's input space. Clearly for this data set, the clusters of pure pixels are well sepa-

Fig. 6.5. Correlation between posterior probability estimates and true sub-pixel proportions

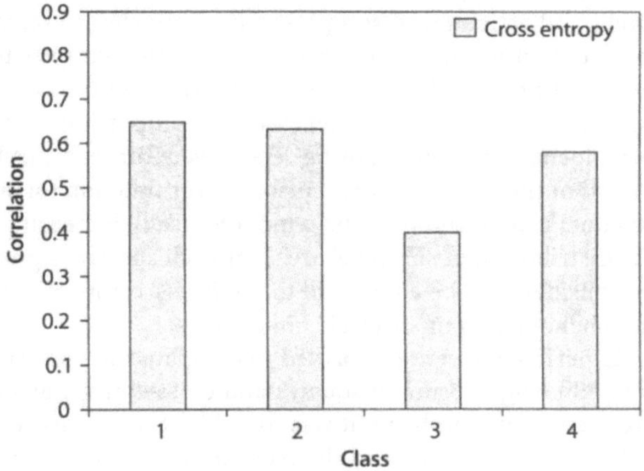

Fig. 6.6. Scatter plot of posterior probability estimates against true sub pixel proportions (◆: cross entropy)

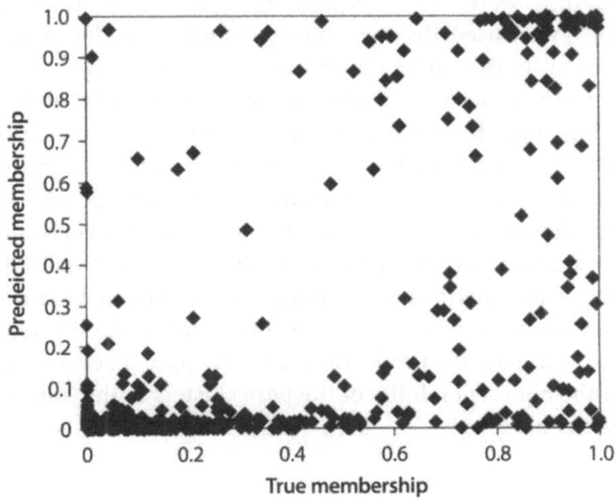

rated in the regions where mixing occurs and hence most mixed pixels in these regions are classified as pure exemplars of one of the target classes with high probability. This may not be so with other data sets: if there is significant spectral confusion between pure exemplars, networks trained on those exemplars will estimate lower posterior probabilities for each of the target classes for much of the input space, and hence produce mixture estimates less tightly clustered around zero and one. Using less flexible networks would produce a similar effect.

6.4 Conclusion

This paper describes the equivalence of sub-pixel area proportions of classes and the posterior probabilities that a sub-pixel point chosen at random from a uniform distribution over the pixel area belongs to each class. This equivalence is used to establish notation and axioms for representing and manipulating sub-pixel area proportions by analogy with the axioms of probability theory, and to show that maximum likelihood estimates of sub-pixel area proportions are, given a set of exemplars, obtained by minimizing the cross entropy function over the set. It is also shown that posterior probabilities of class membership of pixels are not, in general, equal to the optimal estimates of sub-pixel class area proportions if it is assumed that class membership may unambiguously be determined from sub-pixel area proportions. The lower bound of the expected cross entropy error function for sub-pixel area proportion estimators can be shown to be a monotonically increasing function of the variance of the distribution of sub-pixel areas for each spectral signature. This lower bound may be considered to be equivalent to the Bayes' error rate of classifiers, and is related to the phenomenon of spectral confusion.

Experiments were performed to compare the performance of neuronal networks trained using the sum of squares and cross entropy error functions on the sub-pixel area estimation problem. It was found that the measure of correlation between each network's predictions and the true sub-pixel area proportions showed only limited sensitivity to the choice of error function when flexible networks (with ten hidden neurons) were used. However, when less flexible networks (with only two hidden neurons) were used, the cross entropy function produced networks which made sub-pixel area proportion predictions which were more highly correlated with the true proportions than those of the sum of squares trained networks. This effect was particularly pronounced for those classes that formed only a small proportion of the sub-pixel area in a large number of pixels. Networks trained on pure pixels were shown to produce sub-pixel area proportion estimates which, while positively correlated with the true sub-pixel cover proportions, were more weakly correlated than predictions made by networks trained directly on mixed pixels. The distribution of the sub-pixel area proportion predictions on the validation set is related to the relative positions of the clusters of pure pixels in the space of spectral measurements. Specifically, the tendency to over estimate the proportion of nearly pure pixels in the validation set was the result of the high seperability of the pure clusters in the training set with respect to the networks used.

References

Bastin L (1997) Comparison of fuzzy c-means classification, linear mixture modelling and MLC probabilities as tools for unmixing coarse pixels. Int J Rem Sen 18(17):3629–3648

Bishop C (1995) Neuronal networks for pattern recognition. Oxford University Press, Oxford

Canters F (1997) Evaluating the uncertainty of area estimates derived from fuzzy land-cover classification. Photogrammetric Engineering and Remote Sensing 63(4):403–414

Cox RT (1946) Probability, frequency and reasonable expectation. Am J Phy 14(1):1–13

Cracknell AP (1998) Synergy in remote sensing – what's in a pixel? Int J Rem Sen 19(11):2025–2047

DeGroot MH (1989) Probability and statistics. Addison-Wesley Publishing Company, New York

Dunne RA, Campbell NA (1997) On the pairing of the softmax activation and cross-entropy penalty functions and the derivation of the softmax activation function. 8th Australian Conference on Neuronal Networks, pp 181–185

Fisher P (1997) The pixel: a snare and a delusion. Int J Rem Sen 18(3):679–685

Foody GM (1995) Cross-entropy for the evaluation of the accuracy of a fuzzy land cover classification with fuzzy ground data. ISPRS Journal of Photogrammetry and Remote Sensing 50(5):2–12

Foody GM (1996a) Relating the land-cover composition of mixed pixels to artificial neuronal network classification output. Photogrammetric Engineering and Remote Sensing 62(5):491–499

Foody GM (1996b) Approaches for the production and evaluation of fuzzy land cover classifications from remotely sensed data. Int J Rem Sen 17(7):1317–1340

Foody GM (1997a) Fully fuzzy supervised classification of land cover from remotely sensed imagery with an artificial neuronal network. Neuronal Computing and Applications 5:238–247

Foody GM (1997b) Land cover mapping from remotely sensed data with a neuronal network: Accommodating fuzziness. In: Kanellopoulos I, Wilkinson GG, Roli F, Austin J (eds) Neuronal-computation in remote sensing data analysis. Spinger-Verlag, Berlin

Gorte B, Stein (1998) A Bayesian classification and class area estimation of satellite images using stratification. IEEE Trans on Geoscience and Remote Sens 36(3):803–812

Haykin S (1994) Neuronal Networks: A Comprehensive Foundation. Macmillan, New York

Horwitz HM, Nalepka RF, Hyde PD, Morgenstern JP (1971) Estimating the proportions of objects within a single resolution element of a multispectral scanner. Proceedings of the 7th International Symposium on Remote Sensing of Environment, pp 1307–1320

Kent JT, Mardia KV (1988) Spatial classification using fuzzy membership models. IEEE Transactions on Pattern Analysis and Machine Intelligence 10(5):659–671

Maselli F, Rudolfi A, Conese C (1996) Fuzzy classification of spatially degraded thematic mapper data for the estimation of sub-pixel components. Int J Rem Sen 17:537–551

Ripley B (1996) Pattern recognition and neuronal networks. Cambridge University Press, Cambridge

Settle JJ, Drake NA (1993) Linear mixing and the estimation of ground cover proportions. Int J Rem Sen 14(6):1159–1177

Wang F (1990) Fuzzy supervised classification of remote sensing images. IEEE Trans on Geoscience and Remote Sens 28(2):194–201

Wood TF, Foody GM (1989) Analysis and representation of vegetation continua from Landsat Thematic Mapper data for lowland heaths. Int J Rem Sen 10(1):181–191

Part III

Artificial Neuronal Networks in Population, Community and Ecosystem Ecology

Chapter 7

Patterning of Community Changes in Benthic Macroinvertebrates Collected from Urbanized Streams for the Short Time Prediction by Temporal Artificial Neuronal Networks

T.-S. Chon · Y.-S. Park · E.Y. Cha

7.1 Introduction

Patterning temporal development of community is an important topic in ecosystem management as of late. Especially in aquatic ecosystems, where communities are easily affected by disturbances caused by various natural and anthropogenic agents, it is important to know how communities would develop in response to changes in water quality. They would develop either progressively with further disturbances, or regressively in recovery from pollution (Sladecek 1979; Hellawell 1986). Methods for characterizing 'changes' in communities are needed in terms of predicting the future development of the community, detecting mechanism of community differentiation, and assessing ecological status of the target ecosystem.

Data for community dynamics, however, are complex and difficult to analyse since they consist of many species, varying in nonlinear fashion in spatio-temporal domain. Although there have been numerous accounts on community classification through conventional multivariate analyses in ecology (e.g. Bunn et al. 1986; Legendre and Legendre 1987; Ludwig and Reynolds 1988; Quinn et al. 1991), not many studies have been conducted on patterning community dynamics. Legendre et al. (1985) and Legendre (1987) discussed classifying communities in temporal domains, including ordination and segmentation techniques in multivariate data series, and Turchin and Taylor (1992) reviewed time series analysis in analysing dynamic data for populations. Recently, attention has been focused on dynamic neuronal networks for patterning spatio-temporal data in electronics and computer sciences (Kung 1993; Giles et al. 1994). In ecology, artificial neuronal networks have been mainly applied in classifying groups (e.g. Chon et al. 1996; Levine et al. 1996), or patterning complex relationships (e.g. Lek et al. 1996; Huntingford and Cox 1996; Tuma et al. 1996).

In temporal patterning in ecology, artificial neuronal networks were mostly implemented in estimating time development of populations such as the flowering and maturity of soybeans (Elizondo et al. 1994), algal bloom (Recknagel et al. 1997) and changes in the size of animal population (Stankovski et al. 1998). Regarding communities, grassland community changes were predicted by Tan and Smeins (1996). However, these models were essentially applied in static terms; the time of input and output were the same, although the aim of the study was to predict the size of populations or communities. They were mostly based on the back propagation algorithm for patterning the relational effects with environmental factors.

Direct revealing of dynamics of ecological data was conducted by Boudjema and Chau (1996). Sets of univariate time-series data, such as tree ring thickness, were analysed by artificial neuronal network after preprocessing of the data with moving average and linear generation of sequential data. Time-delay effect was considered in training; the previous data were used for predicting future data. It was successful in extracting inherent information of the longitudinal series of complex data, which had been generally short for the time-series analysis and had relatively a high level of noise.

The aim of this study was to pattern and predict multivariate date even in a shorter period. Benthic communities change rapidly in response to natural and anthropogenic agents and show a wide range in response to different pollution impacts. This short term prediction is important in monitoring water quality and setting up strategies for aquatic ecosystem management. After training the multivariate data we would like to test the feasibility of temporal artificial neuronal networks in forecasting community changes in a short time period.

7.2 Methods

7.2.1 Multilayer Perceptron with Time Delay

To pattern relationships between different time events of community changes in this study, initially a simple multilayer perceptron with the back propagation algorithm was used as a nonlinear predictor (Wray and Green 1994; Haykin 1994) (Fig. 7.1a). The architecture consists of the well-known static multilayers; however, input and output data were provided with time delay. The input vector is defined in terms of the past samples, $X(t-1), X(t-2), \ldots, X(t-q)$, where q, prediction order, is the number of the total delays. The current data, $X(t)$, was given as matching output. In our cases, densities of selected 5 genera in sampled communities were provided as data sets for inputs with 1–5 time delays, i.e. $q = 1, 2, \ldots, 5$. With each delay, input nodes were corre-

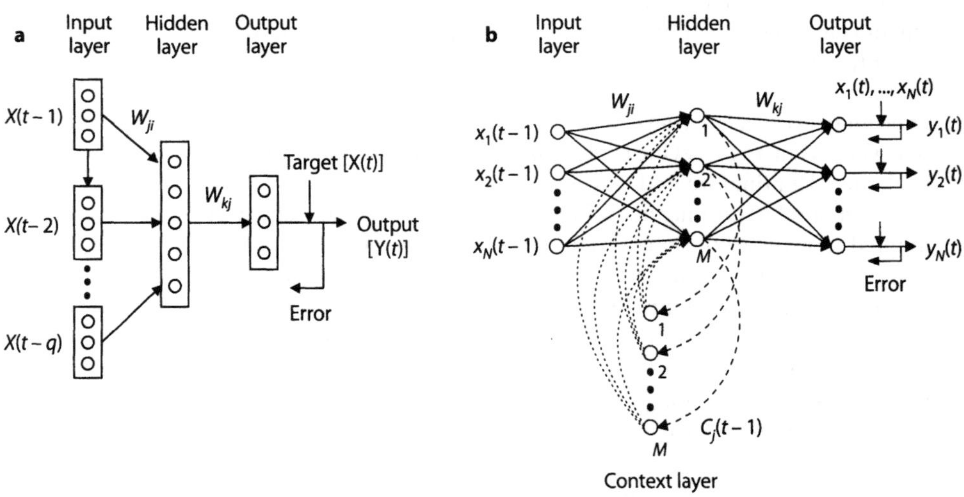

Fig. 7.1. **a** The architecture of the multilayer perceptron with time delay; **b** the Elman type recurrent neuronal network for patterning community changes

spondingly added. For example, if 5 genera were introduced with 2 time delays, $5 \times 2 = 10$ nodes were assigned for each input.

The input layer was subsequently interconnected to the hidden layer. Eight to thirty nodes were used in the hidden layer. The number of nodes in the hidden layer was determined based on experiences on obtaining convergence in training. The number of output nodes was 5, equal to the number of selected genera. The internal state of the network, $NET_{p,j}$, was obtained by linear summation of products of weights and output values of nodes in the hidden layer over time. Subsequently, these values were adjusted in a nonlinear fashion, logistic function in this case, to produce outputs, $Y(t)_{p,j}$, as follows (Wasserman 1989; Zurada 1992):

$$NET_{p,j} = \sum_{i=1} x_{p,i} w_{p,ji} \tag{7.1}$$

$$Y_{p,j} = \frac{1}{1 + \exp(-\lambda NET_{p,j})} \tag{7.2}$$

where $Y_{p,j}$ is activation of neuron j for pattern p, $x_{p,i}$ is output value of the neuron i of the previous layer for pattern p, $w_{p,ji}$ is weight of the connection between the neuron i of the previous layer and the neuron j of the current layer for pattern p, and λ is activation function coefficient (e.g. 1.0 in this study).

The output $Y(t)$ of the multilayer perceptron was produced in response to the input vector, and was equivalent to the one-step prediction for the future development. Subsequently actual data at time t, $X(t)$, were provided as the target and the difference between $Y(t)$ and $X(t)$ was measured and propagated backward for adjusting weights in the usual manner of the back propagation algorithm (Rumelhart et al. 1986). Weights at output neurons were updated as follows (Wasserman 1989):

$$\delta_{p,j} = Y_{p,j}(1 - Y_{p,j})(d_{p,j} - Y_{p,j}) \tag{7.3}$$

$$\Delta w_{p,ji}(t+1) = \eta \delta_{p,j} Y_{p,j} + \alpha \Delta w_{p,ji}(t) \tag{7.4}$$

$$w_{p,ji}(t+1) = w_{p,ji}(t) + \Delta w_{p,ji}(t+1) \tag{7.5}$$

where $d_{p,j}$ is desired output of node j for pattern p, η is training rate coefficient, and α is momentum coefficient. Weight updating at the hidden layers is similar to processes at the neurons of the output layer. Detailed procedures of the learning rules could be referred to Rumelhart et al. (1986) and Zurada (1992).

7.2.2 Recurrent Neuronal Network

An Elman type network (Elman 1990), one of the most well-known dynamic models in artificial neuronal networks, was applied in this study as a recurrent neuronal network. We used one hidden layer for simplicity of the network structure. The architecture of the network is basically similar to the multilayer perceptron, except in the composition of the hidden layer (Fig. 7.1b). Hidden layer embodies another context layer for implementing recurrence. Recurrence implies that the state of network depends on current input and its own internal state on the previous cycle. In this case, the hidden layer has recurrence and its own internal state is represented through the context hidden layer. Similar to the multilayer perceptron, the number of nodes at the input and output layers was 5, and 30 neurons were used for the hidden and context layers.

In the input layer, community data for selected genera, $x_l(t-1)$, were given as external inputs. Concurrently, output values from the hidden layer for the previous cycle are also provided as internal inputs to the hidden layer as $C_l(t-1)$. Initially, some small random numbers are used for the internal inputs. The group of $x_l(t-1)$ and $C_l(t-1)$ consist of the total input for the hidden layer, $z_l(t)$. The sum of linear combination of weights and inputs, $I_j(t-1)$, is subsequently adjusted in a nonlinear function such as $C_l(t-1) = f(I_j(t-1))$. The input process could be summarized as follows (Hecht-Nielsen 1990):

$$z_l(t) = \begin{cases} x_l(t) & \text{if } 1 \le l \le N \\ C_l(t-1) & \text{if } (N+1) \le l \le L \end{cases} \tag{7.6}$$

where $l = 1, 2, \ldots, L$, $L = N$ (number of input nodes) + M (number of hidden nodes), $x(t)$ is external input, and $C(t)$ is context input.

$$I_j(t) = \sum_{l=1}^{L} w_{jl} z_l(t) \tag{7.7}$$

$$f(I_j(t)) = \frac{1}{1+\exp(-\lambda I_j(t))} \tag{7.8}$$

$$C_l(t) = f(I_j(t)) \tag{7.9}$$

The net output in the output layer is determined by the summation of the linear combination of weights and values produced from the hidden layer. As a usual process in artificial neuronal networks, this is subsequently adjusted with a nonlinear function, logistic equation in this case, to produce output values for t as $y_k(t)$. These output values are in turn compared with actual field data, $x_l(t)$. Weight adjustment is conducted in the same way as it is determined in the back propagation algorithm. The difference between desired output and internal output was calculated, and subsequently was backpropagated through the hidden layer down to the context and input layers.

7.2.3
Field Data

Assessment of water quality and prediction of community dynamics in aquatic ecosystems are important in rapidly developing countries in Asia such as Korea, China, etc. For field data, benthic macroinvertebrate communities monthly collected from urbanized streams were used. The streams in the Suyong River, located in the Pusan metropolitan area, on the southern part of the Korean Peninsula, have been urbanized with a wide range of organic pollution mainly from domestic sewage (Fig. 7.2). We selected the sites of Suyong and Soktae streams that formed the Suyong River. The sample sites in the Soktae Stream in the Suyong River showed a wide range of water quality from β-meso-saprobity to iso-saprobity as the water flows down from TSD, TKC to THP. Community compositions generally reflected the water quality status of streams (Kwon and Chon 1993; Kang et al. 1995). Trent Biotic Index (Woodiwiss 1964), a biological index for indicating water quality, correspondingly showed the water quality from seven to two, and BOD also increased downstream (Fig. 7.2b).

In the Suyong Stream, in contrast, all the sites were in slight pollution of β-meso-saprobity. In this state of slight pollution, species richness is relatively high and communities could respond sensitively to environmental disturbances (Hellawell 1986).

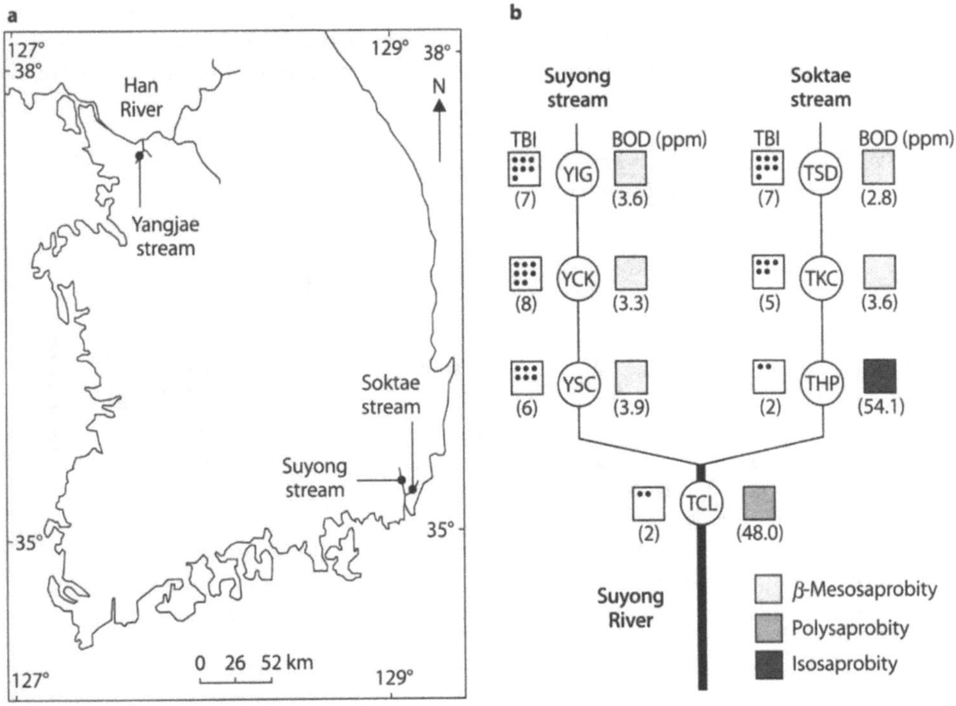

Fig. 7.2 a. Location of streams for collecting benthic macroinvertebrates in the Suyong and Han rivers; **b** water quality of sample sites with saprobity, TBI and BOD in the Suyong and Soktae streams in the Suyong River

The site TCL, where the Suyong and Soktae streams joined, showed poly-saprobity. The sites in the Suyong River were selected such that it would cover a wide range of pollution with an emphasis on slight pollution. Species richness was exceptionally high at relatively clean sites, while only a few species appeared at polluted sites (Kwon and Chon 1993, Kang et al. 1995; Yoon and Chon 1996). This is a typical response of community to pollution (Hellawell 1986). One of the objectives for this study was to investigate how these diverse changes in community could be effectively patterned by the temporal network. Samples collected from March 1992 to March 1995 were used for the learning process in this study in the Soktae Stream, while those from January 1993 to March 1995 were selected in the Suyong Stream. Among these, samples from TKC, THP, YCK, and YSC from September 1994 to March 1995 were set aside for recognition, and the rest were used for training (Table 7.1).

Another set of community data came from relatively homogenous environments. Samples were collected in a relatively short distance within 200 meters in the Yangjae Stream, a tributary of the Han River. The Yangjae Stream is located in the Seoul metropolitan area, on the middle part of the Korean Peninsula (Fig. 7.2a), and is highly polluted with poly-saprobity. This stream, however, is partly in a recovery phase due to the restoration efforts by the city government (Fig. 7.3). The number of species and the water quality indices such as Shannon diversity gradually increased and the species less tolerant to heavy organic pollution reappeared as water quality was slightly improved. Ecological status and water quality in the Yangjae Stream will be reported elsewhere. We attempted to see if the prediction of community change could be possible in this transitional recovery phase of the stream. Data collected from March 1996 to March 1998 were used for the learning process, and a portion of the samples were

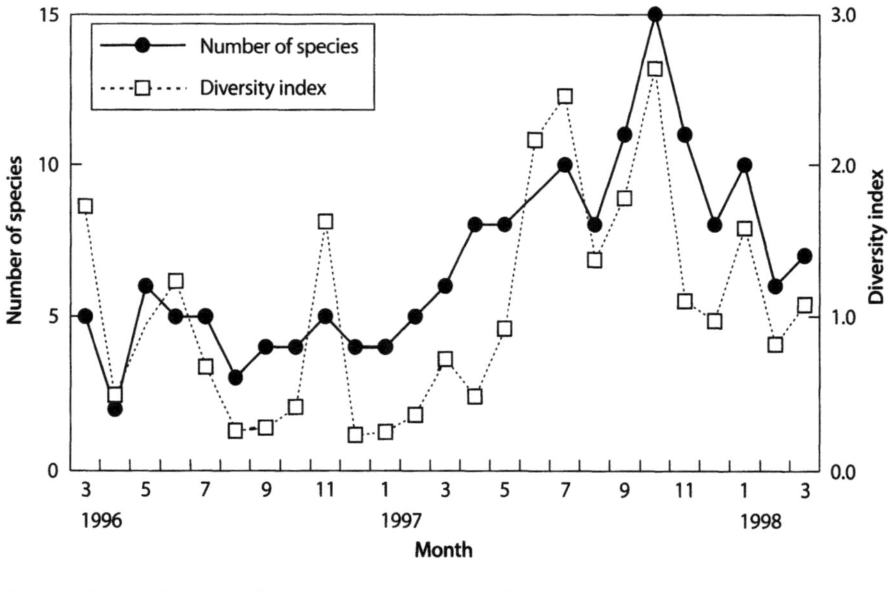

Fig. 7.3. Changes in the number of species and Shannon diversity indices in recovery of water quality in the Yangjae Stream in the Han River from March 1996 for two years

Table 7.1. Number of training and recognition patterns, and convergence results from the training by the multilayer perceptron with time delay when changes in benthic macroinvertebrate communities in streams of the Suyong and Han rivers were given as inputs

Streams (rivers)	Time delay (months)	No. training patterns	No. nodes at input layer	No. weights at input layer	No. nodes at hidden layer	No. nodes at output layer	Iteration	Error term at end of training	No. patterns for recognition
Suyong and Soktae (Suyong)	1	100	5	5	30	5	20000	0.047	28
	2	170	5	10	30	5	30000	0.041	28
	3	170	5	15	30	5	30000	0.012	28
	4	160	5	20	30	5	30000	0.008	28
	5	150	5	25	30	5	30000	0.005	28
Yangjae (Han)	1	100	5	5	30	5	20000	0.098	21
	2	195	5	10	30	5	30000	0.042	21
	3	195	5	15	30	5	30000	0.014	21
	4	195	5	20	30	5	30000	0.008	21
	5	195	5	25	30	5	30000	0.004	21

set aside for recognition, similar to the case of the streams in the Suyong River (Table 7.1).

Figure 7.4 shows examples of input data for the selected genera in the streams in the Suyong and Han rivers. In selecting data, attention was given to taxa more frequently and abundantly collected while the data for rare species were not included for training. For the Suyong and Soktae streams, genera such as *Chironomus* sp., *Conchapelopia* sp., *Orthocladius* sp., *Tanytarsus* sp., and *Limnodrilus* sp. were selected. The first four species belong to Chironomidae, an important indicator family in fresh water (Hellawell 1986), while the last one is an Oligochaeta. In communities of the Yangjae Stream, *Chironomus* sp., *Orthocladius* sp., *Cricotopus* sp., *Limnodrilus* sp. and *Erpobdella* sp. were chosen. The first three species are Chironomidae, while the fourth and fifth species belong to Oligochaeta and Hirudinea, respectively. These selected genera occurred consistently at the study sites during the survey period. The input values with greatly different numerical values in densities were avoided for training. The data were transformed by natural logarithm in order to emphasize the differences in low densities. Subsequently the transformed data were proportionally normalized between 0.01 and 0.99 in the range of the maximum and minimum density for each species collected during the survey period.

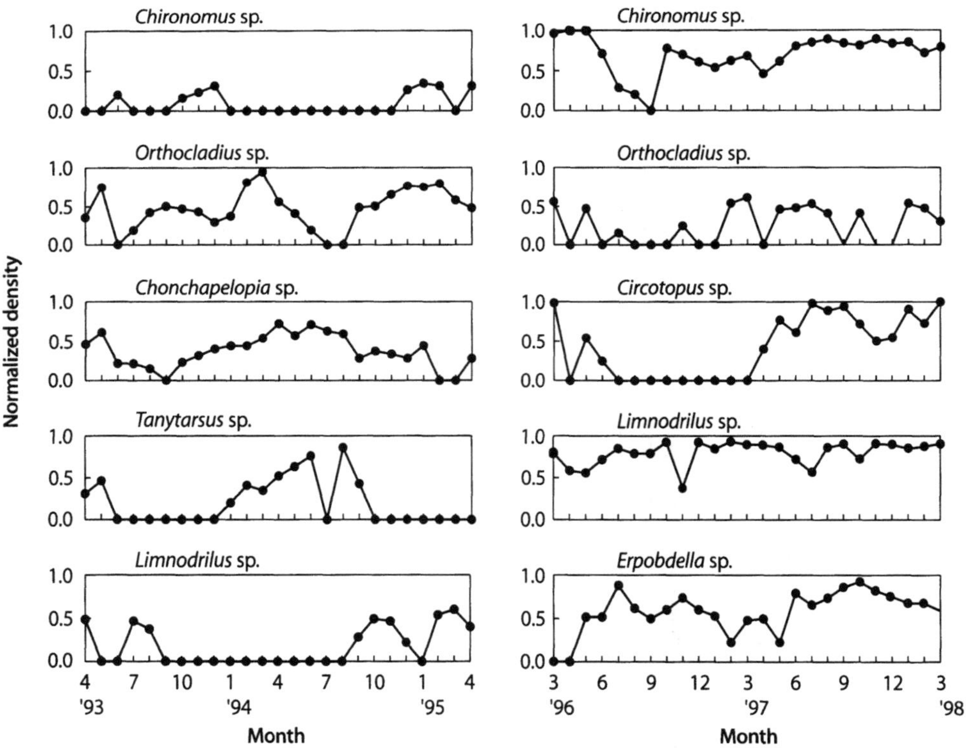

Fig. 7.4. Examples of community dynamics of selected genera in the Soktae (*left*) and Yangjae (*right*) streams as inputs for the training of community changes in artificial neuronal networks

7.3 Training and Recognition

7.3.1 Multilayer Perceptron with Time Delay

When communities were given as inputs to the simple multilayer perceptron with the time delay between one and five months, the convergence was generally reached in the iteration of 20 000–30 000 under the mean error term of 0.05, which is the sum of square terms of difference in output and target values divided by the number of input patterns (Table 7.1). The learning and momentum coefficients were 0.5 and 0.9, respectively. Trained data sets were accordingly matched to the original input data. It appeared that convergence was more easily achieved as time delay of input data was longer. In one month delay, it was generally difficult to obtain a convergence. At the same iteration number of 30 000 except for the one-month delay, the mean error term consistently decreased as the time delay was increased, both in the Han and Suyong rivers (Table 7.1).

When new data were given to the trained network for recognition as mentioned before, the network was able to make one-step predictions for the following community in time. Figure 7.5 shows examples of predicted community data after recognition by the trained multilayer perceptron. In general it appeared that the predicted and actual field data were in accord, although some discrepancies were locally observed. It was relatively difficult to match precisely the density level for each selected taxon. Based on the training experiences with available data (Table 7.1), it appeared that the prediction of community development was slightly more effective in the Yangjae Stream than in streams in the Suyong River.

7.3.2 Recurrent Neuronal Network

When communities were trained with the more sophisticated Elman type recurrent network, convergence was also achieved and its learning efficiency generally appeared to be higher than that by the simple multilayer perceptron. Convergence was usually reached in the 20 000th iteration, and the mean error terms were less than 0.006 in the case of Suyong and Soktae streams and 0.0006 in the Yangjae Stream. These were distinctively lower than the mean error term shown in the training by the previous multilayer perceptron (Table 7.1). The efficiency of learning was also reflected in the frequency of the different size of the error term (Fig. 7.6). When the density difference in each genus in each input community between the predicted and field data was considered individually, there was a higher frequency in the low range of difference in the training by the recurrent neuronal network than by the multilayer perceptron with time delay.

When new data were given to the trained network for recognition, the network was able to predict the status of communities for the next month (Fig. 7.7). It appeared that the predicted and actual field data were generally in accord, better than in the case of the training with the multilayer perceptron. The predicted data from the recurrent network appeared to be in a better accordance in the Yangjae Stream than in the

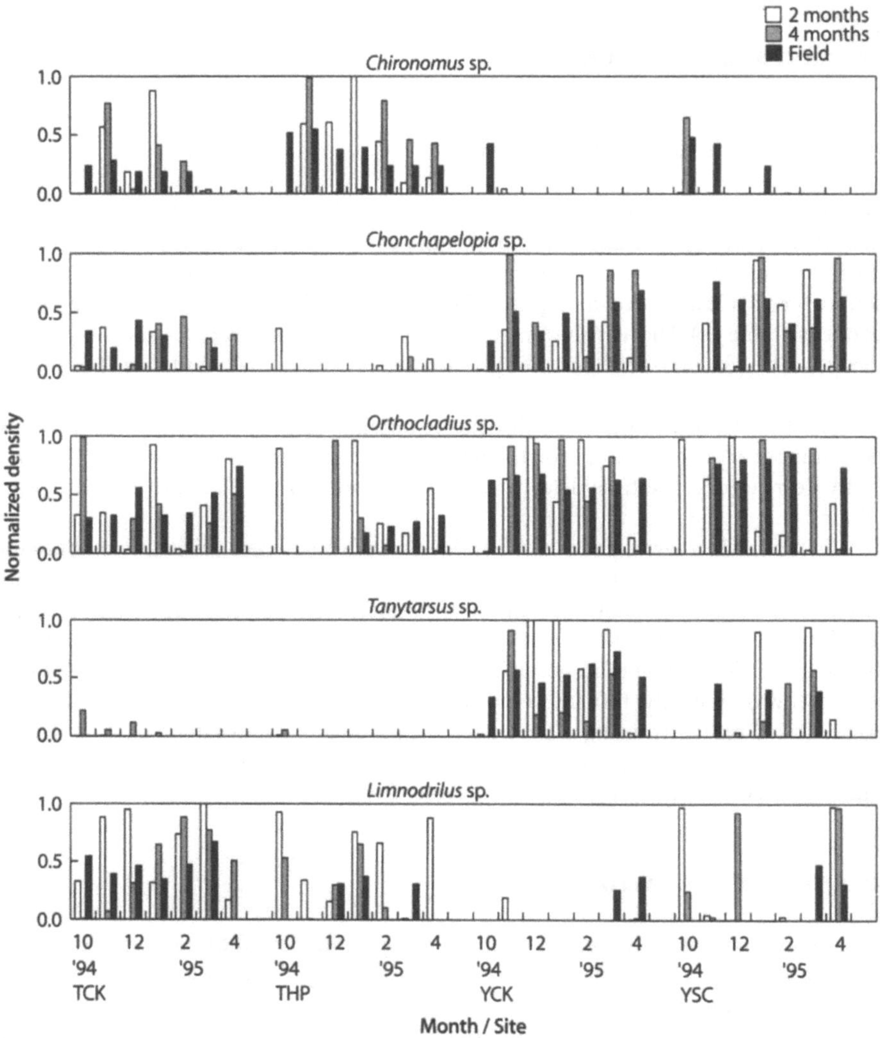

Fig. 7.5 a. Examples of field data and predictions in densities of selected Genera after recognition by the trained multilayer perceptron with time delay (two and four months) when new data for community development were given as inputs in the Soktae and Suyong streams

streams in the Suyong River. This was consistent to the case of training with the multilayer perceptron.

7.4 Discussion and Conclusion

By elaborating the feasibility of temporal artificial neuronal networks for community changes, this study showed that community development could be patterned and fore-

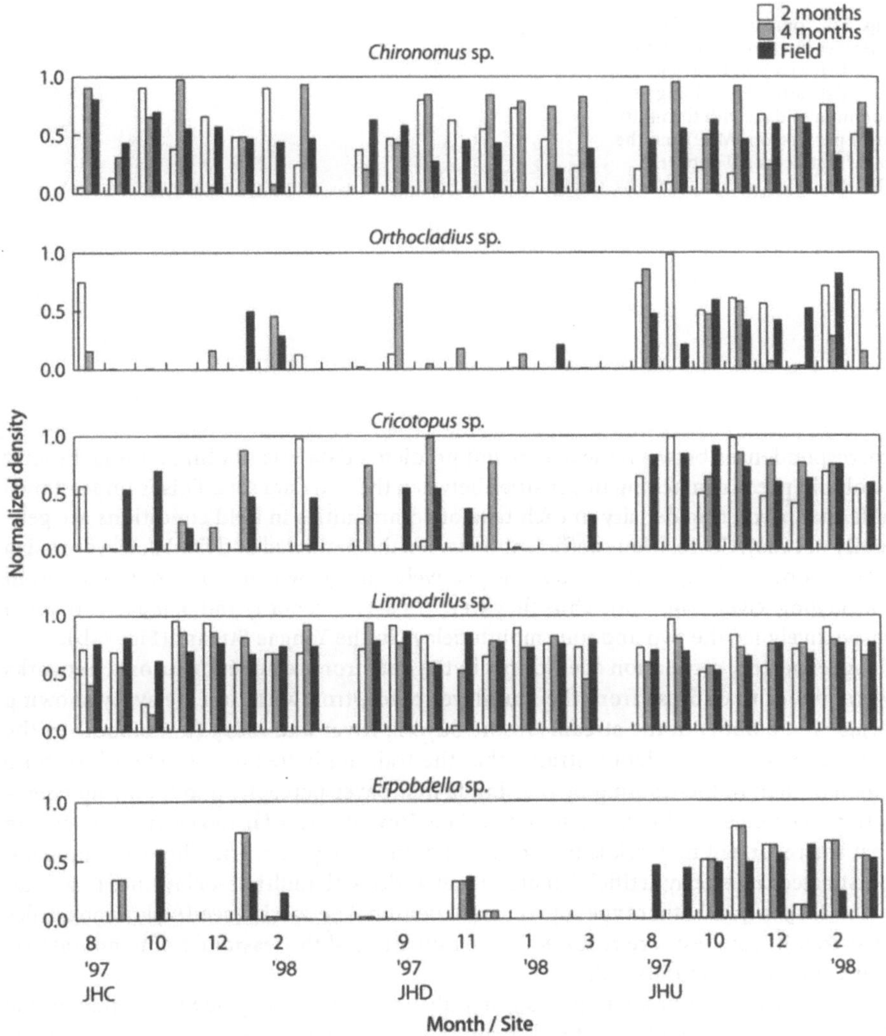

Fig. 7.5 b. Examples of field data and predictions in densities of selected Genera after recognition by the trained multilayer perceptron with time delay (two and four months) when new data for community development were given as inputs in the Yangjae Stream

casted by the temporal artificial neuronal networks in a short period. Boudjema and Chau (1996) effectively utilized artificial neuronal networks to pattern univariate data sets in a relatively short period, to which time series analyses may not be applicable due to the shortness of data. This study demonstrated that artificial neuronal networks could pattern multivariate data sets even in a shorter period.

In general, the trained results correspondingly reflected the expected development of communities under the influence of various degrees of organic pollution in urbanized streams in a certain time span. However, there were occasions that the degree of

Fig. 7.6. Comparison of frequencies in error term in difference between predicted and field data after the training of community data with the multilayer perceptron (MLP) and the recurrent neuronal network (RNN)

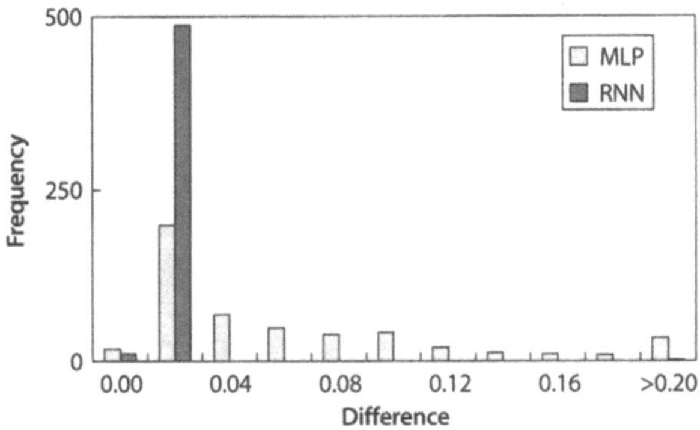

correspondences between the actual and predicted data was not high. It was difficult to obtain precise matching in densities between the two data sets. This is understandable that predicting density in each taxa of communities in field conditions are generally not easy. Correlation coefficients between the predicted and field data were 0.439 ($P < 0.0001$) and 0.481 ($P < 0.0001$) respectively for the two and four month delays in the Suyong River (Fig. 7.5a), while they were 0.556 ($P < 0.0001$) and 0.489 ($P < 0.0001$) respectively for the two and four month delays in the Yangae Stream (Fig. 7.5b).

As expected, correlation coefficients in the data from recurrent neuronal networks were higher than those from the multilayer perceptron with time delay, by showing 0.548 ($P < 0.0001$) in the streams in the Suyong River and 0.675 ($P < 0.0001$) in the Yangjae Stream. This demonstrated that the training by recurrent network is more efficient than in the training by the simple multilayer perceptron in their implementation to changes in diverse community data. However, it is far too early to generalize that the recurrent network is better than the time delay network. There are more sophisticated time delay artificial neuronal networks with multiple-delay lines (e.g. Waibel et al. 1989), and other types of recurrent neuronal networks (see Haykin 1994; Giles et al. 1994). More tests are required in the future to test the feasibility of temporal networks in the community data.

The efficiency of prediction was generally better in the Yangjae Stream than in the streams of the Suyong River. It is difficult to verify why prediction efficiency is different in the two rivers. One possible reason is that, since the sites in the Suyong River came from a wide range of pollution levels (Fig. 7.2), the selected taxa used for input data did not cover fully all the communities at different sites. One taxon may abundantly appear at one site, while it may not appear at the other site. For example, *Tanytarsus* sp. mainly appeared in relatively clean sites, but not much at the polluted sites. In contrast *Chironomus flaviplumus*, an indicator species of organic pollution, abundantly appeared at polluted sites. In the Yangjae Stream, however, all five taxa appeared consistently. As shown in Fig. 7.3, the Yangjae Stream has been in a recovering phase along with the increase in species richness. This study showed that, even in this transition period, the network made it possible to forecast a short-time community change.

Another advantage with the specific forecasting for each taxon is that it could assist to characterize community changes. Even if the predicted data were not in accord

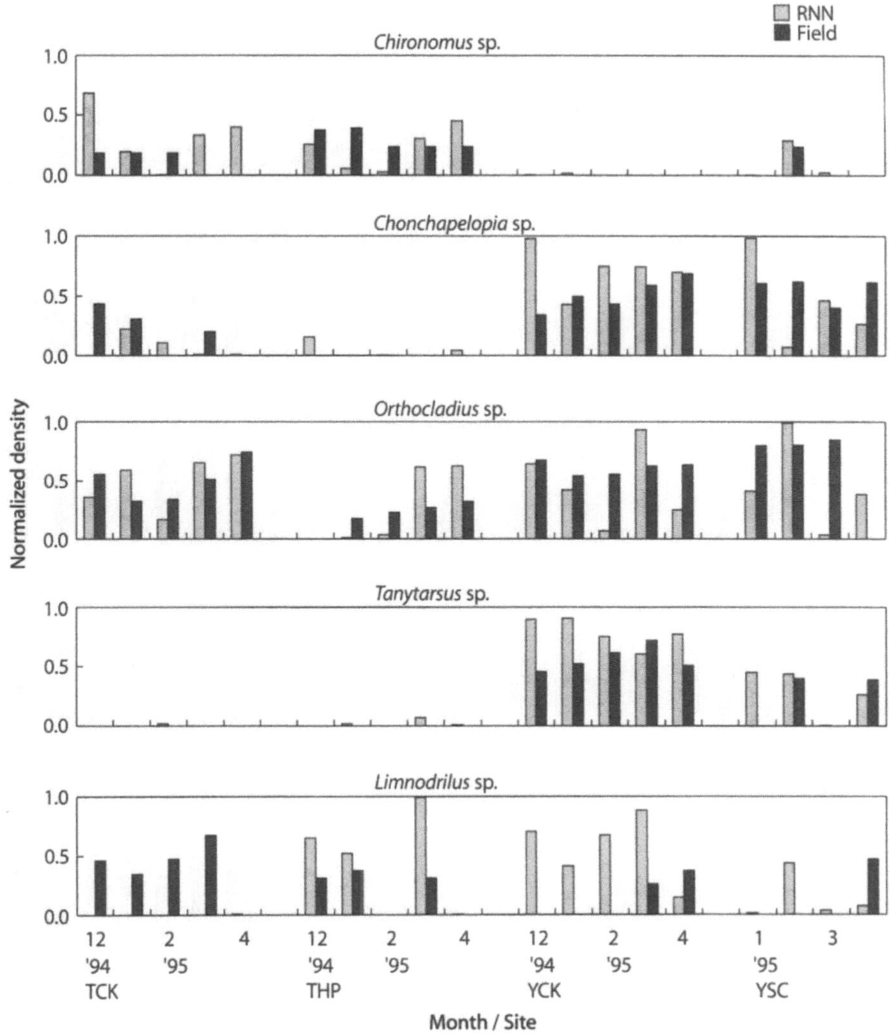

Fig. 7.7 a. Examples of field data and predictions in densities of selected genera in communities of benthic macroinvertebrates after recognition by recurrent neuronal network (RNN) when new data for community change were given as inputs in the Soktae and Suyong streams

with field data, for example, it would still give information to investigate the status of communities. Since the neuronal network represent average effects, more frequently appearing taxa would have more chance to be patterned in the training. Then the mismatching between the actual and predicted data may suggest occurrence of some disturbances in communities of new data. In communities sampled at TKC from December 1994 to March 1995 in Fig. 7.7a, for example, *Limnodrilus* sp., an indicator species of pollution, appeared in field data while there was no density forecasted by the recurrent network. In contrast, *Orthocladius* sp., which generally appears in recovery

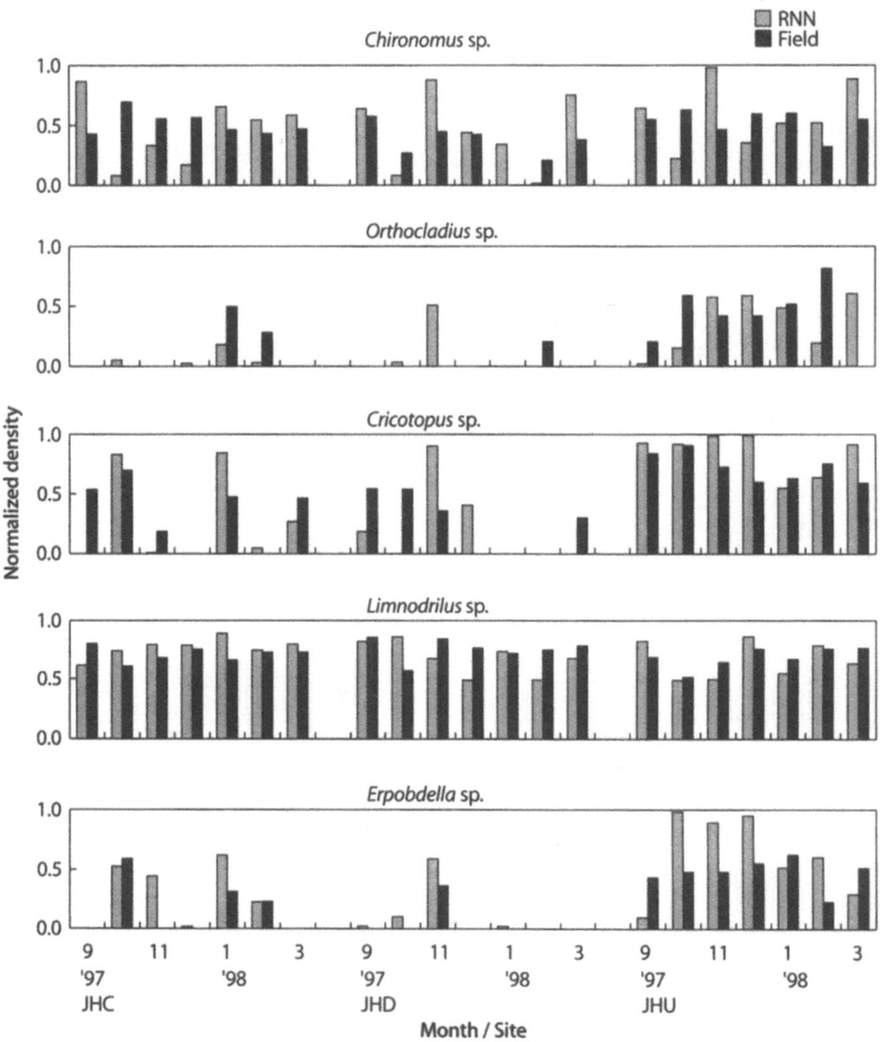

Fig. 7.7 b. Examples of field data and predictions in densities of selected genera in communities of benthic macroinvertebrates after recognition by recurrent neuronal network (RNN) when new data for community change were given as inputs in the Yangjae Stream

state, was collected in a high density in this period. In the original data for training, in contrast, the former appeared in a high level while the latter was rarely collected. Since the recognition period lies in the last part of the survey, this suggested that the water quality be improved. This demonstrated that the forecast of 'changes-in-density' could effectively pinpoint the shifting of ecological status in communities.

In the simple multilayer perceptron with time delay, the increase in the number of nodes in the hidden layer consistently enhanced training efficiency. The number of hidden nodes required were unusually high. Initially, we started with 8 hidden nodes, however, convergence was not satisfactory; the error term was not effectively decreased. As the number of hidden nodes was increased up to 30 nodes, the learning efficiency was accordingly increased (Table 7.1). In contrast it required usually a lower number of hidden nodes (8–10) in the case of training without time delay. This requirement of additional nodes in the hidden layer may be due to complexity of data implementation in time delay. Further study may be required in assessing the role of hidden nodes in the comparative study with a more elaborated time-delayed neuronal network (TDNN) (Waibel et al. 1989) and other related temporal networks.

In conclusion this study demonstrated that temporal artificial neuronal networks could be utilized to forecast and analyse short-period changes in multivariate data sets. The recurrent neuronal network appeared to be effective in patterning development of benthic communities in streams responding in a diverse manner to a wide range of pollution.

Acknowledgements

This paper was supported by Non Directed Research Fund, Korea Research Foundation, 1996.

References

Boudjema G, Chau NP (1996) Revealing dynamics of ecological systems from natural recordings. Ecol Model 91:15–23
Bunn, SE, Edward DH, Loneragan NR (1986) Spatial and temporal variation in the macroinvertebrate fauna of streams of the northern jarrah forest, Western Australia: Community structure. Freshwat Biol 16:67–91
Chon T-S, Park YS, Moon KH, Cha EY (1996) Patternizing communities by using an artificial neuronal network. Ecol Model 90:69–78
Elizondo DA, McClendon RW, Hoongenboom G (1994) Neuronal network models for predicting flowering and physiological maturity of soybean. Transactions of the ASAE 37(3):981–988
Elman JL (1990) Finding structure in time. Cognitive Science 14:179–211
Giles CL, Kuhn GM, Williams RJ (1994) Dynamic recurrent neuronal networks: Theory and applications. IEEE Transactions on Neuronal Networks 5:153–156
Haykin S (1994) Neuronal networks. Macmillian College Publishing Company, New York
Hecht-Nielsen R (1990) Neurocomputing. Addison-Wesley, New York
Hellawell JM (1986) Biological indicators of freshwater pollution and environmental management. Elsevier, London
Huntingford C, Cox PM (1996) Use of statistical and neuronal network techniques to detect how stomatal conductance responds to changes in the local environment. Ecol Model 97:217–246
Kang DH, Chon T-S, Park YS (1995) Monthly changes in benthic macroinvertebrate communities in different saprobities in the Suyong and Soktae streams of the Suyong River. Kor J Ecol 18:157–177
Kung SY (1993) Digital neuronal networks. Prentice Hall, Englewood Cliffs, New Jersey
Kwon T-S, Chon T-S (1993) Ecological studies on benthic macroinvertebrates in the Suyong River. III. Water quality estimations using chemical and biological indices. Kor J Lim 26:105–128
Legendre P (1987) Constrained clustering. In: Legendre P, Legendre L (eds) Developments in numerical ecology. Springer-Verlag, Berlin, pp 289–307
Legendre P, Legendre L (eds) (1987) Developments in Numerical Ecology. Springer-Verlag, Berlin
Legendre P, Dallot S, Legendre L (1985) Succession of species within a community: Chronological clustering, with applications to marine and freshwater zooplankton. Am Nat 125:257–288
Lek S, Delacoste M, Baran P, Dimopoulos I, Lauga J, Aulagnier S (1996) Application of neuronal networks to modelling nonlinear relationships in ecology. Ecol Model 90:39–52
Levine ER, Kimes DS, Sigillito VG (1996) Classifying soil structure using neuronal networks. Ecol Model 92:101–108
Ludwig JA, Reynolds JF (1988) Statistical ecology: A primer of methods and computing. John Wiley and Sons, New York
Quinn MA, Halbert SE, Williams III L (1991) Spatial and temporal changes in aphid (Homoptera: Aphididae) species assemblages collected with suction traps in Idaho. J Econ Entomol 84:1710–1716

Recknagel F, French M, Harkonen, P, Yabunaka K-I (1997) Artificial neuronal network approach for modelling and prediction of algal blooms. Eco Model 96:11–28
Rumelhart DE, Hinton GE, Williams RJ (1986) Learning internal representations by error propagation. In: Rumelhart DE, McCelland JL (eds) Parallel distributed processing: Explorations in the microstructure of cognition, vol I: Foundations. MIT Press, Cambridge, pp 318–362
Sladecek V (1979) Continental systems for the assessment of river water quality. In: James A, Evison L (eds) Biological indicators of water quality. John Wiley & Sons, Chichester, pp 31–32
Stankovski V, Debeljak M, Bratko I, Adamic M (1998) Modelling the population dynamics of red deer (*Cervus elaphus* L.) with regard to forest development. Ecol Model 108:143–153
Tan SS, Smeins FE (1996) Predicting grassland community changes with an artificial neuronal network model. Ecol Model 84:91–97
Tuma A, Haasis H-D, Rentz O (1996) A comparison of fuzzy expert systems, neuronal networks and neuro-fuzzy approaches controlling energy and material flows. Ecol Model 85:93–98
Turchin P, Taylor AD (1992) Complex dynamics in ecological time series. Ecology 73(1):289–305
Waibel, A., Hanazawa, T., Hinton, G., Shikano, K.G. and Lang, K.J. (1989) Phoneme recognition using time delay neuronal networks. IEEE Trans Acoustics, Speech and Signal Process 37:328–339
Wasserman PD (1989) Neuronal computing: Theory and practice. Van Nostrand Reinhold, New York
Woodiwiss FS (1964) The biological systems of stream classification used by the Trent River Board. Chemistry and Industry 14:443–447
Wray J, Green GGR (1994) Calculation of the Volterra kernels of non-linear dynamic systems using an artificial neuronal network. Biol Cybern 71:187–195
Yoon BJ, Chon TS (1996) Community analysis in chironomids and biological assessment of water qualities in the Suyong and Soktae streams of the Suyong River. Kor J Lim 29(4):275–289
Zurada JM (1992) Introduction to artificial neuronal systems. West Publishing Company, New York

Chapter 8

Neuronal Network Models of Phytoplankton Primary Production

M. Scardi

8.1 Introduction

Empirical models of phytoplankton primary productivity have always played an important role in oceanographic research, mainly because direct measurements of this process are difficult, expensive and time-consuming. Moreover, these models are needed to estimate primary production from the large phytoplankton biomass data sets that are obtained by remote sensing. They are also necessary to carry out instrumental estimates of primary production (e.g. by pump and probe fluorometers) and, in general, to post-process phytoplankton biomass data. Many different empirical models have been developed during the last 40 years and several of them have provided very useful results. The most common formulations among these models are based on simple linear relationships, where depth-integrated phytoplankton primary production depends on phytoplankton biomass within the upper layer of the water column (i.e. the upper attenuation length). For instance:

$$\sum PP = 1.254 + 0.728\log(C_k) \qquad \text{(Smith and Baker 1978)} \qquad (8.1)$$

More complex linear relationships have also been used, taking into account simultaneously phytoplankton biomass, surface irradiance and water transparency (represented by the inverse of the light attenuation coefficient or by the euphotic zone depth) as composite variables:

$$\sum PP = a + bBZ_pI_0 \qquad \text{(Cole and Cloern 1987)} \qquad (8.2)$$

or by means of multiple linear regression (e.g. Eppley et al. 1985). Of course, more complex models have also been developed, taking into account other variables, such as, for instance, day length or photosynthetic rates, but the actual significance of the improvement in accuracy that they provided is still to be fully evaluated.

A typical example of an advanced empirical model of phytoplankton primary production is the vertically generalized production model (VGPM) that was developed by Behrenfeld and Falkowski (1997) to estimate annual global primary production in the oceans from remotely sensed biomass data. The VGPM can be summarized as:

$$\sum PP = 0.66125\, P^B_{\text{opt}} \frac{I_0}{I_0 + 4.1} Chl_0 Z_p Dl \qquad (8.3)$$

where ΣPP is the phytoplankton primary production integrated over the euphotic zone (mg C m^{-2} day^{-1}), P^B_{opt} is a photoadaptive variable [maximum carbon fixation rate within the water column, mg C (mg Chl)$^{-1}$ h^{-1}], I_0 is the surface irradiance (E m^{-2} day^{-1}), Chl_0 is the surface phytoplankton biomass (mg Chl m^{-3}), Z_p is the euphotic zone depth (m) and Dl is the day length (decimal hours). For practical model applications P^B_{opt} is approximated by a 7th order polynomial function of the sea surface temperature and Z_p is assessed by means of power functions of the surface phytoplankton biomass.

Even though modern empirical models may provide adequate estimates of phytoplankton primary production on the basis of widely available variables and good generalization capabilities, a further improvement in accuracy can be achieved by means of artificial neuronal networks (Scardi 1996). Even though in the first applications only very simple error back propagation neuronal networks were used, they provided very good results and always performed better than conventional empirical models.

This was not a surprising evidence, since neuronal networks are inherently more flexible than conventional empirical models, as they are able to approximate virtually every multivariable function, provided that enough data are available for their training and that their structure is adequate (see, for instance, Hornik et al. 1989). However, this knowledge is restricted to the connectionist community, while ecologists and oceanographers are still not acquainted with these tools, even though their potentialities in ecological applications have been recently pointed out (Lek et al. 1996). As far as phytoplankton studies are concerned, Recknagel et al. (1996) and Recknagel (1997) used neuronal networks to predict species abundance of blue-green algae.

In this paper a general scheme of neuronal network for modelling phytoplankton primary production is outlined as well as its further development and optimization for small and large spatial scale applications.

8.2
Materials and Methods

The neuronal networks that are presented in this paper were trained using the most common algorithm, i.e. the error back propagation (Rumelhart et al. 1986). A constant unit learning rate and no momentum were used for all the training procedures, since the optimization of these parameters is not critical in primary production modelling, as the shape of the error surfaces is usually very simple.

Conversely, in order to avoid overfitting and to optimize the generalization capabilities of the neuronal networks, which are of paramount importance in ecological applications, several techniques were applied simultaneously. First of all, only a subset of the whole training set was randomly selected for each training epoch. This solution was also needed because a "learning per pattern" strategy had been chosen for the neuronal network training, which required the training patterns to be submitted in a different order at each learning epoch. Moreover, the training procedures were carried out according to an "early stopping" strategy, i.e. the training was stopped as soon as the validation set error started to increase. Finally, a small random amount of gaussian noise ($\mu = 0$ and $\sigma = 0.01$ in all the cases) was added to input patterns at each epoch (Györgyi 1990). This procedure, also known as jittering, contributed to the NN regularization by producing an unlimited number of "artificial" training patterns that were similar, but not identical, to the original ones.

The optimal structure of the neuronal networks was defined after empirical tests, which were carried out by varying the number of hidden layer neurons from one half to the double of the number of input neurons. This range was heuristically set, but the rationale that supports this choice is that the "shape" of the relationships that had to be modelled was not very complex, so that it could be reproduced by a limited number of hidden layer neurons. The optimal structure was the one that provided the smallest mean square error of the phytoplankton primary production estimates. The error distribution was also checked, even though it was not considered as the primary criterion.

All the neuronal network inputs were scaled into the [0, 1] interval, as well as the output (i.e. primary production). All the neurons had sigmoid activation functions, namely:

$$f(a) = \frac{1}{1+e^{-a}} \tag{8.4}$$

Two different data sets were used for the applications that are presented in this paper. The first one is a global data set, which consists of 2 218 phytoplankton biomass, irradiance, temperature and primary production data. The latter variable was integrated within the euphotic zone depth (i.e. within the depth where irradiance is 1% of the surface value), whereas the predictive variables took into account surface measurements only (so that they could be easily substituted by remote sensing data in model applications). The geographical coordinates of the sampling station and the sampling date were also known for every training pattern. The sampling date was "split" into two new variables in order to use it as a neuronal network input. In particular, the day of the year was univocally described by means of two variables ranging from 0 to 1, which correspond to the cartesian coordinates of a point onto a unit diameter circle:

$$date_1 = \frac{1}{2}\left[\cos\left(\frac{2\pi \cdot \mathrm{day}}{365}\right)+1\right] \qquad date_2 = \frac{1}{2}\left[\sin\left(\frac{2\pi \cdot \mathrm{day}}{365}\right)+1\right] \tag{8.5}$$

The second data set was located at the opposite end of the spatial scale, as it was collected at a single sampling station in the Gulf of Napoli (Italy) during a 5-year cycle of fortnight measurements. It consisted of 825 phytoplankton biomass (both as chlorophyll and phaeophytin concentrations) and primary production observations that were collected at different depths, from surface to 60 m, as well as 116 surface irradiance and temperature values. Sampling dates were converted according to the same procedure that has been described for the previous data set. The sampling site is located in the inner part of the Gulf of Napoli, but it is not directly influenced by terrestrial runoff.

8.3 Results

8.3.1 Basic Primary Production Modelling: Neuronal Networks vs. Linear Regressions

Assuming that a phytoplankton species assemblage does not change significantly its composition, the primary production is obviously proportional to the photosynthetically active biomass and is mainly limited by nutrient availability. Light is also an important factor, even though adaptation of phytoplankton cells to different intensities plays a major role in determining the light requirements for optimal primary production. Water temperature also influences primary production, even though to a lesser extent.

Therefore, on a small spatial scale, if a well-defined pelagic ecosystem is taken into account, it is not very difficult to describe and empirically model the relationships between these variables and primary production. This is the reason why simple linear models are quite effective when these conditions are met.

A comparison between a linear model (*crosses*) and a neuronal network model (*circles*) of surface phytoplankton primary production in the Gulf of Napoli is shown in Fig. 8.1. The linear model was based on a composite variable that included surface biomass (B_0, mg Chl m^{-3}), irradiance (I_0, E m^{-2} day^{-1}) and water temperature (T_0, °C):

$$PP_0 = 0.0346\, B_0 I_0 T_0 \tag{8.6}$$

These variables were also used as inputs of a 3-4-1 neuronal network, that was trained on one half of the data set ($n = 58$) and tested on the second half ($n = 58$). The linear model was almost as accurate as the neuronal network ($R^2 = 0.696$ and $R^2 = 0.862$

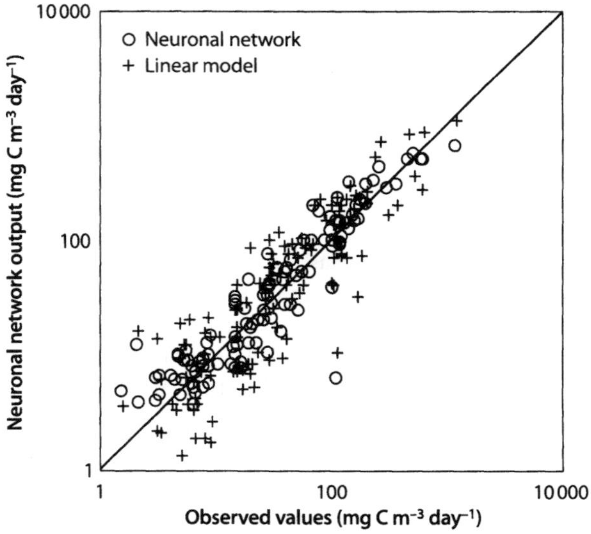

Fig 8.1. A comparison between a linear model of phyto-plankton primary production based on a composite (surface biomass, temperature and irradiance) independent variable (*crosses*, $R^2 = 0.696$) and a 3-4-1 neuronal network model (*circles*, $R^2 = 0.862$)

on log-log scale, respectively) because of the simplicity of the relationship between predictive variables and primary production within the observed range. However, primary production varied according to depth as shown in Fig. 8.2 (only primary production is log-transformed), both because of light attenuation and nutrient availability. The shape of the vertical profiles of primary production that were recorded in the Gulf of Napoli did not vary substantially, even though their magnitude was far from constant. If the integral of the primary production over a 0–60 m interval was taken into account (Fig. 8.3), a 3-4-1 neuronal network model (*circles*, $R^2 = 0.894$) clearly outperformed the linear model based on the aforementioned composite variable (*crosses*, $R^2 = 0.002$), i.e.:

$$\sum PP = 0.1312\, B_0 I_0 T_0 \tag{8.7}$$

It is evident that the poor performance of the linear model was due to the extinction of the primary production with depth, which is strongly nonlinear especially in summer conditions, when high surface phytoplankton biomass is usually observed.

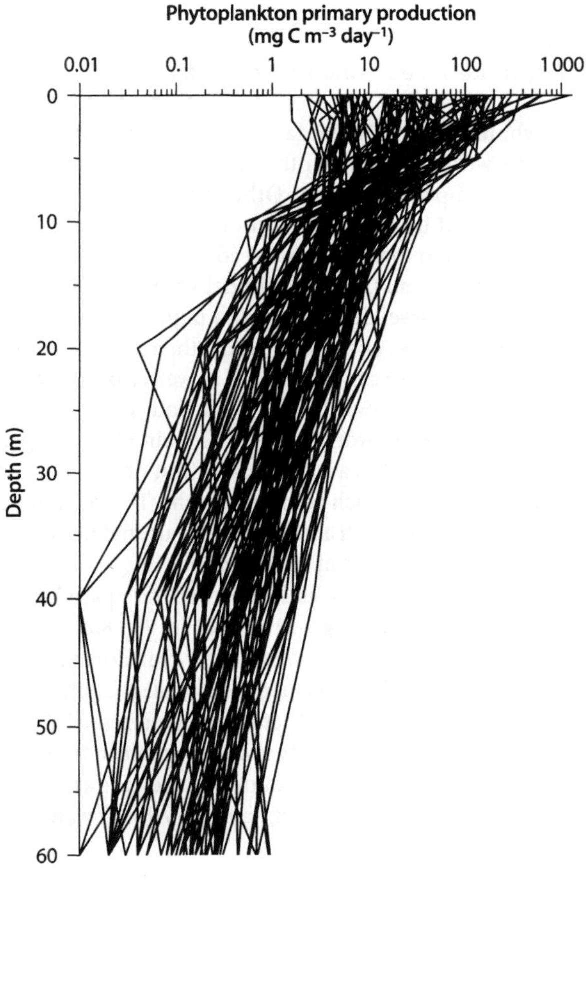

Fig. 8.2. Vertical profiles of phytoplankton primary production in the Gulf of Napoli ($n = 16$, from July 26, 1984 to November 15, 1988)

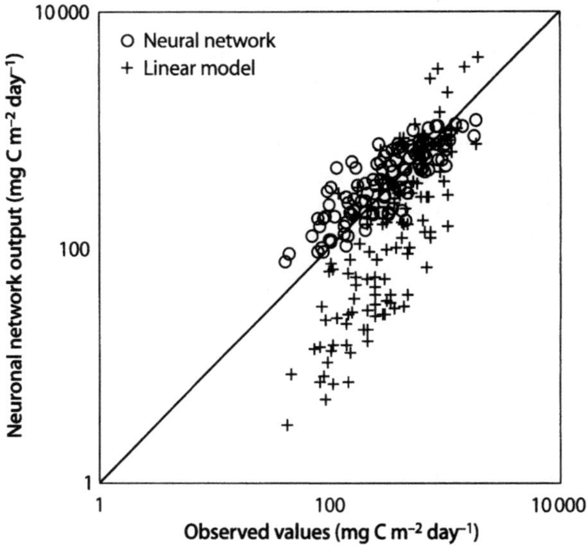

Fig 8.3. A comparison between a depth-integrated linear model of phytoplankton primary production based on a composite (surface biomass, temperature and irradiance) independent variable (*crosses*, $R^2 = 0.002$) and a 3-4-1 neuronal network model (*circles*, $R^2 = 0.894$)

8.3.2
A Depth-Resolved Primary Production Model

The whole Gulf of Napoli data set ($n = 825$) was used to train a more complex neuronal network and to push the connectionist approach beyond the limit of the conventional empirical models. Other predictive variables were added to the basic set in order to obtain primary production estimates at different depths: day of the year (mapped onto a circle as already explained), mean phytoplankton biomass in the upper water column (mg Chl m^{-3}), daily variability of the surface irradiance (see Appendix) and, of course, depth (m). Two pigment concentrations were used to define phytoplankton biomass at the selected depth, namely chlorophyll and phaeophytin (mg m^{-3} in both cases). These concentrations, as well as primary production data, were log-transformed to correctly evaluate the behaviour of the model over a wide range of values.

The neuronal network model was trained on 412 patterns, which were randomly selected out of the 825 available ones (but only 3/4 of the training patterns were randomly submitted at each training epoch). The remaining patterns were used for testing purposes. A 9-7-1 structure of the neuronal network provided the best results and was used in subsequent applications.

The neuronal network estimates were compared to the observed primary production values in the log-log scatter plot in Fig. 8.4. The overall accuracy of these estimates was very good ($R^2 = 0.810$, after transformation to raw data units). A further, although minimal, improvement ($R^2 = 0.822$) could be achieved by means of a linear correction ($PP' = 1.43843PP$). It has to be stressed that this correction, which corresponds to a small vertical shift of all the points in the log-log plot in Fig. 8.4, was needed after the data back-transformation from log to raw units. In particular, it compensates the different role that very large and very small values play with or without log transformation.

Fig. 8.4. Depth-integrated vertical model of the Gulf of Napoli (Italy): observed primary production values vs. neuronal network estimates ($R^2 = 0.810$)

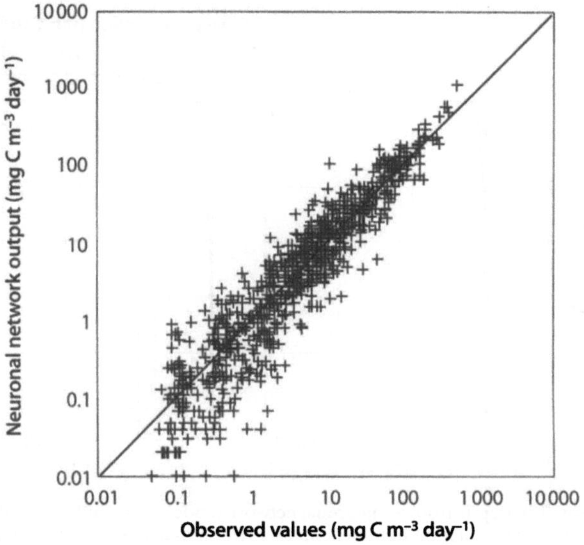

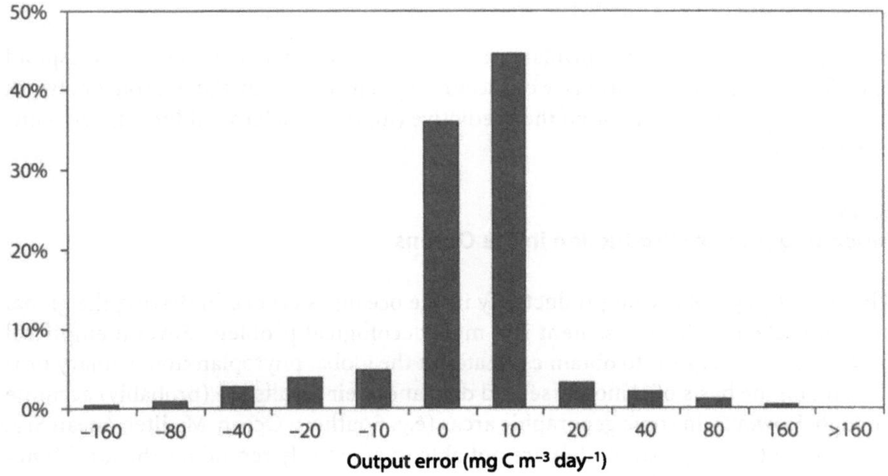

Fig. 8.5. Depth-resolved vertical model of the Gulf of Napoli (Italy): output error distribution. Axis labels indicate the upper limit of each error class

The distribution of the output error, which is shown in Fig. 8.5, provides a further evidence of the good quality of the neuronal network estimates. In fact, 80.6% of the errors were smaller than 10 mg C m^{-3} day^{-1} in absolute value (see bars 0 and 10, i.e. $-10 < \text{error} \leq 0$ and $0 < \text{error} \leq 10$, in Fig. 8.5).

An application of this depth-resolved neuronal network model is summarized in Fig. 8.6, where vertical profiles of primary production were estimated (*solid line*, step = 1 m) and compared to discrete observed data (*black circles*). It is evident that

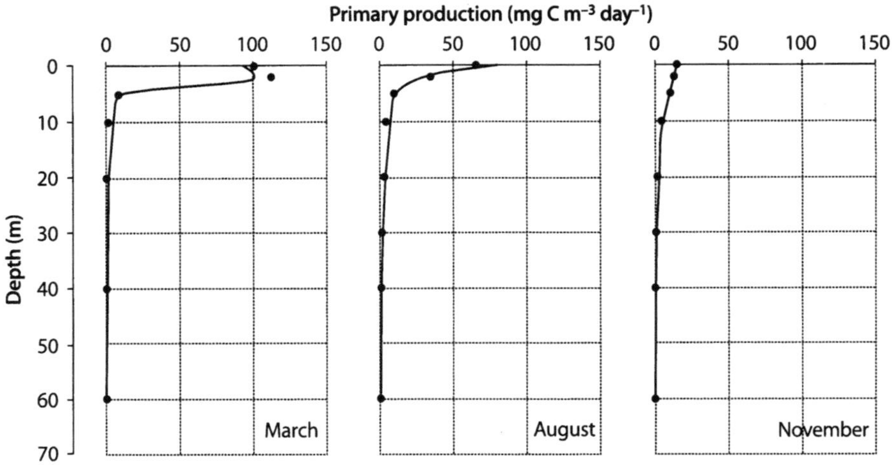

Fig. 8.6. Depth-resolved neuronal network model of the Gulf of Napoli: vertical profiles of phytoplankton primary production obtained. *Black circles* indicate observed values, whereas the *solid line* represents the neuronal network estimates (step = 1 m). A typical spring bloom (March), a summer surface maximum (August) and an almost completely mixed water column (November) are shown

different distributions of the primary production in the water column, that correspond to different seasonal conditions, were accurately reproduced by the neuronal network model using surface data for all the predictive (input) variables but for pigment concentrations.

8.3.3 Modelling Primary Production in the Oceans

The role of phytoplankton productivity in the oceans is crucial in driving the global carbon cycle, and its assessment is a major ecological problem. Several empirical models were developed to obtain estimates of the global phytoplankton primary production on the basis of remotely sensed data and their results are (probably) accurate enough. However, in some geographic areas (e.g. Southern Ocean, Mediterranean Sea, etc.) conventional global models are not able to effectively reproduce the local functioning of the water column, because they were poorly calibrated or because they lack the necessary degree of flexibility. The first problem obviously depends on the scarcity of data, but a neuronal network approach may be very effective in solving the second one.

Two neuronal network models were trained on 1 109 patterns (i.e. one half of the available global data set), using only 554 (i.e, one half of the training set) at each epoch. The first one was a very simple 3-4-1 model, which used only surface biomass (Chl_0, mg Chl m^{-3}), irradiance (I_0, E m^{-2} day^{-1}) and water temperature (T_0, °C) as predictive variables. The second one was a 7-7-1 model and used the same predictive variables, but also considered geographical coordinates (latitude and longitude) and day of the year. The latter variable was mapped onto a circle as previously described. Even though log-transformation of biomass and primary production data was tested, the best results were obtained using raw data in the training patterns and computing the mean square error on log-transformed output and target data.

In order to reduce the effect of outliers on the training procedure, the mean square error was computed only on 90% of the training patterns, excluding the larger errors. The rationale for this choice is that primary production measurements may be affected by severe errors, and that the training data are not to be taken for granted.

The results of the simpler model are shown in Fig. 8.7a. It is evident that they are not satisfactory, especially with respect to low primary production values. The second model successfully exploited the information contained in the additional input (co)variables and performed considerably better, as shown in Fig. 8.7b ($R^2 = 0.573$ with untransformed data).

The output error distribution was almost symmetrical, and 3 out of 4 output values were within 500 mg C m^{-2} day^{-1} from the observed value (Fig. 8.8). This is not a negligible error in the framework of a small spatial scale application, but it is almost insignificant at global scale, when, for instance, equatorial regions are modelled together with polar regions.

8.3.4
Neuronal Network Models vs. Conventional Models

These neuronal network models compared favourably to conventional models in several applications, providing better results not only with respect to data fitting, but also when the meaning of the "shape" of the modelled relationships was taken into account.

The primary production estimates obtained from a very simple 3-4-1 neuronal network model trained on Mediterranean Sea data (a subset of the global data set, $n = 97$) under different combinations of surface irradiance and biomass values are shown in Fig. 8.9a. These estimates refer to a Western Mediterranean station (43° N, 8° E) in mid-

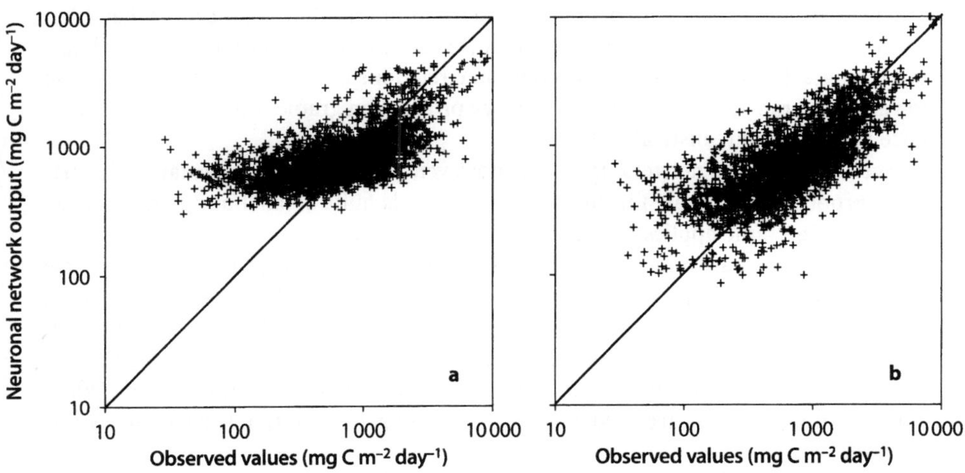

Fig. 8.7. Global neuronal network models of phytoplankton primary production; **a** estimates were based on surface biomass, irradiance and temperature only; **b** estimates were based on the whole set of predictive variables

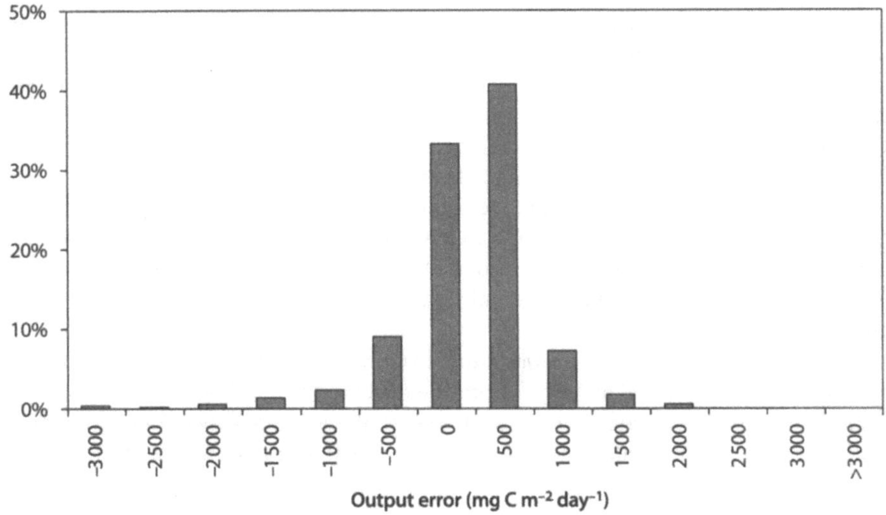

Fig. 8.8. Output error distribution of the 7-7-1 global neuronal network model. Axis labels indicate the upper limit of each error class

June surface temperature conditions (18.94 °C). The primary production estimates obtained from the VGPM (Behrenfeld and Falkowski 1997) under the same conditions are shown in Fig. 8.9b.

The neuronal network model shows a more complex behaviour than the VGPM model, and includes features that are not represented in the latter. For instance, the neuronal network model correctly reproduces the effect of self-shading on phytoplankton productivity, as the primary production estimates do not increase monotonically with surface chlorophyll concentration.

The neuronal network model also mimics photoinhibition that occurs when high surface irradiance conditions are observed, except in case of very high phytoplankton biomass. It also describes only monotonic relationships and is much too sensitive to very low irradiance values, as the primary production gradient in the [0, 10] irradiance interval is unrealistically steep.

Even the neuronal network approach, however, is not perfect. For instance, the primary production estimate for null biomass values is quite small, but it is not null, as expected. The same applies to null irradiance conditions.

Of course, these are minor problems in real world applications, but they clearly show that neuronal network models completely depend on training data, even on those data that are unlikely to be actually measured (as, for instance, a null biomass values). In other words, neuronal networks are modelled after the "shape" of the training data sets, so they tend to retain all (and only) the features that are found in them. When small training data sets are used, the need for good generalization is particularly important.

The surface in Fig. 8.9c shows the primary production estimates provided by a 3-4-1 neuronal network similar to the one presented in Fig. 8.8a, but for the lack of adequate generalization. In this case the neuronal network was overtrained and acted

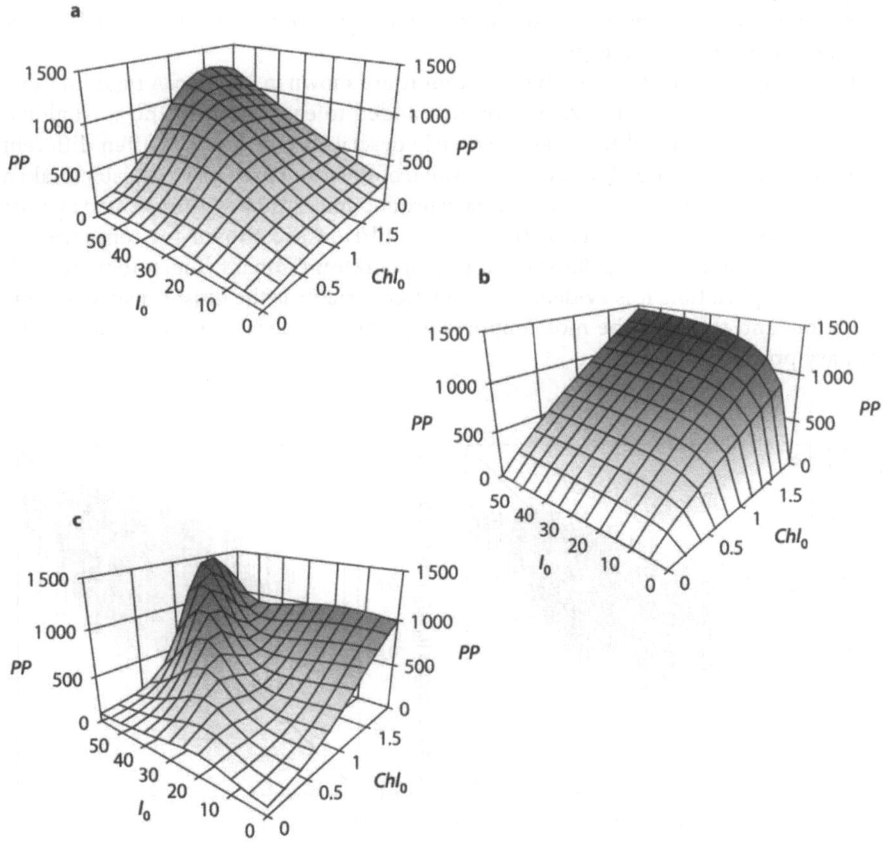

Fig 8.9. Primary production estimates (*PP*) obtained from different combinations of surface irradiance (I_0) and phytoplankton biomass (Chl_0). The primary production surfaces were obtained for a Western Mediterranean station (43° N, 8° E) in mid-June surface temperature conditions (18.94 °C); **a** from a 3-4-1 Mediterranean Sea neuronal network model; **b** from the VGPM (Behrenfeld and Falkowski 1997); **c** from an overtrained 3-4-1 neuronal network

as a memory rather than as a model. It accurately reproduces the training data, even though its shape does not make sense from a more general point of view. High values are attained, for instance, in the rightmost corner of the primary production surface, even when irradiance is null or very low.

This problem, of course, cannot affect conventional models, but can seriously hinder neuronal network models if it is not carefully taken into account.

8.3.5 Sensitivity Analysis

A better understanding of the modelled processes can be achieved by means of sensitivity analysis, which can provide very useful information about the role of the predictive variables in determining the neuronal network outputs. Of course, only first-order interactions can be easily considered and therefore the sensitivity analysis results are to be carefully interpreted.

The results of a sensitivity analysis procedure are shown in Fig. 8.10. A random value from a [−0.5, 0.5] uniform distribution was added to each input of the 3-4-1 global neuronal network model that was previously described (see Fig. 8.7a). Ten different random values were added to each input value of the three predictive variables, taken one at a time, and the ranges of the primary production estimates were plotted against observed values. The effects of this noise addition are shown for temperature (Fig. 8.10a), irradiance (Fig. 8.10b) and phytoplankton biomass (Fig. 8.10c). According to the expectation, it is evident that the latter variable is the most sensitive to perturbation, and that it is the most important predictive variable in determining the primary production values.

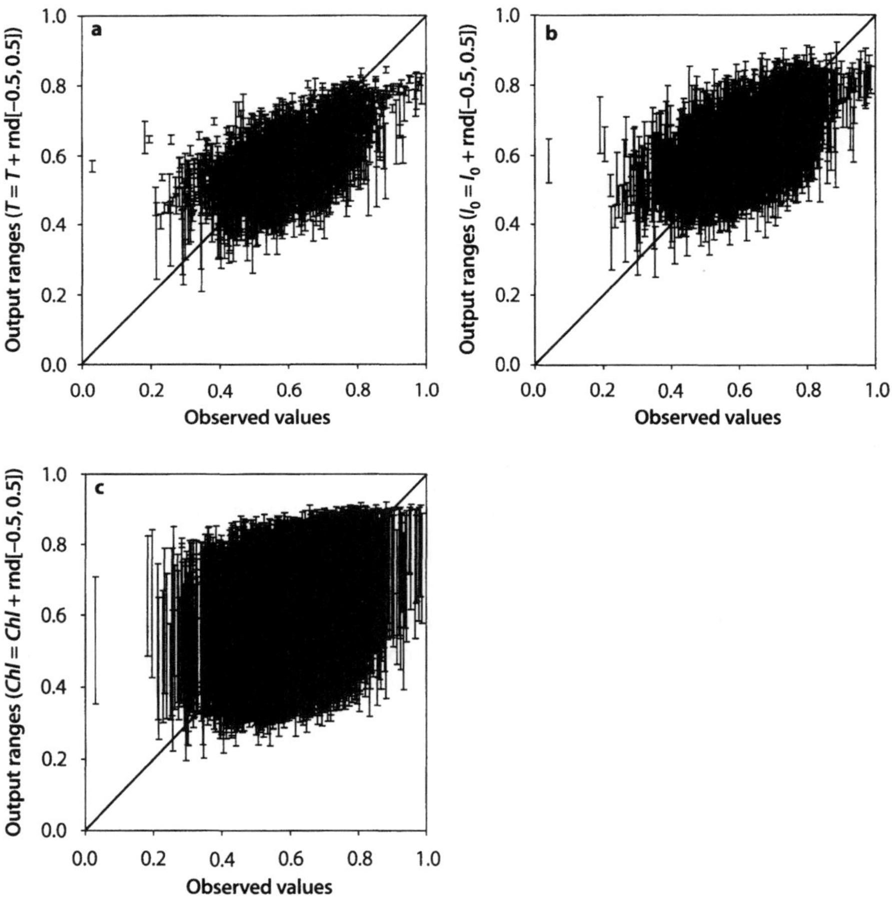

Fig. 8.10. Sensitivity analysis of the 3-4-1 global neuronal network model. A random value in the [−0.5, 0.5] range was added to each input variable, one at a time. The range of the primary production estimates obtained after perturbation of 10 input values are shown for; **a** surface temperature (T_0); **b** surface irradiance (I_0); **c** surface phytoplankton biomass (Chl_0)

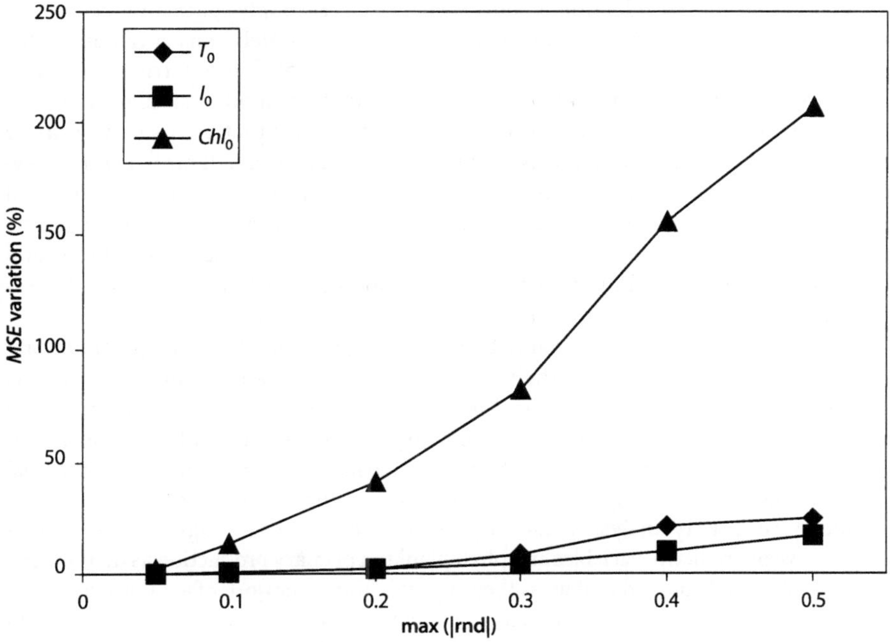

Fig. 8.11. Mean square error variation caused by different levels of random perturbation of the input variables in the 3-4-1 global neuronal network model

It can be also interesting to try how much noise can be added to each input with no significant modification of the neuronal network output. The results of this test on the same 3-4-1 global model are shown in Fig. 8.11. It can be noticed that a [−0.05, 0.05] random noise addition did not cause perceivable effects on the mean square errors. This evidence implies that this neuronal network model is very robust with respect to small errors in predictive variables (e.g. sampling errors).

When greater amounts of white noise are added to the neuronal network inputs, the changes in output values become more evident. It is not surprising that phytoplankton biomass is the most sensitive predictive variable, but it is interesting to notice that irradiance was the least sensitive one. This result probably depends on the fact that temperature, which is usually the least sensitive variable at small spatial scale, is strongly related to latitude at global scale.

8.4 Discussion

Neuronal networks can be very effective tools in ecological modelling. As demonstrated by the sample applications that have been briefly presented in this paper, they are able to reproduce complex nonlinear relationships and to provide better accuracy than conventional models do. They are also very flexible tools, which can be applied to a number of different problems.

As far as primary production is concerned, there is certainly space for an enhancement of the results that have been already presented. Remotely sensed radiances, for instance, could be directly used as predictive variables, or fluorometric data can be used in the depth-resolved model of the Gulf of Napoli or in other similar models.

In general, neuronal network models provide estimates that are usually accurate (low mean square error) and unbiased (small deviation from a null mean error). However, one of the main problems in the application of neuronal networks to ecological modelling is their tendency to overfit the training data patterns. This problem is particularly relevant in an ecological context, where data sets are usually scanty. Several strategies can be applied in order to obtain a good generalization, but none can replace adequate data sets for training and testing procedures.

A major advantage of neuronal network models, not only for phytoplankton primary production modelling, is their capability of incorporating information that is difficult to manage with conventional models (e.g. geographical coordinates, binary or nominal data, etc.), but that tends to co-vary with the modelled (i.e. output) variables. The covariation does not have to be linear, because neuronal networks can deal with nonlinear relationships more easily than other empirical models.

Finally, sensitivity analysis makes it possible to assess the strength of the relationships between predictive variables and phytoplankton primary production as they emerge from the observed data rather than as they appear in the framework of an a priori simplified theoretical model. Therefore, sensitivity analysis can provide a deeper insight into the dynamics of phytoplankton primary production under real world conditions.

Acknowledgements

I wish to thank Vincenzo Saggiomo and Maurizio Ribera of the Stazione Zoologica "A. Dohrn" in Napoli, Italy for providing the primary production data set of the Gulf of Napoli and the Ocean Primary Productivity Working Group (OPPWG) for its data base (ftp://warrior.rutgers.edu/pub/Database/Database2.html), which significantly contributed to the training and validation sets of my global model.

References

Behrenfeld MJ, Falkowski PG (1997) Photosynthetic rates derived from satellite-based chlorophyll concentration. Limnol Oceanogr 42(1):1–20

Cole BE, Cloern JE (1987) An empirical model for estimating phytoplankton productivity in estuaries. Mar Ecol Prog Ser 36:299–305

Eppley RW, Stewart E, Abbott MR, Heyman U (1985) Estimating ocean primary production from satellite chlorophyll. Introduction to regional differences and statistics for the Southern California Bight. J Plankton Res 7:57–70

Györgyi G (1990) Inference of a rule by a neuronal network with thermal noise. Phys Rev Lett 64:2957–2960

Hornik K, Stinchcombe M, White H (1989) Multilayer feedforward networks are universal approximators. Neuronal Networks 2:359–366

Lek S, Delacoste M, Baran P, Dimopoulos I, Lauga J, Aulagnier S (1996) Application of neuronal networks to modelling nonlinear relationships in ecology. Ecol Model 90(1):39–52

Recknagel F (1997) ANNA – artificial neuronal network model predicting blooms and succession of blue-green algae. Hydrobiol 349:47–57

Recknagel F, French M, Harkonen P, Yabunaka K-I (1996) Artificial neuronal network approach for modelling and prediction of algal blooms. Ecol Model 96(1–3):11–28

Rumelhart DE, Hinton GE, Williams GE (1986) Learning representations by back-propagating errors. Nature 323:533–536

Scardi M (1996) Artificial neuronal networks as empirical models of phytoplankton production. Mar Ecol Prog Ser 139:289–299
Smith RC, Baker KS (1978) The bio-optical state of ocean waters and remote sensing. Limnol Oceanogr 23 (2):247–259

Appendix

The daily variability of the surface irradiance caused by cloudiness was considered as a neuronal network input to account for the effect of photoadaptation of phytoplankton cells on primary production.

Assuming that under constant weather conditions the surface irradiance may be heuristically described by a gaussian curve, the differences between log-transformed irradiance values at time t and $t+1$ should be placed along a straight line. The determination coefficient (R^2) of the linear regression of these differences against time indicates how much variance is explained by the linear regression (i.e. by the gaussian curve). Therefore, $1 - R^2$ is a measure of the unexplained variance due to deviations from the gaussian model because of cloudiness. Only data from 8:00 A.M. to 4:00 P.M., i.e. from hours with daylight conditions during the whole year, were taken into account.

Since high variability of the irradiance is likely to stress phytoplankton cells more severely under high irradiance conditions than under low irradiance conditions (e.g. during winter), the $1 - R^2$ term was multiplied by the total daily irradiance scaled into the [0, 1] interval.

The resulting index varies within the [0, 1] range and summarizes the variability of the irradiance during the day. Small values indicate constant conditions, whereas large values indicate variable cloudiness. An example of the index calculations for a sunny day, with small deviations from maximum irradiance, is shown in Fig. A8.1.

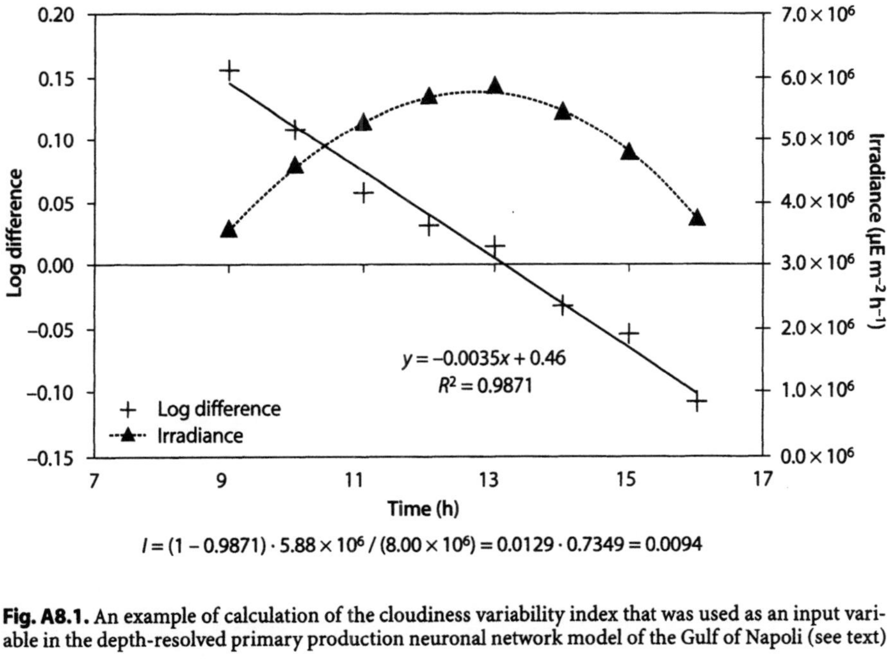

$I = (1 - 0.9871) \cdot 5.88 \times 10^6 / (8.00 \times 10^6) = 0.0129 \cdot 0.7349 = 0.0094$

Fig. A8.1. An example of calculation of the cloudiness variability index that was used as an input variable in the depth-resolved primary production neuronal network model of the Gulf of Napoli (see text)

Chapter 9

Predicting Presence of Fish Species in the Seine River Basin Using Artificial Neuronal Networks

P. Boët · T. Fuhs

9.1 Introduction

Fish communities are the expression of fundamental biological processes (reproduction, feeding, and shelter) at different scales in time and space. They can be considered to be good indicators of the health of aquatic ecosystems (Fausch et al. 1990). This paradigm is the basis for using biological monitoring of fish to assess environmental degradation (Karr 1987).

Such a community-based approach for monitoring aquatic ecosystems, however, requires a sound understanding of the nature of the major factors that cause, or at least explain, the patterns of a fish community's structure and composition among water bodies (Lyons 1996). The identification of these factors, their evaluation, and their ranking according to a hierarchical system, would be therefore an essential tool for the conservation, or the restoration, of both the populations and the aquatic environments.

In the Seine River basin, an initial study has evidenced the main factors that contribute to the current organization of the fish assemblages on the scale of the entire hydrographic network (Belliard 1994). The characteristics of the environment associated with the longitudinal and regional organization of the basin proved to be determining factors (Belliard et al. 1997).

This work is based on the results of in situ fish sampling made by electrofishing. These data are by nature extremely heterogeneous. The reasons are twofold: on the one hand, they come from sampling aimed at various objectives, and, on the other hand, the catch technique presents unavoidable biases, especially in the case of large watercourses.

The determination of the relationships between the habitat characteristics and the presence of the fish species would be of great interest. First, they would quantify the comparative importance of the environmental variables in the structuring mechanisms of the fish communities. Second, they would allow researchers to test the impact of perturbations of those variables.

For these reasons we attempted to use connectionist networks, capable of solving nonlinear problems, with robustness with respect to noised or incomplete data, to predict the composition of a fish community according to the environmental characteristics.

9.2 Description and Selection of Data

Our work is based on the exploitation of a large database covering the whole Seine River basin. These data include the results of over 700 fish catches at 583 sampling sta-

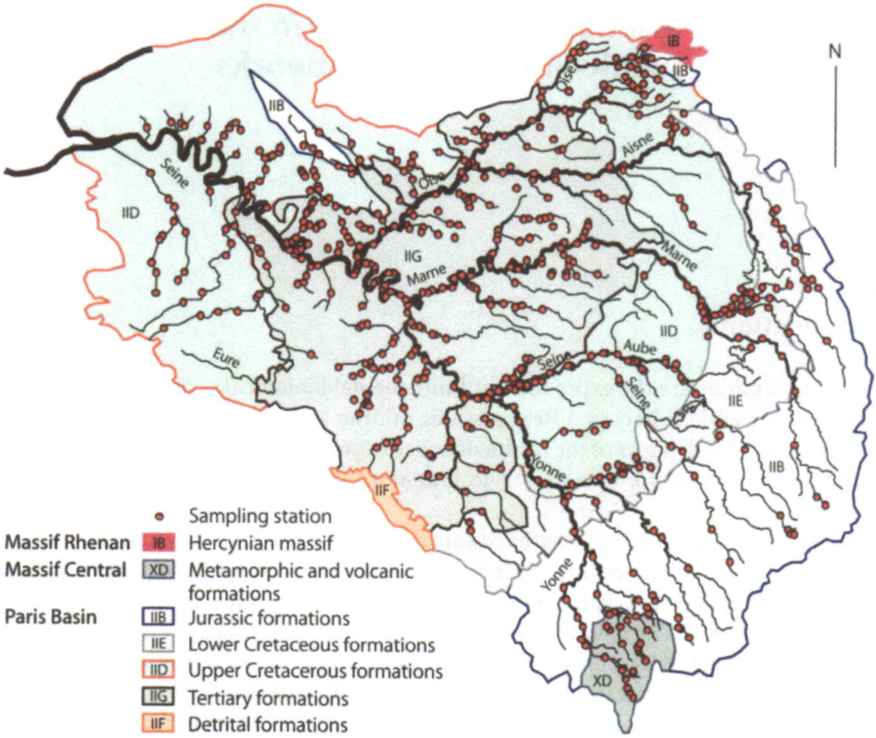

Fig. 9.1. Map of the Seine River basin and localisation of the sampling stations

tions (Fig. 9.1). They consist of fish species abundance, representing more than 200 000 fish belonging to 39 species.

In addition, for each station, about 15 covariates describe the characteristics of the river reach and its nearby environment. Among them, we select the most relevant input data on the basis of the results of our previous studies (Belliard 1994; Belliard et al. 1997). We keep six of them.

- The *stream order* (Strahler 1957) is a parameter of the position of the fishing station within an upstream-downstream gradient in the hydrographic network. Given its summary nature, it accounts for numerous physical and functional variables in the waterway. The lower stretches of the Seine River are order eight.
- The *ecoregion* is the natural region based on homogeneous environmental characteristics to which the station belongs. We used the first two hierarchical levels of the map of the phytoecological zones of France (Dupias and Rey 1985), which are determined by climatic differences and by the geological nature of the substrate. This yielded seven regions for the whole basin (Belliard et al. 1997).
- The *slope* and *width* are classic morphological descriptors in hydrobiology (Huet 1949; Illies and Botosaneanu 1963; Schlosser 1982; Zalewski and Naiman 1985; Rahel and Hubert 1991).

- The *quality of the water* is a score provided by the Seine-Normandy Water Agency (AREA 1992). It ranges from 1 to 5 according to the degradation of the water quality. This index uses 5 qualitative values (excellent, good, fair, mediocre, bad) to integrate 12 variables (temperature, dissolved oxygen, pH, % oxygen saturation, 5-days biological demand, nitrates, phosphates, ammonia, heavy metals, turbidity, chlorophyll, and faecal coliforms).
- The *quality of the habitat* is a summary index taking into account the degree of degradation of five components of the physical habitat: the major and the minor bed, the nature of the banks and of the substrate, and the degree of regulation of the flow (Souchon and Trocherie 1990). It ranges from 0 to 3 with the increase of the artificialisation of the habitat.

Each component is graded 0, 1 or 2, according to its situation, i.e. pseudonatural, influenced or artificial. The global quality of the habitat is equal to the worst note obtained by one of components. When several components obtain the poorest grade, the final grade is decreased by 1. Finally, 4 classes of habitat quality are distinguished:

0. very good, all components are pseudonatural (0) ;
1. good, only one component is influenced (1) ;
2. mediocre, at least two components are influenced (1), or one is artificial (2) ;
3. bad, at least two components are artificial (2).

As the values of the descriptors have not always been collected at the time of the fishing investigations, we reduced our database to the 507 more recent fish catches, i.e. since 1980.

Finally, we considered as output the only 26 most representative fish species (frequency >9%, Table 9.1) out of the 39 species present over the basin. Since the ANN method is known to be sensitive to greatly unbalanced samples, it could not yield an accurate prediction for rare species despite of their likely biological significance.

9.3 Methodology

9.3.1 Choice of Implementation and Tuning of Parameters

The objective of our study is therefore to predict the presence or absence of fish based on the environmental characteristics of the river. The problem is one of classification, in which the use of multilayer connectionist networks has revealed its value in a broad field of studies (Rumelhart et al. 1986).

We chose the so-called MASS implementation of ANN, provided by Ripley and described in Ripley (1996) and Venables and Ripley (1997). This implementation is freely available as an add-on library package for S-PLUS and R software packages. It is based on the BFGS method, a quasi-Newton optimizer which estimates iteratively the Hessian matrix. As a second-order estimation method, it is free of any rule-of-thumb tuning parameters, like learning rate or momentum, contrary to the classical back propagation algorithm (Dennis and Schnabel 1983).

Table 9.1. List of the 26 most frequent fish species in the database

Family	Species	Common name	Species code
Petromyzonidae	*Lampetra planeri*	Brook lamprey	Lap
Anguillidae	*Anguilla anguilla*	Eel	Ana
Cyprinidae	*Alburnus alburnus*	Bleak	Ala
	Barbus barbus	Barbel	Bab
	Abramis brama	Common bream	Abb
	Blicca bjoerkna	White bream	Blb
	Cyprinus carpio	European carp	Cyc
	Chondrostoma nasus	Nase	Chn
	Gobio gobio	Gudgeon	Gog
	Leuciscus cephalus	Chub	Lec
	Leuciscus leuciscus	Dace	Lle
	Phoxinus phoxinus	Common minnow	Php
	Rutilus rutilus	Roach	Rur
	Scardinius erythrophtalmus	Rudd	Sce
	Tinca tinca	Tench	Tit
Cobitidae	*Barbatula barbatula*	Stone loach	Baa
Esocidae	*Esox lucius*	Pike	Esl
Salmonidae	*Oncorhynchus mykiss*	Rainbow trout	Onm
	Salmo trutta fario	Brown trout	Sat
Gadidae	*Lota lota*	Burbot	Lol
Gasterosteidae	*Gasterosteus aculeatus*	Three-spined stickleback	Gaa
	Pungitius pungitius	Nine-spined stickleback	Pup
Cottidae	*Cottus gobio*	Sculpin	Cog
Percidae	*Gymnocephalus cernua*	Ruffe, pope	Gyc
	Perca fluviatilis	Perch	Pef
	Stizostedion lucioperca	Pike-perch	Stl

The parameters to be tuned are then restricted to *(i)* the maximum number of iterations, *(ii)* the tolerance for the error criterion and *(iii)* the weight decay parameter. For the first one, we decided to always wait for the convergence of the algorithm, increasing the maximum number of iterations if needed. The second one equals 0.0001, which is a sensible value for our data. We finally followed Ripley's recommendations for the weight decay, choosing it to be 0.005.

However, the effective use of this technique is nonetheless a delicate operation, since it still requires a correct sizing of the network, for which there is no straightforward methodology.

9.3.2
Architecture of the Network and Error Criterion

The inputs of the network are closely related to the predictive variables we selected. Only one of them, the Ecoregion, is purely qualitative, and must be split in as many binary variables as its different values (6 in this case) minus 1. The quantitative variables being 5, the input layer includes 10 units. In order for the weight decay regularization to make sense, Ripley's implementation normalizes the continuous variables between 0 and 1.

Our first idea was to build a network able to predict in one shot each of the 26 species we selected. Doing so, we hoped to take into account the correlation between species. However, this would have required a 26 units output layer. If N is the number of hidden units, the total number of parameters would have raised $37N + 26$. This number has a questionable statistical relevance as soon as N is moderately high (say 6–8) in regard to our 507 catches. We have eventually chosen in this first attempt to focus on the hidden layer size and consequently we trained a different network for each of the 26 species. Together with the need to train 26 different networks instead of one, the major drawback of this approach is to neglect the possible interactions between species despite of their biological evidence.

Since our problem is a binary classification, we need a sole output unit whose value is the probability of presence of the studied species. This implies that each unit has a logistic transfer mode, and allows the use of a maximum likelihood error criterion, i.e. the entropy fit criterion (Ripley 1996).

9.3.3
Weighting of Data

ANN models, like many classification methods, are biased towards the majority class. This may yield unsound prediction quality for unbalanced data. For example, initial experiments on the common carp (10% of the catches show this species) often give a 0-valued response network, unable to predict any presence. Even if such a network shows a "good" prediction ratio, it is totally useless! This is the reason why we chose to weight the minority class to get a balanced sample. We uniformly assigned the ratio of the two class sizes to any observation belonging to the minority class.

9.3.4
Prediction Error Assessment

As any statistical inference model, ANN are subject to a bias-variance dilemma (Geman et al. 1992). This dilemma stems from the finiteness of the catch sample we own. The bigger ANN model we select, the lower the bias, but the higher the between-sample variability and vice versa. The degrees of freedom of ANN model are the size of the hidden layer, i.e. the number of its units. Then, selecting a good ANN model is closely related to the choice of this number.

A standard method to select this number consists of separating the available data in a training set and a validation set. The prediction error rate is then computed over the validation set for each different number of hidden units. Eventually, the network with the lowest generalization error is kept. Although the method is applicable in our case, the rather small number of observations of data (507 catches) led us to prefer a computer intensive but statistically better founded method, the cross-validation introduced by Stone (1977).

For each number of hidden units, we performed a 5-fold cross-validation to assess an unbiased prediction error rate. Because of the known high variability of cross-validation, we conducted the experiment at least ten times for each combination of a species and a number of hidden units. We finally kept the mean and standard deviation of these (at least ten) experiments. Obviously, each experiment saw a complete redrawal of the 5 data folds to be used in cross-validation.

To sum up, for each species and for each hidden layer size between 1 and 10 (or more), we developed the following protocol.

1. Repeat 10 times
   - 5-fold breakdown of the weighted sample
   - Train ANN for each 4/5 fold
   - Build confusion matrix by adding the errors on each corresponding 1/5 validation fold
   - Compute the total classification error, and the differentiated error ratio for presence and absence
2. Compute mean and standard deviation of the prediction errors
3. Select the "best" hidden layer size

9.4 Results

As stated in the previous section, our aim was to select the best network architecture with respect to the cross-validation prediction error. This proved to be a more difficult task than we expected. We plotted on Fig. 9.2 the error curve against the number of hidden units for 8 different species. Each graph shows the three errors (also appearing on Table 9.2 for 5-unit hidden layers): *(i)* bad prediction of absence ratio, *(ii)* bad prediction of presence ratio, *(iii)* overall bad prediction ratio. The problem is that none of the curves appeared to be J-shaped, as is classical in ANN prediction. They rather involve an initial decrease followed by a plateau behaviour. It is then not straightforward to select the number of units realizing the minimum of the curve, since the error decrease between successive number of hidden units has a magnitude similar to the standard deviation of the error. To achieve a mode unquestionable choice of the ANN architecture, we plan for future work the use of penalization criteria such as Akaike and/or Bayesian information criteria (Schwarz 1978; Sakamoto et al. 1986). These criteria amount to selecting a model with not too many parameters (e.g. the number of weights in an ANN), but achieving a satisfactory accuracy.

However, our results are rather good in general as it appears in Table 9.2, where we summarized the results for 5-unit hidden layer networks. This number appears to be a good trade-off between the network size and the quality of the prediction errors.

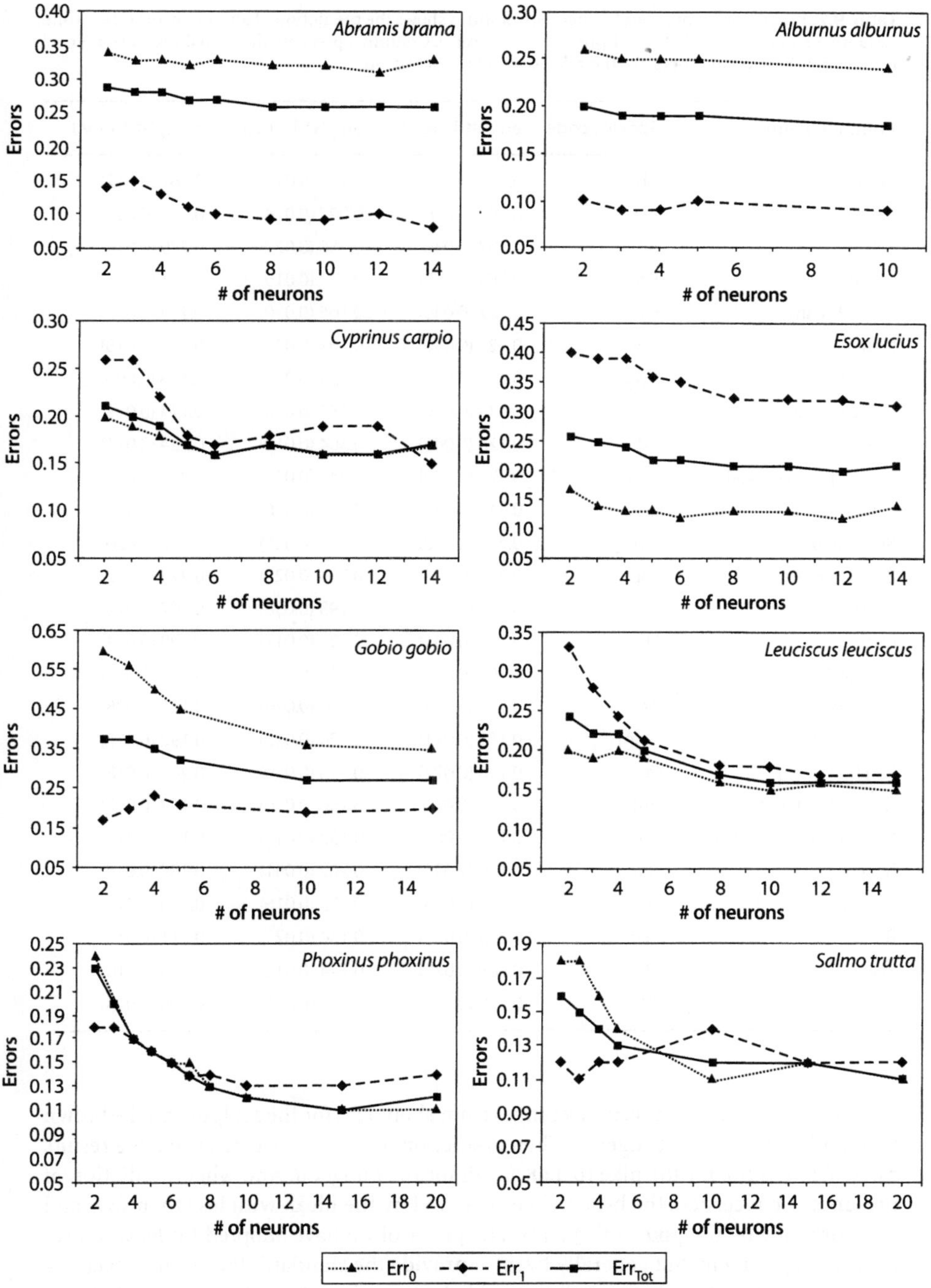

Fig. 9.2. Prediction error ratios behaviour against the number of hidden units for 8 species

Table 9.2. Prediction error ratios by species. Column *3* shows the prediction of absence error ratio, while Column *4* shows the prediction of presence error ratio. Column *5* presents the global prediction error ratio. Parenthesised numbers are the related standard deviations

Common name	Species code	err_0 (std. dev.)	err_1 (std. dev.)	err_{Tot} (std. dev.)
Common bream	Abb	0.324 (0.028)	0.107 (0.015)	0.267 (0.012)
Bleak	Ala	0.247 (0.016)	0.097 (0.014)	0.190 (0.008)
Eel	Ana	0.297 (0.036)	0.256 (0.027)	0.279 (0.016)
Barbel	Bab	0.189 (0.027)	0.147 (0.019)	0.181 (0.012)
White bream	Blb	0.227 (0.040)	0.165 (0.030)	0.214 (0.017)
Nase	Chn	0.229 (0.030)	0.162 (0.024)	0.217 (0.018)
Sculpin	Cog	0.132 (0.013)	0.123 (0.009)	0.128 (0.006)
European carp	Cyc	0.173 (0.030)	0.182 (0.016)	0.174 (0.013)
Pike	Esl	0.126 (0.046)	0.364 (0.019)	0.224 (0.011)
Three-spined stickleback	Gaa	0.227 (0.050)	0.247 (0.027)	0.230 (0.018)
Gudgeon	Gog	0.448 (0.047)	0.214 (0.056)	0.324 (0.016)
Ruffe, pope	Gyc	0.226 (0.058)	0.266 (0.025)	0.230 (0.020)
Brook lamprey	Lap	0.221 (0.048)	0.254 (0.027)	0.225 (0.022)
Chub	Lec	0.231 (0.021)	0.195 (0.038)	0.207 (0.016)
Dace	Lle	0.194 (0.031)	0.206 (0.017)	0.198 (0.014)
Burbot	Lol	0.167 (0.027)	0.115 (0.017)	0.157 (0.013)
Stone loach	Baa	0.131 (0.039)	0.166 (0.026)	0.145 (0.008)
Rainbow trout	Onm	0.189 (0.071)	0.224 (0.022)	0.192 (0.017)
Perch	Pef	0.196 (0.027)	0.213 (0.050)	0.207 (0.018)
Common minnow	Php	0.164 (0.041)	0.156 (0.026)	0.162 (0.017)
Nine-spined stickleback	Pup	0.175 (0.060)	0.137 (0.017)	0.167 (0.011)
Roach	Rur	0.146 (0.026)	0.202 (0.041)	0.189 (0.014)
Brown trout	Sat	0.135 (0.024)	0.123 (0.015)	0.130 (0.010)
Rudd	Sce	0.245 (0.029)	0.134 (0.022)	0.211 (0.012)
Pike-perch	Stl	0.189 (0.026)	0.144 (0.012)	0.185 (0.010)
Tench	Tit	0.328 (0.037)	0.252 (0.040)	0.308 (0.023)

The global prediction accuracy ranges from about 13% for the sculpin and the brown trout, till 32.4% for the gudgeon. When prediction of absence is concerned, the results range from 12.6% for the pike to 44.8% still for the gudgeon. And when prediction of presence is concerned, the best ratio is achieved by the bleak with less than 10% and the worst one by the pike with 36.4%. The protocol we have adopted (at least ten repeats of experiment, but generally twenty) provide the standard deviation of each ratios. These deviations show a rather good behaviour: less than 3% for the global prediction errors, whereas the mean for all species is about 1.5. But they are a little larger for the by-class prediction: from above 1 to 7% for the absence prediction while from 1 to less than 6% for the presence prediction. This means that the different networks obtained after training show similar prediction accuracy, even if this accuracy is estimated by cross-validation which is known to be variable.

According to the error ratios, the studied species may be classified in four subsets:

- Species with good prediction ratios (<19%), and with presence and absence prediction errors of the same magnitude. This group includes barbel, sculpin, carp, burbot, stone loach, minnow, nine-spined stickleback, brown trout and pike-perch.
- Species with average prediction ratios (<25%), still with equivalent presence and absence prediction ratios. This subset is made of white bream, nase, three-spined stickleback, ruffe, lamprey, chub, dace, rainbow trout, perch and roach.
- Species with not so bad global prediction ratio (<23%), but with a dissymmetry between presence and absence prediction. This consists of bleak, pike and rudd.
- Species with at least a bad prediction ratio (>25%), often combined with the dissymmetry mentioned above. This group is made of common bream, eel, gudgeon and tench.

We must notice that this hierarchy does not change with the increase of the number of units in the hidden layer. This reinforces the usefulness of raising ecological reasons to the observed differences among species. The Section 9.5.1 of the following discussion is devoted to this attempt.

9.5 Discussion

Prediction of a species in terms of presence or absence using connectionist multilayer neuronal networks, tried and tested on the scale of the Seine River basin, and according to very global descriptors of the quality of the aquatic medium (six input summary variables) proves to be relevant.

Whereas the input data suffer from considerable noise, the success rates in generalization vary from 67.6 to over 87% according to the species. That represents a highly appreciable performance, as a measurement error in the order of 10 to 20% is common in this type of data.

These results confirm the capacity of neuronal networks to accurately predict fish communities as already shown by Mastrorillo et al. (1998) or Guégan et al. (1998), who dealt with species richness at a large scale according to few physical variables, and Mastrorillo et al. (1997), who studied three species of small-bodied fish at the river scale.

9.5.1 Ecological Soundness

The best levels of success were obtained for four species: the trout *Salmo trutta fario*, the sculpin *Cottus gobio*, the stone loach *Barbatula barbatula*, and the minnow *Phoxinus phoxinus*. These are headwater fish species, where stream habitats still remain relatively preserved from human disturbances, and whose ecological profiles are clear, shown, for example, in Fig. 9.3.

On the other hand, the poorest result is observed for the gudgeon *Gobio gobio* and may be explained by its ecological particularities. Indeed, in the Seine River basin, the

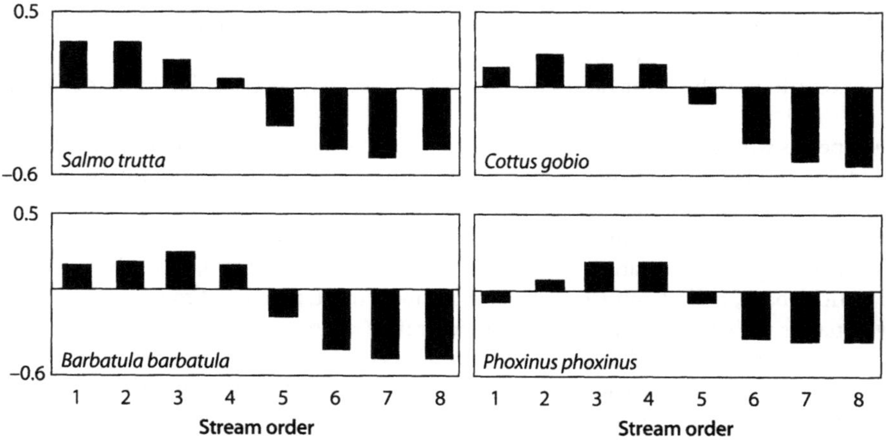

Fig. 9.3. Ecological profile of the trout (*Salmo trutta fario*), the sculpin (*Cottus gobio*), the stone loach (*Barbatula barbatula*), and the minnow (*Phoxinus phoxinus*) according to the stream order in the Seine River (from Belliard, 1994). The profile value $V_i = F_i - F_{tot}$, where F_i is the relative frequency of the species in the sets of class *i*, and F_{tot} is the relative frequency of the species over the whole data sets. All these profiles are statistically significant χ^2 test; $p < 0.001$)

distribution of this species exhibits two distinct groups (Fig. 9.4). In the upstream part, populations composed of small individuals are commonly linked to good quality running water flowing over sand or gravel substrates. On the contrary, downstream, in areas richer in organic matter, are found large gudgeons sometimes measuring up to 25 cm long. Obviously, it will be wise to carefully separate these two subsets to drive the networks better and to improve the quality of prediction.

Likewise, the ecology of the eel *Anguilla anguilla* may explain the poor results observed. Because of the heavy pollution of the river in the sixties, this species tends to decline from the basin. Now, with the general improvement of the water quality due to the wastewater treatment plants, this migratory species is coming back again, but its spread in the upper reaches of the river is restricted by the navigation locks. Its distribution on the river basin depends therefore on other factors.

The in-depth examination of the results is also interesting. For example, in the case of the pike *Esox lucius* (Table 9.2), the total error appears relatively low, but in fact the prediction of the presence of this fish is bad. In most cases, the network predicts the absence of the species where it is present (see also Fig. 9.2). This species is involved in numerous restocking operations, and therefore we believe that its presence is partly independent of the aquatic environment characteristics.

Conversely, in the case of the common bream *Abramis brama*, the total error appears high, but it is essentially due to the prediction of the presence of this fish by the network, whereas in fact it is not included in the catch record (Table 9.2 and Fig. 9.2). Because of its morphology and its deep water behaviour, this species is difficult to catch efficiently by electrical fishing, even when the bream is present in the environment. The situation is similar with the tench *Tinca tinca*. This illustrates the importance of the representativeness of the sampling, which raises the problem, already widely discussed, of the ecological significance of the absence of a species in the fauna records.

Fig. 9.4. Ecological profile of the gudgeon (*Gobio gobio*) according to the stream order in the Seine River (from Belliard, 1994). The profile value $V_i = F_i - F_{tot}$, where F_i is the relative frequency of the species in the sets of class i, and F_{tot} is the relative frequency of the species over the whole data sets. This profile is statistically significant χ^2 test; $p < 0.001$)

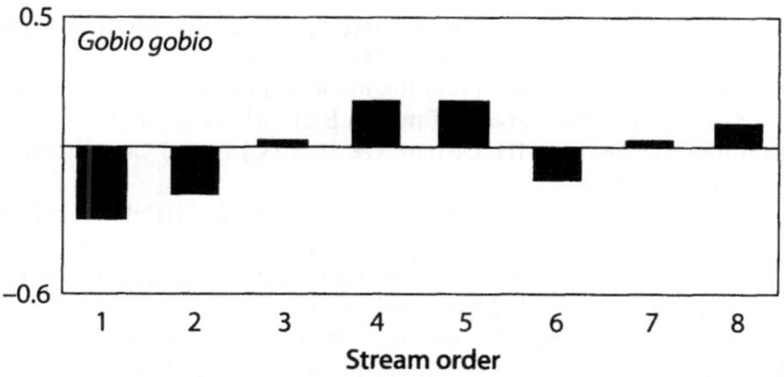

Obviously, these poorly predicted species might also be influenced by other ecological variables than the ones we kept in our model.

9.5.2
On Methodology

Finally, these results focus on the prediction of the presence or absence of a given species, whereas our initial aim was to consider directly the entire community. As this could not be done with all the species at the same time, we think that the catch data would have to be reprocessed to identify the different types of fish stock in place. Then, multilayer networks could be driven, output of which would no longer be a particular species but a given type of fish community.

Nonetheless, given the nature of the data processed and the very summary character of the input variables, these exploratory results prove to be satisfactory. Indeed, they are already very close to those obtained using discriminant analyses or multiple regressions (Pouilly 1994; Capra 1995). But these authors, working at the scale of the microhabitat, have obviously obtained extremely reliable data on the description of the habitat and the sampling of the in situ fauna.

Oberdorff et al. (1998) modelled statistically the occurrence of the 34 most common fish species of France using logistic regression models applied to presence/absence data, in relation to regional and local environmental factors in order to adapt and calibrate a multimetric index which could serve as a practical technical reference for conducting biological assessments of lotic systems.

Our initial trials are therefore encouraging, considering that there are currently very few predictive models of fish on the scale of a whole river basin (Huet 1959; Verneaux 1977; Oberdorff et al. 1998). In future work, we plan to extend this initial study to the comparison with other nonlinear classification methods like generalized additive models (McCullagh and Nelder 1989), and decision trees (Breiman et al. 1984). Our long term objective is to simulate the consequences of natural or human disturbances on the composition of a fish community at the scale of the hydrographic basin, thanks to such predictive models.

References

AREA (1992) Cartographie de synthèse des schémas départementaux de vocation pisicole et des ZNIEFF humides du bassin Seine-Normandie. Rapport DIREN Ile-de-France Délégation de bassin Seine-Normandie, Agence de l'eau Seine-Normandie

Belliard J (1994) Le peuplement ichtyologique du bassin de la Seine: Rôle et signification des échelles temporelles et spatiales. Thèse Doct., Paris VI
Belliard J, Boët P, Talès E (1997) Regional and longitudinal patterns of fish community structure in the Seine River basin, France. Environ Biol Fish 50:133–147
Breiman L, Friedman JH, Olshen RA, Stone CJ (1984) Classification and regression trees. Chapman & Hall, New York
Capra H (1995) Amélioration des modèles prédictifs d'habitat de la truite fario : échelles d'échantillonnage ; intégration des chroniques hydrologiques. Thèse Doct., Univ. Claude Bernard, Lyon I
Dennis JE, Schnabel RB (1983) Numerical methods for unconstrained optimization and nonlinear equations. Prentice-Hall, Englewood Cliffs
Dupias G, Rey P (1985) Document pour un zonage des régions phyto-écologiques. CNRS, Février
Fausch KD, Lyons J, Karr JR, Angermeier PL (1990) Fish communities as indicators of environmental degradation. In: Adams SM (ed) Biological indicators of stress in fish. American Fishery Society Symposium 8:123–144
Geman S, Bienenstock E, Doursat R (1992) Neuronal networks and the bias/variance dilemma. Neuronal Computation 4:1–58
Guégan J-F, Lek S, Oberdorff T (1998) Energy availability and habitat heterogeneity predict global riverine fish diversity. Nature 391(22):382–391
Huet M (1949) Aperçu des relations entre la pente et les populations piscicoles des eaux courantes. Schweiz Z Hydrol 11(3–4):332–351
Huet M (1959) Profiles and biology of western European streams as related to fish management. Trans Am Fish Soc 88(3):155–163
Illies J, Botosaneanu L (1963) Problèmes et méthodes de la classification et de la zonation écologique des eaux courantes, considérées surtout du point de vue faunistique. Mitt Intern Verh Limnol 12:1–57
Karr R (1987) Biological monitoring and environmental assessment: A conceptual framework. Environmental Management 11:249–256
Lyons J (1996) Patterns in the species composition of fish assemblages among Wisconsin streams. Environ Biol Fish 45:329–341
Mastrorillo S, Lek S, Dauba F, Belaud A (1997) The use of artificial neuronal networks to predict the presence of small-bodied fish in a river. Freshwat Biol 38:237–246
Mastrorillo S, Dauba F, Oberdorff T, Guégan J-F, Lek S (1998) Predicting local fish species richness in the Garonne River basin. C R Acad Sci Paris, Life Sciences 321:423–428
McCullagh P, Nelder JA (1989) Generalized linear models, 2nd edn. Chapman & Hall, London
Oberdorff T, Pont D, Hugueny B, Boët P, Porcher J-P, Chessel D (1998) A probabilistic model characterizing riverine fish communities of French rivers: A framework for the adaptation of a fish based index. In: Jungwirth M, Schmutz S, Kaufmann M (eds) Assessing the ecological integrity of running waters. Intern. Symposium, Vienna, November 9–11, 1998
Pouilly M (1994) Relations entre l'habitat physique et les poissons des zones à cyprinidés rhéophiles dans trois cours d'eau du bassin rhodanien: vers une simulation de la capacité d'accueil pour les peuplements. Thèse Doct., Univ. Claude Bernard, Lyon I
Rahel FJ, Hubert WA (1991) Fish assemblages and habitat gradients in a Rocky Moutain-Great Plains stream: biotic zonation and additive patterns of community change. Trans Am Fish Soc 120:319–332
Ripley BD (1996) Pattern recognition and neuronal networks. Cambridge University Press, Cambridge
Rumelhart DE, McClelland JL, the PDP Research Group (1986) Parallel distributed processing. MIT Press/ Bradford Books, Cambridge
Sakamoto Y, Ishiguro M, Kitagawa G (1986) Akaike Information Criterion Statistics. Reidel Publishing Company
Schlosser IJ (1982) Fish community structure and function along two habitat gradients in a headwater stream. Ecol Monogr 52(4):395–414
Schwarz G (1978) Estimating the dimension of a model. Ann Stat 6:461–464
Souchon Y, Trocherie F (1990) Technical aspects of French legislation dealing with freshwater fisheries (June 1984): "Fisheries orientation schemes" and "fishery resources management plans". In: Van Densen WLT, Steinmetz B, Hughes RH (eds) Management of freshwater fisheries, Proceeding of EIFAC Symposium, Göteborg, May 31 – June 3, 1988, Pudoc Wageningen, pp 190–214
Stone M (1977) Cross-validation: a review. Math Operationforsch Statist Ser Statist 9:127–139
Strahler AN (1957) Quanitative analysis of watershed geomorphology. Trans Am Geophysi Union 38:913–920
Venables W, Ripley BD (1997) Modern applied statistics with S-Plus, 2nd edn. Springer-Verlag, New-York
Verneaux J (1977) Biotypologie de l'écosystème "eau courante". Déterminisme approché de la structure biotypologique. C R Acad Sc Paris 284:77–79
Zalewski M, Naiman RJ (1985) The regulation of riverine fish communities by a continuum of abiotic-biotic factors. In: Alabaster JS (ed) Habitat modification and freshwater fisheries. Butterworths, London, pp 3–9

Elucidation and Prediction of Aquatic Ecosystems by Artificial Neuronal Networks

F. Recknagel · H. Wilson

10.1 Introduction

Models in aquatic ecology are needed for hypothesis testing (elucidation) and management (predictions) of changing properties in estuaries, lakes, wetlands, and rivers. Two modelling approaches are distinguished to achieve these aims: inductive and deductive modelling. Inductive modelling is considered to be the result of structuring, aggregation, or pattern extraction of ecological data (see Fig. 10.1). The most common techniques available for inductive modelling are regression analysis and neuronal network training. Deductive modelling goes much further towards integration of structured and aggregated ecological data into relevant ecological theory (see Fig. 10.1). Deductive modelling is normally based on physical mass balances for food webs and nutrient cycles, or heuristic rule sets.

While both inductive and deductive models are supposed to be suitable for predictions of ecosystems (Peters 1986; Rigler and Peters 1995; Livingstone and Imboden 1996), there is a dispute about the potential of inductive models for elucidation. Livingstone and Imboden (1996) conclude that only deductive models based on all available information, including knowledge of relevant processes obtained in other contexts, are likely to be elucidative.

The aim of this paper is to introduce artificial neuronal networks as a new generation of inductive models that have potential for ecosystem elucidation as well as for ecosystem prediction. Two examples are chosen to demonstrate the superiority of neuronal networks over alternative models: prediction and elucidation of phytoplankton abundance, and prediction of density of brown trout redds.

10.2 Phytoplankton Abundance in Lakes and Rivers

10.2.1 Prediction

Blooms of toxic cyanobacteria and dinoflagellates are considered globally a threat to drinking water supply, aquaculture, fishery and tourism. Predictions of phytoplankton abundance with high resolution in time and species composition are therefore of a high concern for water quality management (NRA 1990). Even though there is a broad variety of phytoplankton models in the literature, there are big differences in the predictive validity and capacity of these models.

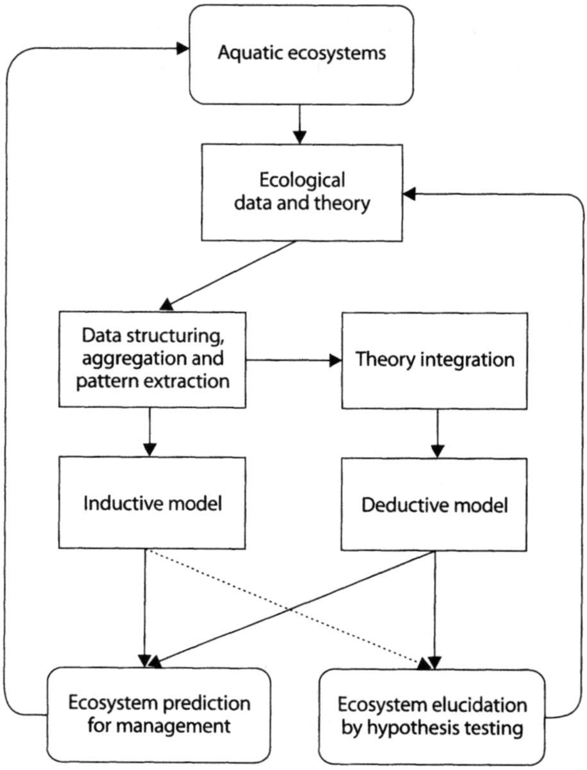

Fig. 10.1. Paradigm on inductive and deductive modelling of aqua-tic ecosystems

In Figs. 10.2 and 10.3, six prototypes of inductive and deductive models for phytoplankton are represented. Regression models (see Fig. 10.2a) can predict steady-state abundance of chlorophyll a for seasons or years but fail to predict dynamics in abundance. Long-term predictions by regression models are widely applied for the trophic status classification and eutrophication management of lakes (Sakamoto 1966; Vollenweider 1976; Dillon and Rigler 1975). Time series models (see Fig. 10.2b) can predict weekly abundance of chlorophyll a but fail to predict species abundance and succession. It was successfully implemented for short-term predictions of water quality in the River Thames (Whitehead and Hornberger 1984).

Artificial neuronal networks trained by time series of physical, chemical and biological water quality data proved to be predictive for dynamics of the most abundant algae species in lakes (French and Recknagel 1994; Recknagel et al. 1997), and rivers (Meier et al. 1998). Recknagel (1997) and Recknagel et al. (1998) have trained the neuronal network model ANNA with eight years of limnological data from Lake Kasumigaura, Japan. In this model, cell numbers of 5 blue-green algae species had been chosen as model outputs, and light, temperature, nutrient and zooplankton data were used as inputs. The authors used the trained model ANNA to predict abundance and succession of the algae species for two independent years not used for training. The

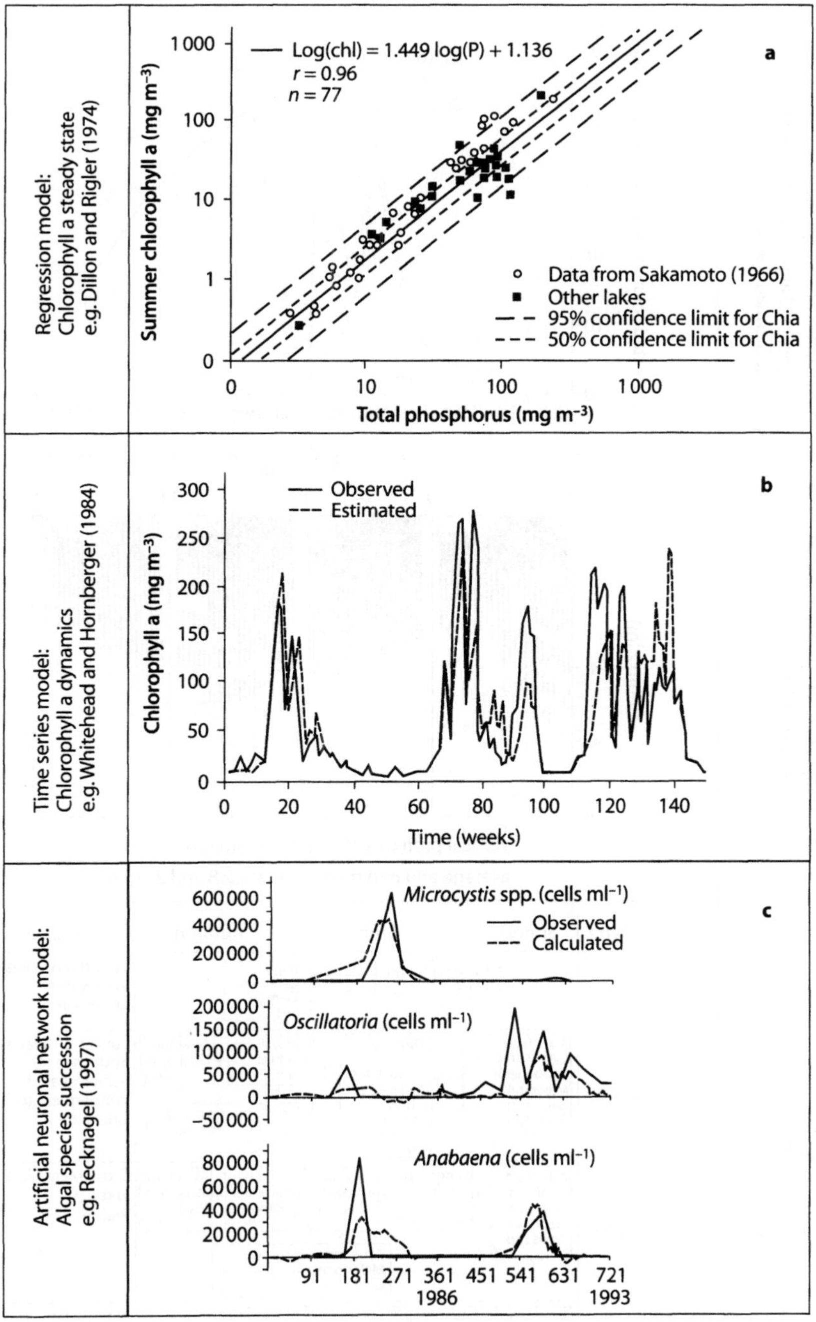

Fig. 10.2. Inductive prediction of phytoplankton abundance

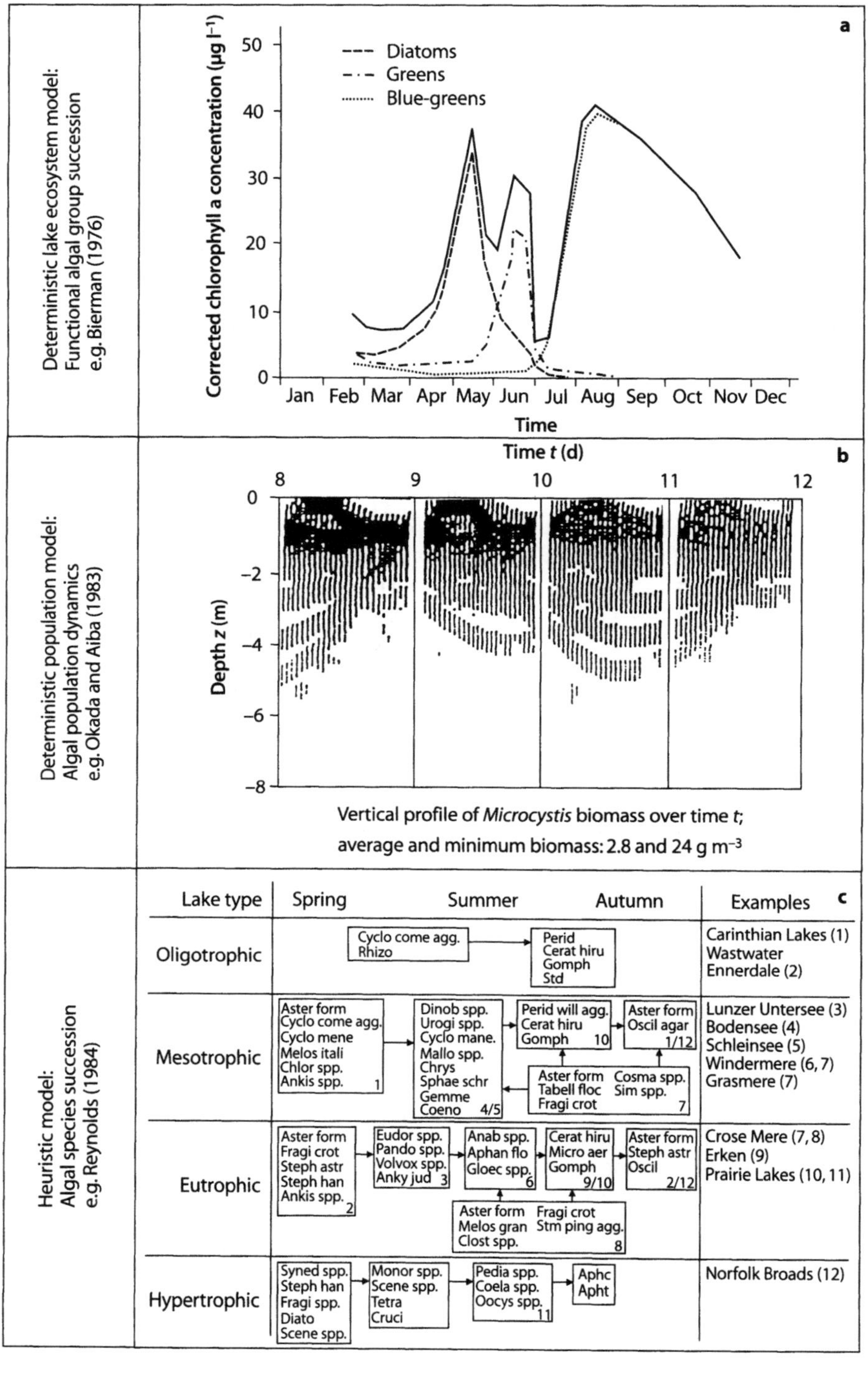

Fig. 10.3. Deductive prediction of phytoplankton abundance

model achieved reasonable accuracy in timing and magnitudes (see Fig. 10.2c). These results are encouraging in that short-term forecasting, and control of algae species abundance is achievable in the near future using artificial neuronal networks.

Deterministic ecosystem models (see Fig. 10.3a) predict daily abundance and succession of functional algal groups but fail to predict varieties of algal species. This type of model is widely used for scenario analyses to predict efficiency of measures for lake eutrophication control in medium- or long-term modes (Recknagel 1989; Recknagel et al. 1995). Deterministic population models (see Fig. 10.3b) predict vertical abundance of a single algal population on a daily basis but do not predict species succession. Okada and Aiba (1983) and Kromkamp and Walsby (1990) developed population models for *Microcystis* which can be used to predict horizons of low concentrations of algal cells for optimum withdrawal of raw water or for simulation of the effects of artificial mixing on algal growth. Heuristic models (see Fig. 10.3c) predict seasonal succession of algal species but fail to predict species abundance. Reynolds (1984) constructed a heuristic model to qualitatively predict seasonal succession of 49 algal species in temperate lakes of different trophic states. It is used as a rapid assessment guide for likely occurrence of certain algae species in lakes.

It can be summarized that currently only artificial neuronal network models enable us to predict timing and magnitudes of species dynamics and therefore species succession in lakes (Fig. 10.2c).

10.2.2
Elucidation

Underwater light, limiting nutrients, zooplankton grazing, and mixing conditions are known as environmental factors which drive phytoplankton dynamics in aquatic ecosystems (Reynolds 1984; Harris 1986). But little is known about temporal and spatial interconnections of these factors causing constellations for instantaneous abundance and succession of algae species. In an attempt to gain insight into these questions, the neuronal network model ANNA (Recknagel 1997; Recknagel et al. 1998) has been applied to scenario and sensitivity analyses. The model ANNA explicitly considers all these environmental factors driving phytoplankton dynamics mentioned above as input variables and the five dominating phytoplankton species as output variables.

10.2.2.1
Scenario Analysis: Succession of Microcystis by Oscillatoria

A succession from *Microcystis* dominance to *Oscillatoria* dominance was observed in Lake Kasumigaura in 1987. Takamura et al. (1992) measured a significant increase in the TN (total nitrogen)/TP (total phosphorus) ratio from 10 to approximatly 20 at the time of the species succession. Even though *Microcystis* is considered as a non N_2-fixing Cyanophyceae, it occurred as abundantly as N_2-fixing Cyanophyceae during times of nitrate deficiency and PO_4-P sufficiency in Lake Kasumigaura before 1986 (Takamura et al. 1987). Therefore, Takamura et al. (1992) proposed the consideration of the changed nutrient conditions as the possible reason for the species succession. They concluded that *Microcystis* tolerates nitrate deficiency at phosphorus sufficiency as observed in Lake Kasumigaura before 1987, while *Oscillatoria* depends on nitrate sufficiency as observed from 1987 afterwards.

A scenario analysis was carried out with the neuronal network model ANNA (Recknagel 1997) to test which limiting factor for algal growth may have triggered species succession in Lake Kasumigaura. The following four scenarios have been defined and tested: *(1)* swap of phosphorus and nitrogen data between 1986 and 1993 , *(2)* swap of zooplankton data between 1986 and 1993, *(3)* swap of light, temperature and Secchi depth data between 1986 and 1993, *(4)* swap of chlorophyll a data between 1986 and 1993.

Figure 10.4 shows the results of the scenario analysis. While the swap of the nutrient conditions between 1986 and 1993 did not influence the behaviour of *Microcystis*, *Oscillatoria* experienced a shift of its maximum peak from 1993 to 1986. The swap of zooplankton data in scenario two produced similar effects on the behaviour of *Microcystis* and *Oscillatoria* as scenario one. The only case where the maximum peak of *Microcystis* was shifted from 1986 to 1993 was in scenario three, where light, temperature and Secchi depth data were swapped between 1986 and 1993. But these conditions in scenario three caused significant decreases of peaks of *Oscillatoria* in both years. Finally, scenario four shows that chlorophyll a is obviously the most sensitive forcing function of the model for *Oscillatoria* by strengthening maximum peaks in 1993, while chlorophyll a did not affect significantly the behaviour of *Microcystis*.

The scenario analysis allows one to conclude that *Microcystis* seems less to be affected by the change in nitrate conditions between 1986 and 1993 but by the change in light, transparency and temperature conditions. By contrast, *Oscillatoria* seems to be sensitive to changes in nitrate conditions as well as to changes in zooplankton conditions.

10.2.2.2
Sensitivity Analysis: Input Sensitivity of Phytoplankton Species

A sensitivity analysis was carried out with the artificial neuronal network model ANNA to further clarify findings from the scenario analysis. The trained and validated neuronal network was again fed with independent input data for the years 1986 and 1993 of Lake Kasumigaura. Each input variable was separately changed once by +20% and once by −20% in independent experiments. These experiments were repeated for all 8 input variables, where absolute values of the resulting daily outputs for the blue-green algae species *Microcystis*, *Anabaena* and *Oscillatoria* from the two experiments were averaged between 1986 and 1993. As both years 1986 and 1993 appeared to be significantly different as indicated by the abundances of *Microcystis* and *Oscillatoria*, this approach of sensitivity analysis allowed the investigation of a reasonable spectrum of diverse conditions.

Results in Fig. 10.5a show that changes in global solar radiation by ±20% influence significantly seasonal dynamics of *Oscillatoria*, especially in winter and early and mid summer. At the same time *Oscillatoria* appears to be relatively insensitive against changes in water transparency (Fig. 10.5b) and water temperature (Fig. 10.5c). These results correspond with findings of Takamura et al. (1992), who observed several *Oscillatoria* peaks under winter conditions at low temperatures obviously mainly driven by light and nutrients. Its insensitivity against water transparency

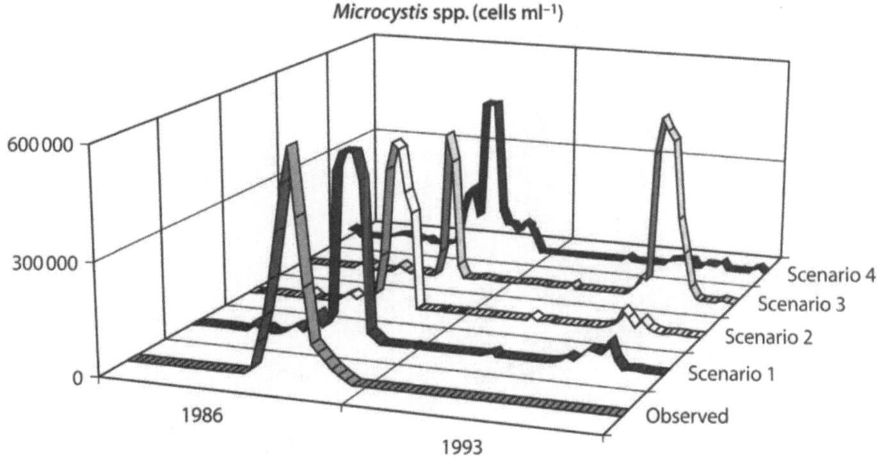

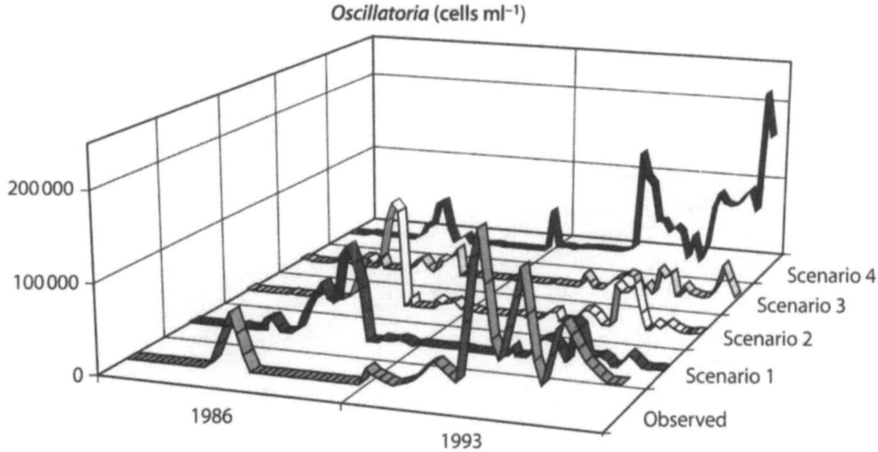

Scenario 1: Swap phosphorus and nitrogen data between 1986 and 1993
Scenario 2: Swap zooplankton data between 1986 and 1993
Scenario 3: Swap light, temperature, Secchi depth and depth data between 1986 and 1993
Scenario 4: Swap chlorophyll a data between 1986 and 1993

Fig. 10.4. Scenario analysis on succession of *Microcystis* by *Oscillatoria* at the time of the species succession

(see Fig. 10.5b) can be explained by its ability to perform buoyancy. *Microcystis* and some taxa of *Anabaena* are known as buoyant algae as well, and therefore are almost insensitive against water transparency under calm conditions in summer (Fig. 10.5b). *Anabaena* behaves differently under mixed conditions in spring by showing high sensitivity against water transparency. This may be correlated with the timing of the log-growth phase of *Anabaena* that is indicated by its high temperature sensitivity in spring

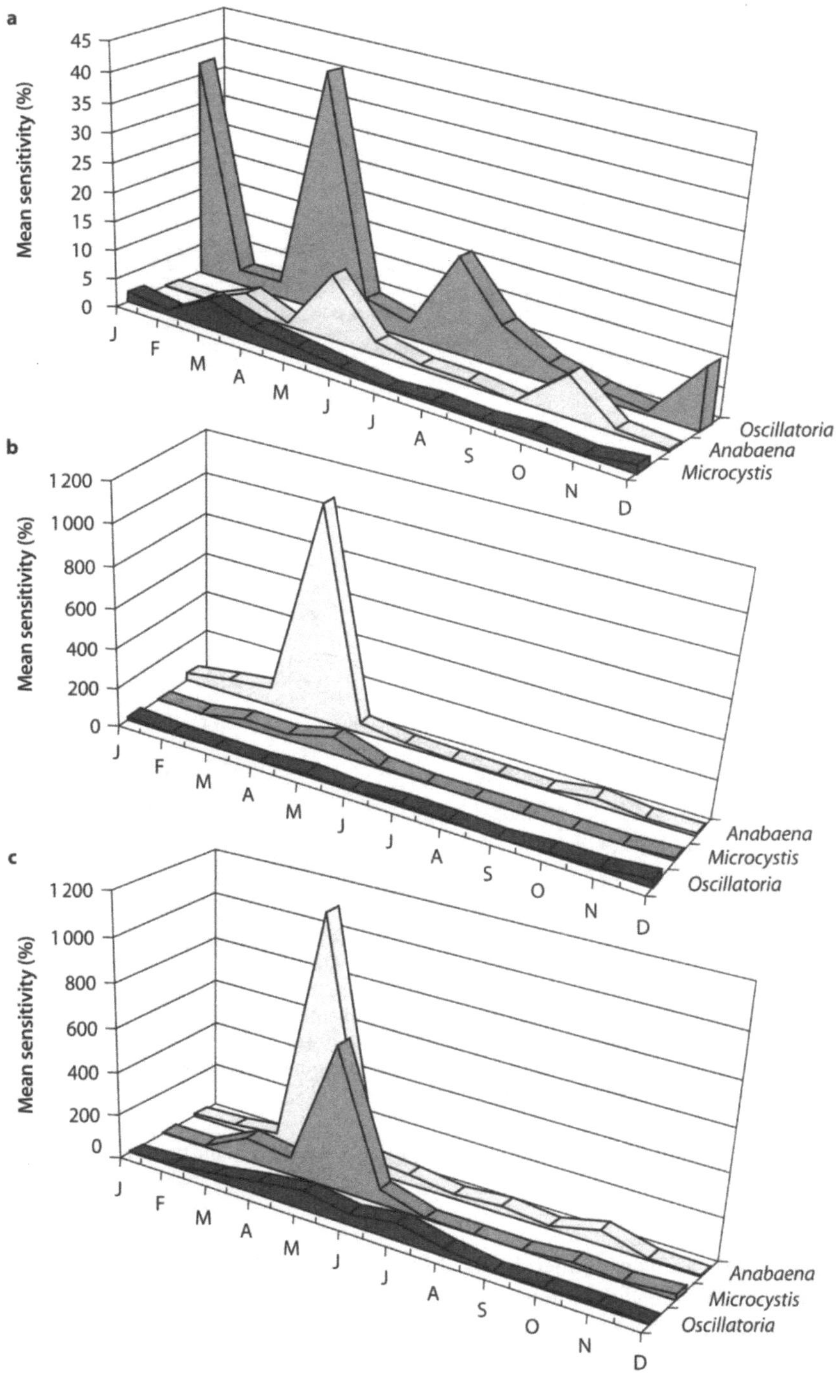

Fig. 10.5. Sensitivity analysis of *Microcystis*, *Anabaena* and *Oscillatoria*; **a** on changes of solar radiation by ±20%; **b** on changes of water transparency by ±20%; **c** on changes of water temperature by ±20%

too. *Microcystis'* known preference of high water temperature is confirmed by the high temperature sensitivity in early summer (Fig. 10.5c).

Results in Fig. 10.6a show high sensitivity of *Microcystis* and *Anabaena* in spring, late summer and autumn against changes in orthophosphate concentrations that indicates their known phosphorus limitation. As *Anabaena* is considered to be a N_2-fixing Cyanophyceae, results in Fig. 10.6b show the expected low sensitivity against changes in nitrate only during the growing season in summer, but not in spring and autumn. Further results in Fig. 10.6 suggest that *Oscillatoria* seems to cope well with

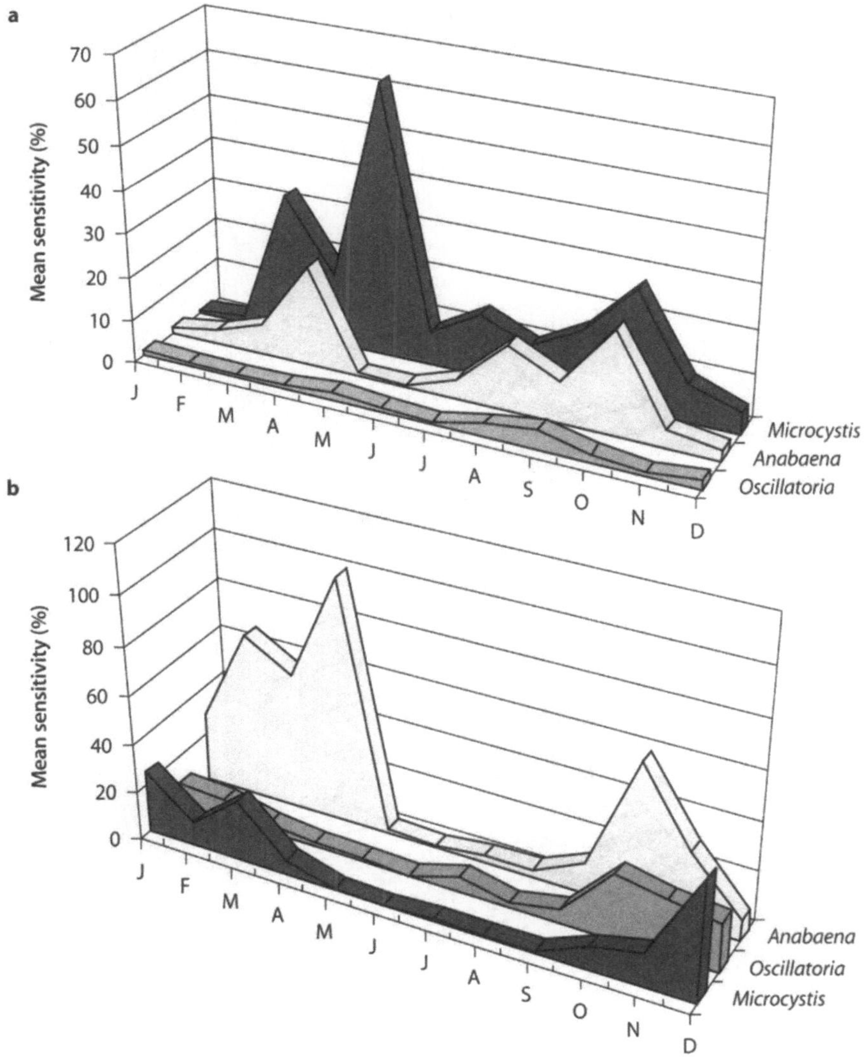

Fig. 10.6. Sensitivity analysis of *Microcystis*, *Anabaena* and *Oscillatoria*; **a** on changes of orthophosphate (PO_4-P) by ±20%; **b** on changes of nitrate (NO_3-N) by ±20%

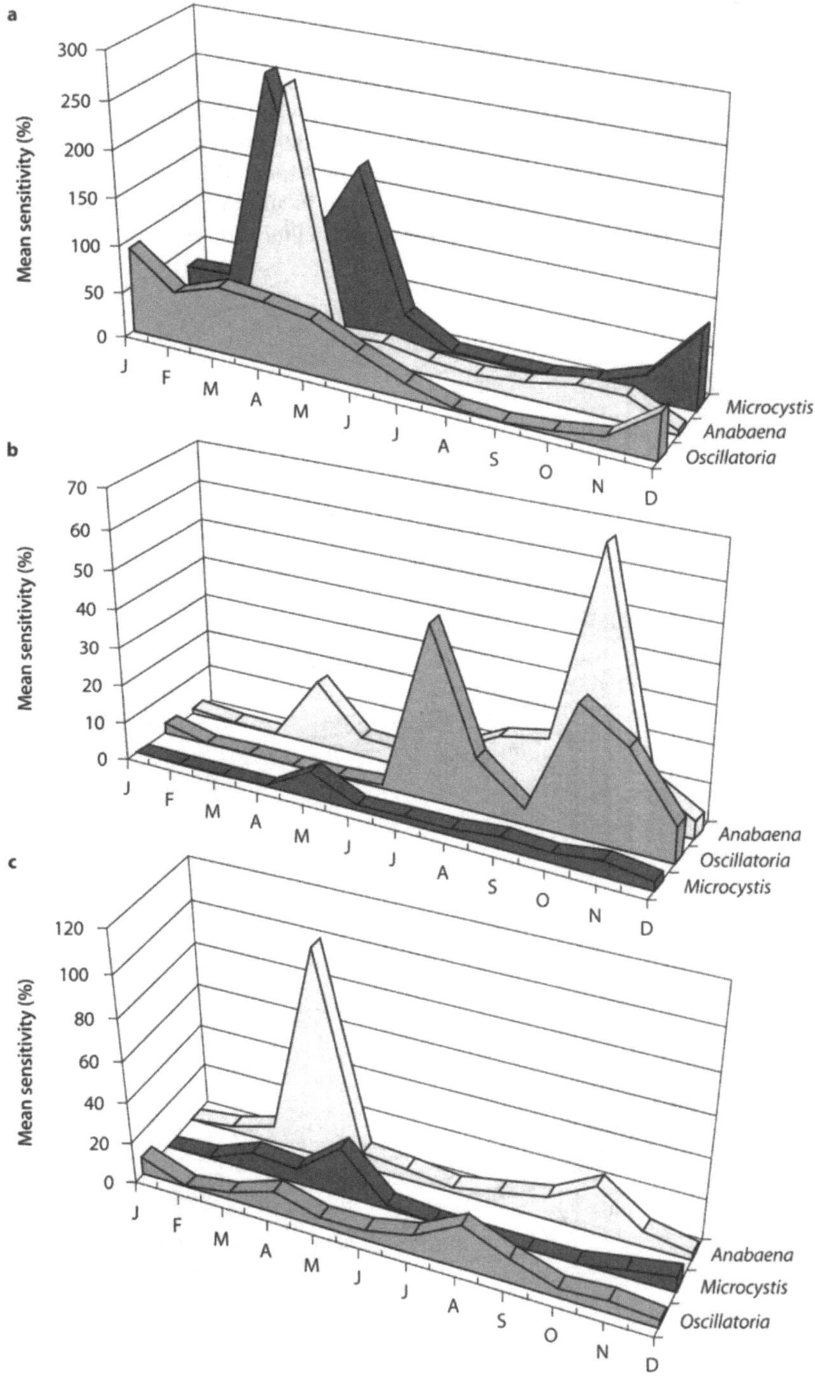

Fig. 10.7. Sensitivity analysis of *Microcystis*, *Anabaena* and *Oscillatoria*; **a** on changes of Copepoda by ±20%; **b** on changes of Cladocera by ±20%; **c** on changes of Rotifera by ±20%

changing PO_4-P concentrations, as discussed in the context of scenario analysis, but shows a relatively high sensitivity against nitrate in summer and autumn.

Results in Fig. 10.7 show a high sensitivity of blue-green algae against crustacean copepods in spring and early summer (see Fig. 10.7a). Although copepods are known as raptorial feeders, allowing them to feed on large particles of organic matter or zooplankton in summer and autumn, only juvenile live stages of copepods can be expected in spring and early summer which feed on smaller particles such as single algae cells. Single blue-green algae cells are mainly available in spring and early summer before they build large colonies (*Microcystis*) or filaments (*Anabaena* and *Oscillatoria*) in summer. Therefore, the high sensitivity, especially of *Microcystis* and *Anabaena*, against changes in copepods in spring and early summer (Fig. 10.7a) may be explained by grazing pressure from nauplius and copepodite larvae. Crustacean Cladocera tend to be abundant in summer and autumn and are known as filtration feeders utilizing small organic particles and algae cells. As *Microcystis* tends to build large cell colonies in summer, little grazing is expected by Cladocera as proven by the low sensitivity in Fig. 10.7b. But much higher sensitivity was calculated for *Anabaena* and *Oscillatoria* against Cladocera in summer and autumn (Fig. 10.7b). These results indicate that Cladocera might cope quiet well with filamentous algae for grazing. Results in Fig. 10.7c show that rotifera as suspension feeders influence *Anabaena* abundance mainly in spring.

Even though the present documentation of the sensitivity results may have limited the analysis of specific effects by using only absolute values and averaging them between 1986 and 1993, it can be concluded that artificial neuronal network models enable us to study interconnections and sensitivities of species dynamics and environmental control factors.

10.3
Prediction of Density of Brown Trout Redds in Streams

The understanding of organism-habitat relationships becomes increasingly important for the design of sustainable landscape use practices. The investigation of relationships between physical characteristics of streams and the abundance of trout is one interesting research area in this context where numerous modelling efforts have been applied. Commonly used techniques include multiple regression analysis, principal component analysis, and discriminant analysis. Most of these techniques failed to overcome statistical constraints and to cope with distinct nonlinearities in ecological data.

Only recently, genetic algorithms (d'Angelo et al. 1995) and artificial neuronal networks (Lek et al. 1996) were successfully applied to predict abundance of trout depending on physical characteristics of streams. Both applications have demonstrated the ability of so-called machine learning methods to overcome previous constraints and improve predictions.

In Fig. 10.8 results of a multiple regression model and a neuronal network model are compared by means of data from six mountain streams in France (Lek et al. 1996). Both models predicted the density of brown trout redds per linear meter of stream bed using stream habitat characteristics as input variables, such as: wetted width, area with suitable spawning gravel for trout per linear meter of river, surface velocity, bottom velocity, flow, water gradient, and mean depth. The results in Fig. 10.8a and 10.8b

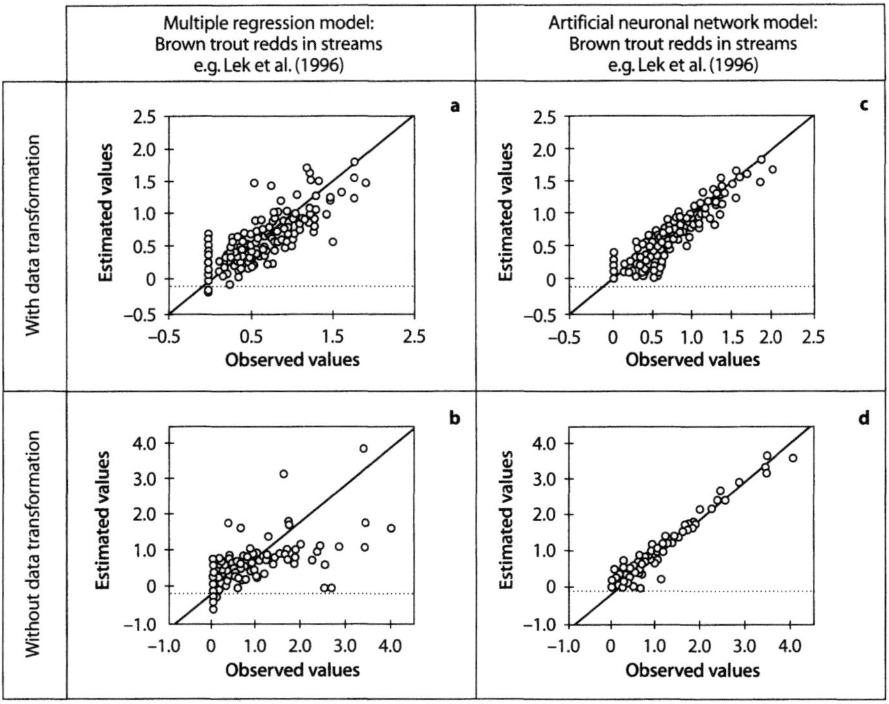

Fig. 10.8. Prediction of density of brown trout redds; **a,b** by a multiple regression model; **c,d** by an artificial neuronal network model

show that the multiple regression model performs better after linearisation of the data by transformation, where the determination coefficient is $R^2 = 0.643$ instead of $R^2 = 0.444$. But both applications can not overcome a distinct overestimation of low values and underestimation of high values. By contrast, the neuronal network does not require linearisation of data for optimum performance with a determination coefficient of $R^2 = 0.96$ (Fig. 10.8d).

10.4 Conclusions

Prediction and elucidation in aquatic ecology requires models which can cope with the distinct complexity, nonlinearity and interconnectance of ecosystems. Artificial neuronal networks appear to suit these purposes very well. The phytoplankton abundance and trout abundance examples have shown that artificial neuronal network models show up well in terms of predictive accuracy and elucidative capacity compared to alternative models. There seems to be great potential in exploring further machine learning techniques for ecological modelling, such as genetic algorithms, either in combination with neuronal networks or as exclusive applications.

Acknowledgements

The authors wish to thank Takehiko Fukushima, Noriko Takamura and Takayuki Hanazato of the National Institute of Environmental Studies in Tsukuba, Japan, for providing data of the Lake Kasumigaura. Additionally the authors thank William Lewis, Jr., Centre for Limnology at Boulder University of Colorado, and two anonymous reviewers for valuable comments on the manuscript.

References

d'Angelo DJ, Howard LM, Meyer JL, Gregory SV, Ashkenas LR (1995) Ecological uses for genetic algorithms: Predicting fish distribution in complex physical habitats. Can J Fish Aquat Sci 52:1893–1908
Dillon PJ, Rigler FH (1975) The phosphorus-chlorophyll relationship in lakes. Limnol Oceanogr 19:767–773
French M, Recknagel F (1994) Modeling of algal blooms in freshwaters using artificial neuronal networks. In: Zanetti P (ed) Computer techniques in environmental studies V, vol II: Environmental Systems. Computational Mechanics Publications, Southampton, Boston
Harris GP (1986) Phytoplankton ecology: Structure, function and fluctuation. Chapman and Hall, London
Kromkamp J, Walsby AE (1990) A computer model of buoyancy and vertical migration in cyanobacteria. J Plankton Res 12:161–183
Lek S, Delacoste M, Baran P, Dimopoulos I, Lauga J, Aulagnier S (1996) Application of neuronal networks to modelling nonlinear relationships in ecology. Ecol Model 90:39–152
Livingstone DM, Imboden DM (1996) The prediction of hypolimnetic oxygen profiles: A plea for a deductive approach. Can J Fish Sci 53:924–932
Meier HR, Dandy GC, Burch MD (1998) Use of artificial neuronal networks for modelling cyanobacteria *Anabaena spp.* in the River Murray, South Australia. Ecol Model 105:257–272
NRA (1990) Toxic blue-green algae: A report by the National Rivers Authority. NRA Water Quality Series Report No. 2, London
Okada M, Aiba S (1983) Simulation of water blooms in a eutrophic lake. Modeling the vertical migration in a population of *Microcystis aeruginosa*. Wat Res 20(4):485–490
Peters RH (1986) The role of prediction in limnology. Limnol Oceanogr 31(5):1143–1159
Recknagel F (1989) Applied systems ecology approach and case studies in aquatic ecology. Akademie Verlag, Berlin
Recknagel F (1997) ANNA – artificial neuronal network model for predicting species abundance and succession of blue-green algae. Hydrobiologia 349:47–57
Recknagel F, Hosomi M, Fukushima T, Kong D-S (1995) Short- and long-term control of external and internal phosphorus loads in lakes: A scenario analysis. Wat Res 29(7):1767–1779
Recknagel F, French M, Harkonen P, Yabunaka K-I (1997) Artificial neuronal network approach for modelling and prediction of algal blooms. Ecol Model 96(1–3): 11–28
Recknagel F, Fukushima T, Hanazato T, Takamura N, Wilson H (1998) Modelling and prediction of phyto- and zooplankton dynamics in Lake Kasumigaura by artificial neuronal networks. Lakes & Reservoirs 3(2):123–133
Reynolds CS (1984) The ecology of freshwater phytoplankton. Cambridge University Press, Cambridge
Rigler FH, Peters RH (1995) Science and limnology. In: Kinne O (ed) Excellence in ecology, vol VI. Ecological Institute, Oldendorf
Sakamoto M (1966) Primary production by phytoplankton community in some Japanese lakes and its dependence on lake depth. Arch Hydrobiol 62:1–28
Takamura N, Iwakuma T, Yasuno M (1987) Uptake of ^{13}C and ^{15}N (ammonium, nitrate and urea) by *Microcystis* in Lake Kasumigaura. J Plankton Res 9:151–165
Takamura N, Otsuki A, Aizaki M, Nojiri Y (1992) Phytoplankton species shift accompanied by transition from nitrogen dependence to phosphorus dependence of primary production in Lake Kasumigaura, Japan. Arch Hydrobiol 124(2):129–148
Vollenweider R (1976) Advances in defining critical loading levels for phosphorus in lake eutrophication. Mem Ist Ital Idrobiol 33:33–83
Whitehead P, Hornberger G (1984) Modelling algal behaviour in the River Thames. Wat Res 18(8):945–953

Chapter 11

Performance Comparison between Regression and Neuronal Network Models for Forecasting Pacific Sardine (*Sardinops caeruleus*) Biomass

M.A. Cisneros-Mata · T. Brey · A. Jarre-Teichmann

11.1 Introduction

Forecasting is particularly important for the management of harvested marine fish populations. Unfortunately, random and deterministic factors are pervasive characteristics which can undermine one's ability to conduct accurate forecasts. In some cases the span or resolution of available data can limit development or use of a particular kind of model.

Examples of such situations are fisheries management of small pelagics like sardines and anchovies, the most important group of aquatic species in volume captured. Despite vast work on variability experienced by these populations, regulation of their fisheries continues to be particularly difficult. The population dynamics of small pelagics have been investigated using statistical, empirical and mechanistic modelling approaches. Studies indicate that variability in these species results from a combination of intrinsic factors such as density-dependence, and environmental forcing of survival rates (Huato-Soberanis and Lluch-Belda 1987; Jacobson and MacCall 1995; Jarre-Teichmann et al. 1995; Cisneros-Mata et al. 1996), although no clear-cut explanation has been provided. Again, however, when dealing with management of these resources, one needs to explore alternative research approaches, and use available data/resources to come up with prescriptions.

The Pacific sardine population (*Sardinops caeruleus*), an inhabitant of the California current, has experienced high variability in abundance over the past several decades (Butler et al. 1993). Maximum spawning biomass (fish of age 2 yr. and older) was about 3.6×10^6 metric tons (t) in 1932, declined to 6 000 t in 1975, then increased to about 353 000 t in 1995 (Table 11.1). Scale deposition in anaerobic sediments revealed that this population's abundance also underwent oscillations of several orders of magnitude in the past 20 centuries (Baumgartner et al. 1992). By virtue of being closely related to the environment as primary and secondary consumers, annual recruitment of the Pacific sardine is highly variable. Temperature is of foremost importance for this species; for example, incubation period of their pelagic eggs, as well as maturation time due to high temperatures, might result in decreased mortality rates of the vulnerable first life stages, with a consequent increase in population biomass (Smith 1995).

In the present work, we develop models to forecast one year in advance the spawning biomass of Pacific sardine of the California current. For management purposes, one year would be a minimum necessary time to take provisions in the fishery, both administrators and fishers as well. We use regression and artificial neuronal network models and compare results. Our goal is to increase our understanding of the dynamics of Pacific sardine to develop appropriate management schemes, using available time series which have meant years of effort and resources.

Table 11.1. Pacific sardine spawning biomass and sea surface temperature index (ΔT). Biomass (metric tons) corresponds to the year indicated. ΔT indicates a deviation of average temperature over the 3 years immediately before the year indicated, as compared to the overall mean of the time period considered

Year	Biomass	ΔT (°C)	Year	Biomass	ΔT (°C)
1932	3524000	0.71	1960	88000	0.82
1933	3415000	0.15	1961	54000	0.41
1934	3625000	−0.36	1962	27000	−0.21
1935	2845000	−0.46	1963	21000	−0.56
1936	1688000	−0.15	1964	11000	−0.45
1937	1207000	0.13	1965	3000	−0.39
1938	1201000	0.15	1983	8024	0.27
1939	1608000	0.01	1984	19609	0.59
1940	1760000	0.17	1985	21107	0.81
1941	2458000	0.32	1986	31117	0.88
1942	2065000	0.52	1987	54047	0.72
1943	1677000	0.23	1988	68807	0.57
1944	1260000	−0.04	1989	90364	0.26
1945	720000	−0.27	1990	101628	0.21
1946	566000	−0.34	1991	204686	0.36
1947	405000	−0.26	1992	191544	0.63
1948	740000	−0.22	1993	166532	0.86
1949	793000	−0.47	1994	238127	0.98
1950	780000	−0.67	1995	353000	not used
1951	277000	−0.57			
1952	136000	−0.43			
1953	202000	−0.56			
1954	239000	−0.59			
1955	170000	−0.57			
1956	108000	−0.48			
1957	90000	−0.33			
1958	177000	0.05			
1959	122000	0.63			

11.2 Materials

We used annual spawning biomass data of the Pacific sardine stock off the coasts of California, USA and Baja California, México. Biomass data up to 1987 were taken from Jacobson and MacCall (1995) and the rest were provided by L. Jacobson and T. Barnes (SWFSC, La Jolla, CA, pers. comm.). The 46-year series used runs from 1932 to 1965, and from 1983 to 1995, with data from 1966 to 1982 missing due to unavailability of reliable estimates (Barnes et al. 1992).

As an indicator of the environment for sardine, in our models we used temperature data (°C) recorded at Scripps Pier, in Southern California. We averaged temperature for the 3 years prior to a given year, then for each year we computed deviations from the overall mean over the time period considered (e.g. Hilborn and Walters 1992). Jacobson and MacCall (1995) found a significant relation between annual recruitment of Pacific sardine as a function of past temperature and spawning biomass. That study also used temperature measured at Scripps Pier and recruits were number of adult sardines of age 2 yr. that enter the fishery every year.

11.3 Methods

We performed three kinds of analyses comparing results from regressions and artificial neuronal network (ANN) models. ANNs are computer programs, which, by means of massive parallel computations, can identify patterns within and between series of data. This modelling approach is starting to find application in ecology (e.g. Tan and Smeins 1996) and fisheries research (e.g. Komatsu et al. 1994).

First, we investigated the possibility of forecasting biomass using only past biomass and tested models with the last data points of the series. Then we added temperature and tested performance of models using randomly selected southsets of data. Finally, we incorporated results of the two previous analyses and conducted biomass forecasts for the last three years of the series, which represent a particular challenge (see below).

A preliminary analysis showed high positive auto-correlation ($r^2 \geq 0.92$) of spawning biomass at lag 1 yr. Such auto-correlation is likely to result from the age-structure and even survival rates of adult sardines in the spawning population. Thus we tested how next year's biomass can be forecasted using the simplest model possible, based on this year's biomass only.

The regression models were of the form:

$$B_{t+1} = \psi + \mu B_t + \varepsilon \tag{11.1}$$

where B_t is biomass at year t, ψ and μ are parameters of the linear model, and ε is an error term. The general ANN models constructed for this first analysis followed:

$$B_{t+1} = f(B_t) \tag{11.2}$$

This notation indicates that biomass this year is "related," or is a function of only biomass last year. Using the two modelling approaches, biomass was forecasted for the years 1990, 1992, 1994 and 1995, and results were compared. We chose those years for no other reason than to test forecasted values for every other year (1990, 1992, and 1994), plus for the last year of the series, when an important increase in biomass occurred. We analysed our forecasted values in 15 bootstrap trials, using the same data southsets for both regressions and ANNs. We caution that although 15 bootstrap trials sufficed for our comparison purposes, such low number may not be adequate for management models. Southsets used for bootstrapping included data from 1932 up to the year previous to that for which the forecast was made; for example, up to 1992 to forecast biomass in 1993.

In the second analysis, we used mean annual temperature deviations (anomalies) over the past three years with respect to the overall mean, in addition to current biomass, to predict next year's sardine biomass level. We chose temperature over three years because such time span will cover recruitment variability due to environmental changes. Recruitment in this species may cause annual biomass to fluctuate widely, and three years will suffice to represent environmental influence on this species (Jacobson and MacCall 1995; Cisneros-Mata et al. 1996).

To test how well the models performed, the three series: B_t, $\Delta T_{t-3,\,t-2,\,t-1}$ and B_{t+1}, were randomly ordered, and a southset of data from 10 years was left of both regressions and ANNs. Then the 10 expected values of B_{t+1} were compared to values forecasted by the models. This procedure of randomly selecting southsets of data to fit and test models was repeated 10 times to compute statistics for analysis and comparison of approaches.

The general regression models used in this instance were of the form:

$$B_{t+1} = \alpha + \beta B_t + \chi \Delta T_{t-3,\,t-2,\,t-1} + \varepsilon \tag{11.3}$$

where α, β and χ are parameters of the linear model, B_t is biomass at year t, and $\Delta T_{t-3,\,t-2,\,t-1}$ is annual temperature anomaly corresponding to the 3-year mean prior to year t. This series was computed as follows: to the overall mean of the series, we subtracted the 3-year running average over the three years prior to the year of interest.

The ANN models in this analysis were of the form:

$$B_{t+1} = f(\Delta T_{t-3}, \Delta T_{t-2}, \Delta T_{t-1}, B_t) \tag{11.4}$$

In other words, biomass level relates to temperature anomalies over the average of the previous three years, and biomass level the previous year. In the formal ANN argot, this means there were two input (temperature and biomass), and one-output (biomass) nodes, respectively. Our third analysis was similar to the previous one just described, except that the three last years in the series were forecasted, following the same bootstrapping procedure as in the first analysis. The same general models, as expressed in Eqs. 11.3 and 11.4, were tested in this last instance.

Regression models were fit using the Simplex algorithm in SYSTAT 5 and, due to the exploratory nature of this work, we did not try alternative fitting techniques. ANNs were constructed using the Professional Works II+ software by NeuronalWare. All ANNs were constructed using the commonly used back propagation algorithm (Dayhoff 1990), using the sigmoid transfer function; all networks were trained for 10 000 cycles, and had one intermediate layer. Criteria used to decide on performance of models were coefficients of determination (r^2), root mean square (*RMS*) and, for comparison purposes only, approximate 95% confidence intervals (*ACI*) for the mean biomass values.

To construct the *ACI* we computed the bootstrap standard error (*SE*) of biomass as described in Efron and Tibshirani (1993). We then obtained the *ACI* as ($\hat{O} \pm 1.96\ SE$), where $\hat{O}$ is the mean of the bootstrap values of predicted biomass. To determine performance of a given kind of model in our second analysis, we compared the ratios RMS_{test}/RMS_{fit} within both modelling approaches. Here, RMS_{fit} and RMS_{test} are the *RMS*s between expected data southsets used to fit the regression models or train the ANNs, and the corresponding forecasted values. Given that a good model fit is observed, then RMS_{test} should be equal or smaller than RMS_{fit} and, consequently, *RMS* ratios should be ≤1, and otherwise would indicate the inability to extrapolate (forecast) using the models developed.

In the third analysis we computed the ratio $RMS_{test\ ANN}/RMS_{test\ regression}$ to determine the relative performance of ANN and regression models. A value near or equal to 1 would indicate even performance. A ratio >>1 would indicate better performance of regression than ANN models, and vice versa.

11.4 Results and Discussion

Our results showed that past sea temperature is a good indicator of Pacific sardine abundance. Also, with the database we used, performance of regressions was in general better than that of artificial neuronal networks. We elaborate on these results below and make an effort to interpret them. Two potential problems were considered in our analyses: extrapolating and over-parameterisation of models. The former was explicitly tested and the latter was implicit when we kept the number of parameters as low as possible.

The very high positive auto-correlation at one-year lag in the biomass series of sardine suggested that good extrapolations could be made by means of simple linear regression and ANN models using only past biomass. Our first analysis showed this to be true, except that after 1992, both types of models underestimated the expected values. Using ANNs the mean forecasted values were lower and the 95% *ACI* were wider, hence less precise and more biased low as compared to those from regression models (Table 11.2). The low bias shown by both types of models could be explained by a change

Table 11.2. Performance of regression models and ANNs in the first analysis. Expected is the actual sardine biomass in the indicated year; *LL* and *UL* are the lower and upper limit of a 95% approximate confidence interval; *Mean* is the average value of 15 bootstrap trials

	Year	Expected	*LL*	Mean	*UL*
Regression models	1990	101628	72979	96493	120006
	1992	191544	169698	198030	226362
	1994	238127	113406	150546	187687
	1995	353000	99089	224545	350000
ANNs	1990	101628	27466	100213	172960
	1992	191544	54357	138789	223222
	1994	238127	46258	123894	201530
	1995	353000	146646	179750	212855

in the dynamics of Pacific sardine in recent years related to a change in its coastal pelagic habitat. Studies indicate that an extended warming period in the California current system could have boosted the currently increasing abundance trend of Pacific sardine (Smith 1995). Such sudden changes in the trends of abundance of small pelagics have been termed the regime problem (Lluch-Belda et al. 1989). It is therefore not surprising that inclusion of temperature improved performance of all models for the last years of the series.

In the second analysis, both regression and ANN models yielded good forecasts of data values left out for testing (Table 11.3). The mean ratio RMS_{test}/RMS_{fit} of ANNs was not different from 1 ($0.05 < p < 0.10$), but that of regression models was smaller than 1 ($0.001 < p < 0.01$). Paired t-tests did not reveal significant differences between both modelling approaches in terms of r^2 ($p = 0.37$), although the ratio RMS_{test}/RMS_{fit} was greater ($p = 0.014$) for ANNs than for regression models. This suggests a potential bias of extrapolations using ANNs. Previous results indicated the possibility that ANNs underestimate biomass when expected values are very high (Jarre-Teichmann et al. 1995), but this topic needs further exploration.

The third analysis showed how inclusion of sea temperature improved the forecasting performance for the critical last years of the series, by both modelling approaches. For all three years, 95% *ACI* contained the expected biomass; for 1993, ANNs performed well, although forecasts deteriorated for 1994 and 1995 (Table 11.4). The ratio $RMS_{test\ ANN}/RMS_{test\ regression}$ were 0.53, 1.83 and 2.52 for 1993, 1994 and 1995; that is, ANNs performed better for 1993, yet regression models worked better than ANNs in the subsequent years (Table 11.4).

The reasons why ANNs were outperformed by regression models are not clear. One possibility is that the time series used were not long enough for the networks to learn possibly existing patterns. Consequences of the shortness of the series became evi-

Table 11.3. Performance of ANN and regression models in the second analysis. r^2_{tes} = coefficient of determination between expected and forecasted values; Ratio = RMS_{tes} / RMS_{fit}, where tes refers to test of forecasted values and fit to values used to build the models

Trial	ANNs r^2_{tes}	ANNs Ratio	Regression models r^2_{tes}	Regression models Ratio
1	0.96	1.21	0.95	0.78
2	0.97	1.75	0.99	0.72
3	0.94	0.62	0.74	0.74
4	0.98	0.70	0.97	0.67
5	0.80	1.51	0.98	0.88
6	0.83	1.03	0.96	0.46
7	0.94	1.11	0.92	0.60
8	0.94	1.54	0.98	0.62
9	0.94	1.22	0.97	1.25
10	0.79	2.39	0.96	0.41

Table 11.4. Performance of regression models and ANNs in the third analysis. Expected is actual sardine biomass (metric tons) in the year indicated; *LL* and *UL* are the lower and upper limit of a 95% approximate confidence interval; *Mean* is the average value of 15 bootstrap trials

	Year	Expected	*LL*	Mean	*UL*
Regression models	1993	166532	119001	285945	452889
	1994	238127	178864	253951	329038
	1995	353000	235019	319351	403684
ANNs	1993	166532	63505	201832	340159
	1994	238127	75464	220374	365284
	1995	353000	59039	263481	467923

dent in some instances when we considered more series of parameters that were judged pertinent. For example, when we included an index of annual zooplankton abundance, or annual wind speed cube for the California current, we observed overtraining of the networks. In these instances the trained networks yielded nonsensical forecasts even if they had low *RMS* and high r^2, computed with data used for training and the corresponding values predicted by the networks. This indicates that there is a minimum, or threshold, length of the series to be used by networks in order to obtain reasonably good forecasts, yet this investigation was beyond our present goal. Future work with enough data points (e.g. Jarre-Teichmann et al. 1995) might address this issue by analysing performance of ANNs with increasing length of data series.

Acknowledgements

We thank Larry Jacobson and Tom Barnes for updating the biomass series, and Greg Hammann (CICESE, Ensenada, México) for facilitating temperature data. Two anonymous reviewers and the editor provided useful comments; errors remain solely our responsibility.

References

Barnes JT, Jacobson LD, MacCall AD, Wolf P (1992) Recent population trends and abundance estimates for sardine (*Sardinops sagax*). Calif Coop Ocen Fish Invest Rep 33:60–75
Baumgartner TR, Soutar A, Ferreira-Bartrina V (1992) Reconstruction of the history of Pacific sardine and Northern anchovy populations over the past two millennia from sediments of the Santa Barbara Basin, California. Calif Coop Ocen Fish Invest Rep 33:24–40
Butler J, Smith PE, Lo NC (1993) The effect of natural variability of life-history parameters on anchovy and sardine population growth. Calif Coop Ocen Fish Invest Rep 34:104–111
Cisneros-Mata MA, Montemayor-López G, Nevárez-Martínez MO (1996) Modeling deterministic effects of age-structure, density-dependence, environmental forcing and fishing on the population dynamics of *Sardinops sagax caeruleus* in the Gulf of California. Calif Coop Ocen Fish Invest Rep 37:201–208
Dayhoff JE (1990) Neuronal network architectures: An introduction. Van Nostrand Reinhold, New York
Efron B, Tibshirani RJ (1993) An introduction to the bootstrap. Chapman and Hall Inc., New York
Hilborn R, Walters CJ (1992) Quantitative fisheries stock assessment: Choice, dynamics and uncertainty. Chapman and Hall Inc., New York

Huato-Soberanis L, Lluch-Belda D (1987) Mesoscale cycles in the series of environmental indices related to the sardine fishery in the Gulf of California. Calif Coop Ocen Fish Invest Rep 28:128–134
Jacobson LD, MacCall AD (1995) Stock-recruitment models for Pacific sardine (*Sardinops sagax*). Can J Fish Aquat Sci 52:566–577
Jarre-Teichmann A, Brey T, Haltof H (1995) Exploring the use of neuronal networks for biomass forecasts in the Peruvian upwelling system. NAGA, the ICLARM Quarterly, October 1995, pp 38–40
Komatsu T, Aoki I, Mitani I, Ishii T (1994) Prediction of the catch of Japanese sardine larvae in Sagami bay using a neuronal network. Fish Sci 60:385–391
Lluch-Belda D, Crawford RJM, Kawasaki T, MacCall AD, Parrish RH, Shwartzlose RA, Smith PE (1989) Worldwide fluctuations of sardine and anchovy stocks: The regime problem. S Afr J Mar Sci 8:195–206
Smith PE (1995) A warm decade in the southern California bight. Calif Coop Ocen Fish Invest Rep 36:120–126
Tan SS, Smeins FE (1996) Predicting grassland community changes with an artificial neuronal network model. Ecol Model 84:91–97

A Comparison of Artificial Neuronal Network and Conventional Statistical Techniques for Analysing Environmental Data

G.R. Ball · D. Palmer-Brown · G.E. Mills

12.1 Introduction

The use of artificial neuronal networks (ANNs) to model environmental problems is increasing. They have been used to model systems as diverse as algal distributions in oceans (Simpson et al. 1992), grassland community changes (Tan and Smeins 1996), and in the recognition of birdsong (McIlraith and Card 1997). They are well suited to modelling complex nonlinear systems which are inherently 'noisy,' a characteristic that makes them suited to modelling environmental systems. They have been used in studies in this and related papers, to model environmental influences on the impact of tropospheric ozone pollution on plants (Balls et al. 1995; Balls et al. 1996; Roadknight et al. 1997; Ball et al. 1998).

It is important that statistical or ANN approaches should be able to produce accurate generalized predictions which are applicable to real world systems. If this is the case, the model can then be interpreted in a wider sense. One of the main criticisms of ANNs is that they are black boxes which give no indication of the processes involved in the modelled system (Benitez et al. 1997). This criticism is waning as techniques for the analysis of the mechanism of trained ANN models are being developed. In particular, methods for the identification of the importance of input variables are valuable in the development of simplified models of complex environmental systems. Using methods developed by the authors and others, the importance of variables can be identified by analysis of the performance of models or by analysis of the weights of the trained ANN model (Balls et al. 1995; Roadknight et al. 1997).

Linear regression, linear regression using exponential, logarithmic and power transformations, and multiple regression can be used to create mathematical models of data. These models may be of a predictive form that combines multiple factors in the equation to predict an effect, using a mathematical equation. In the case of simple linear regression and linear regression using transformations, the importance of individual influences upon a measured parameter may be determined by regressing each of the influences against the measured parameter and determining the r^2 value of each. This method was employed by Colls et al. (1993) to determine the most appropriate ozone dose parameter for use in the yield response of beans to ozone. When using multiple regression the strength of influences are determined from the regression equation and from probability values produced within the analysis, for each independent variable.

Principal components analysis is known as a method of data reduction because it reduces the data into a number of principal components based on the eigenvectors. The principal components produced are linear combinations of the original variables (Fry 1993). They are found by multiplying by the eigenvectors which are established during the PCA. The eigenvectors are the principal directions in the data. Principal components analysis can be combined with a least squares regression method to create a predictive model. As principal components analysis acts as a data reduction method, the least squares method employed will work efficiently on relatively complex data which has been previously converted into principal components.

A number of studies have compared the performance of back propagation ANNs with conventional statistical methods (Urquidimacdonald and Macdonald 1994; Wise et al. 1995; Timofei et al. 1997). Although relatively few have compared them for analysis of nonlinear environmental data. For example, Lek et al. (1996), Comrie (1997) and Paruelo and Tomasel (1997) all found that ANNs showed superior performance to conventional statistical techniques. There has been little comparison of the ability of both methods to identify the importance of influences.

The database used in this study is from the exposure of plants to controlled ozone levels. Visible injury (off white chlorotic flecking) is thought to occur in response accumulated ozone doses above a concentration of 40 ppb (a dose parameter known as *AOT*40), as low as 200 ppb.h. depending on the conditions at the time of exposure (Sanders et al. 1995). The sensitivity of the plant is strongly influenced by microclimatic conditions at the time of exposure both singly and in combination (Grantz and Meinzer 1990; Aphalo and Jarvis 1991; Leuning 1995). These act primarily on the stomata, the main route of ozone entry to the plant (Kerstiens et al. 1992). When measured under controlled conditions, humidity (vapour pressure deficit) is one of the primary influences on stomatal conductance and thus ozone flux into the plant (Weiser and Havranek 1995; Leuning 1995). Other influences from photosynthetically active radiation (PAR) and temperature have also been identified when measured singly (Aphalo and Jarvis 1991; Herbst 1995). Gutiérrez et al. (1994) concluded that interactions between these variables could mask the ozone dose response of the plant, making simple analysis using ozone as a single parameter inappropriate. Clearly the interactions between microclimatic conditions and ozone uptake by the plant are complex, and include variables that interact with each other as well as influencing ozone uptake. These will ultimately determine the extent of visible ozone injury on the plant.

The aims of the work were to assess the ability of a range of statistical methods to analyse and model an environmental data set; to compare the accuracy and performance of these techniques with those of a back propagation ANN model; and to assess the ability of conventional and ANN techniques to indicate the importance of parameters in the data.

12.2 Methods

12.2.1 Database Development

The data set used for analysis in this chapter was produced from the exposure of subterranean clover plants to controlled ozone doses under co-varying ambient environmental conditions in acrylic chambers (Balls et al. 1995). Ozone was produced by ultra-violet light from oxygen (Light-O3-clean AS, Denmark). During exposure to ozone concentration, temperature, relative humidity and photosynthetically active radiation

(*PAR*) were measured and logged to a PC (Viglen, UK), producing 30-minute mean values for each parameter. One week after exposure, visible ozone injury (characterized by off white chlorotic lesions) was assessed using a visible injury score key, representing 6 injury classes between 0 and 5. Mean ozone and climatic conditions were calculated for the exposure period and the mean visible injury score for 20 plants was determined. Exposures were repeated on a regular basis to create a data set containing 256 points.

A random number variable was inserted into the database to represent a factor that had no influence upon the extent of visible injury. This helps to assess the performance of the various modelling and analysis approaches since ideally a low importance should be identified for this 'dummy' variable. In all cases the performance of the various statistical approaches were compared to the performance of the ANN approach.

12.2.2 Development and Analysis of ANN Models

12.2.2.1 *Training*

Prior to training, validation and test data southsets, each comprising 20% of the data, were extracted randomly from the database. The remaining data was used to train an ANN model, within the Neuroshell 2 package (Ward Systems Group Ltd), using all of the available input variables. These were ozone dose (*AOT*40), leaf age (*leaf*), photosynthetically active radiation (*PAR*), temperature (*temp*), relative humidity (*RH*) and the random number (*Rand*). The model was retrained and tested using a range of numbers of hidden nodes, a momentum of 0.9 and a learning rate of 0.05. Training was finished when the error of the validation data failed to improve for 20 000 epochs. The trained models were then tested by presenting them with the validation data southset and the training data. The optimum model was the one with the best performance on validation data. The ability of this model to make generalized predictions was then tested by presenting it with the test data subset.

12.2.2.2 *Weightings Analysis of the ANN Model*

The weights of the trained optimized ANN model were determined. The relative importance of the input variables to the best model were then determined by taking the sum of the products of absolute weight values leading from each input.

12.2.2.3 *ANN Modelling Using Combinations of Input Variables*

Another approach used to determine the importance of input variables was to train multiple ANN models using all possible input combinations. The modelling methods described earlier were used to train each model. The performance of each combination was assessed (optimized using a number of hidden nodes). Comparison of the performance for selected combinations gave an indication of their importance and of the importance of inputs singly and in combination. In this paper results for the input combinations not significantly different from the best model, all of the single input models, and a model with all inputs with the exception of *AOT40* are presented.

12.2.3 Conventional Statistical Methods

A number of statistical methods were used to analyse the same database as used in the development of the ANN models.

12.2.3.1 *Simple Linear Regression Analysis*

Simple linear regression analysis was carried out on all of the data using the Microsoft Excel package (version 5.0). Regression analysis was carried out for *AOT40*, *PAR*, *temp*, *RH* and *Rand*, against the visible injury score for each *leaf* using standardized values to prevent the magnitude of the values having an influence. The regression coefficients for each of these variables were calculated, giving a measure of the strength of their influence on the visible injury score.

12.2.3.2 *Linear Regression Analysis Using Exponential, Logarithmic and Power Transformations*

Regression analysis was carried out in a similar way to the simple linear regression analysis using the logarithmic, exponential and power transformations within the Microsoft Excel package and the training data set. Ozone dose *AOT40*, *PAR*, temperature, relative humidity and the random number were regressed against the visible injury score. This was repeated for each of the three leafs, using logarithmic, power and exponential functions. The regression analysis produced r^2 values for each parameter (based on the Pearson correlation coefficient), for each leaf using each function. Average r^2 values were calculated for each parameter for all of the leafs. These average values were used to indicate the linear strength of the correlations produced between each parameter and as a comparison of the linear strength of correlation using different transformation functions.

12.2.3.3 *Multiple Regression*

Multiple regression analysis (Fry 1993) was carried out on the training data, using the Statmost (Datamost corporation) statistical analysis package. The independent variables used were the same as those used as inputs to the ANN model. After analysis, the equation produced was applied to the validation data subset and the complete data set. The actual injury score was then plotted against predictions made for the validation data. This produced an r^2 value for the predictions of the multiple regression analysis equation based on the validation data, and allowed comparison of the performance of the multiple linear regression analysis with the ANN analysis.

12.2.3.4
Stepwise Multiple Regression

Stepwise multiple regression serves to determine the variables for inclusion into multiple regression. Forward and backward stepwise multiple regression was carried out on the whole data set using the Unistat data analysis package (version 1.12). Consecutive multiple regression analyses were run to determine the variables, which influenced ozone injury formation. Variables were included or discarded based upon the strength of their correlation with the visible injury score. The output from the analysis indicated which variables should be included within the regression equation and the r^2 value when each variable was included. The r^2 values produced were compared with the performance of ANNs and other modelling methods for the whole data set, and the ranking of variables was compared with the ranking obtained from other methods.

12.2.3.5
Principal Components Analysis (PCA)

PCA (Fry 1993) was carried out in the Unistat package. Again the training data set was used in the analysis having been transformed within Unistat. Each value was multiplied by the mean value for the variable involved and divided by its standard deviation. This served to put each parameter into the same range, producing values which were given an equal bias and which were not affected by exceptionally large or small units. Once transformation was complete the PCA was run, and eigenvalues, eigenvectors and principal components generated. The output produced was saved as a Microsoft Excel worksheet. Regression analysis was carried out between the actual data for individual variables and the principal components to identify the source of the principal components. The regression that produced the highest r^2 value for a principal component indicated its most likely source.

12.2.4
Combination of PCA with Other Techniques

12.2.4.1
Combination of PCA with Least Squares Regression (LSR)

The principal components of the training data generated by PCA were regressed with their corresponding visible injury scores. This regression was carried out using a back propagation program written in C. The model was trained with no hidden nodes to simulate a multiple least squares regression using an iterative process. The method could only model linear functions within the data, because the network had no hidden nodes. Analysis was carried out using a momentum of 0.9 and a learning rate of 0.05, for 1 000, 5 000 and 10 000 epochs. Learning was convergent in this case, the network having reached a minimum error by 10 000 epochs.

After the regression was complete, the principal components of the validation data set used in earlier analysis methods were generated from the eigenvectors of the PCA. These validation principal components were used to test the principal components least squares model to generate predicted injury scores. The predicted and actual values produced were then plotted within Microsoft Excel to determine the accuracy of the PCA combined with LSR approach, an r^2 value was generated for the approach by linear regression.

12.2.4.2
Combination of PCA with ANNs

To assess the performance of ANNs combined with PCA, the C program was used to retrain an ANN model with the principal components of the data. A range of hidden nodes between 2 and 10 was used, simulating an ANN approach rather than LSR. Again training was carried out for 1 000, 5 000, and 10 000 epochs, with a learning rate of 0.05 and momentum rates of 0.9 and 0.5. The same procedure was used to test the predictions of the model, on the validation data set, as for the PCA combined with LSR analysis, i.e. by regression of actual and predicted values. The performance of PCA and ANNs was compared with PCA combined with LSR.

12.3
Results

12.3.1
ANN Analysis and Testing

12.3.1.1
ANN Optimization

Results of the optimization of the ANN model using all inputs, within the Neuroshell 2 package, indicated that the best performance was produced when the model was trained using 10 hidden nodes, a momentum of 0.9 and a learning rate of 0.05. This training resulted in a r^2 value of 0.742 for training data, an r^2 of 0.84 for the validation data set and an r^2 of 0.68 for the test data set. Actual versus predicted values were plotted for the validation data (Fig. 12.1). The line of best fit for this data had an equation:

$$\text{Actual} = 0.82\ \text{Predicted} - 0.07$$

This equation shows that the network tends to slightly underestimate low levels of injury, and slightly overestimate high levels of injury.

12.3.1.2
Analysis of Weightings

Analysis of the weightings of the trained model using all inputs (Table 12.1) indicated that the input variables could be ranked in the order *AOT40* > *RH* > *leaf* > *Rand* = *temp* > *PAR*. Temperature and *Rand* were of similar importance.

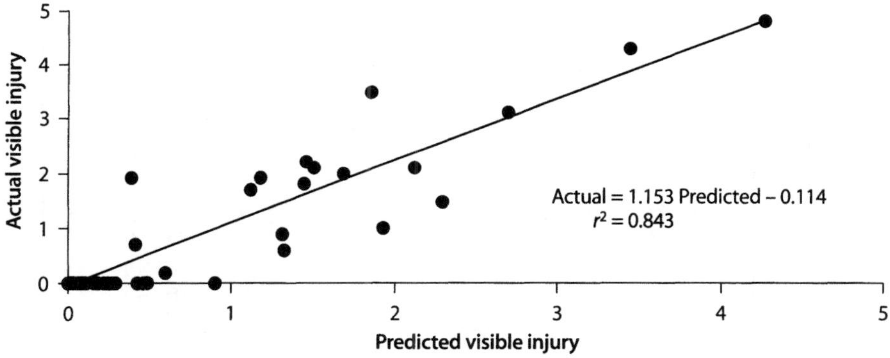

Fig. 12.1. Actual versus predicted plot for validation data applied to the ANN model using all inputs, indicating the equation of the line of best fit and the r^2 of the line

Table 12.1. Results of weightings analysis of the ANN model

Input variable	Relative importance / Arbitrary units
AOT40	0.412
RH	0.214
Leaf	0.187
Rand	0.075
Temp	0.072
PAR	0.039

12.3.1.3
Analysis of Input Combinations

Analyses of the performance of multiple models using different combinations of input variables are presented in Table 12.2. The model with the best performance for validation data ($r^2 = 0.850$) used the input combination *leaf, AOT40, temp* and *RH*. This model had an r^2 of 0.7894 for training data and an r^2 of 0.668 for test data. These inputs could be ranked in the order *AOT40* > *leaf* > *RH* > *temp* based on changes in the performance of models using combinations of these inputs.

The next 14 models produced r^2 values that were not significantly different from the above result, having r^2 values from 0.850 to 0.831.

Of the single input models only *AOT40* produced a good performance ($r^2 = 0.655$ for validation data and 0.436 for test data). All other single input models produced a performance of less than 0.071 for validation data and 0.108 for test data. This indicated that these other inputs were secondary influences to the ozone response. This was also confirmed by the poor performance of the model using all inputs apart from the *AOT40* input ($r^2 = 0.304$ for validation data and 0.179 for test data).

Table 12.2. Results of training multiple models using all combinations of input variables

Combination	Trn r^2	Valid r^2	Test r^2	Combination	Trn r^2	Valid r^2	Test r^2	Combination	Trn r^2	Valid r^2	Test r^2
PAR	0.044	0.001	0.031	*Leaf, PAR, RH*	0.154	0.114	0.149	*Rand, AOT40, Temp, RH*	0.628	0.723	0.508
Rand	0.000	0.004	0.002	*Rand, PAR, RH*	0.148	0.124	0.039	*AOT40, Temp*	0.634	0.726	0.533
Temp	0.038	0.018	0.088	*Rand, Leaf PAR*	0.185	0.132	0.170	*Rand, AOT40, RH*	0.627	0.727	0.485
PAR, Temp	0.096	0.034	0.094	*Rand, Leaf, Temp*	0.164	0.136	0.070	*AOT40, Temp, RH*	0.643	0.736	0.570
Rand, Temp	0.044	0.042	0.067	*Leaf, Temp, PAR, RH*	0.170	0.146	0.055	*AOT40, PAR, Temp, RH*	0.651	0.753	0.587
Rand, RH	0.000	0.047	0.007	*Rand, AOT40, Temp*	0.172	0.148	0.254	*Leaf, AOT40, Temp*	0.661	0.779	0.602
RH	0.000	0.049	0.006	*Rand, Leaf, PAR, RH*	0.257	0.153	0.056	*Leaf, AOT40*	0.657	0.788	0.599
Rand, PAR, Temp	0.082	0.067	0.053	*Rand, Leaf, Temp, RH*	0.082	0.158	0.217	*Rand, Leaf, AOT40, PAR, RH*	0.767	0.831	0.657
PAR, RH	0.076	0.068	0.157	*Rand, Leaf, RH*	0.090	0.159	0.080	*Leaf, AOT40, PAR, RH*	0.809	0.832	0.676
Leaf	0.075	0.071	0.108	*Leaf, Temp, RH*	0.138	0.161	0.147	*Rand, Leaf, AOT40, RH*	0.730	0.834	0.670
Leaf, RH	0.084	0.079	0.069	*Rand, Leaf, PAR, Temp, RH*	0.179	0.187	0.012	*Rand, Leaf, AOT40*	0.739	0.836	0.573
Rand, PAR, Temp, RH	0.074	0.087	0.155	*Leaf, PAR, Temp, RH*	0.375	0.304	0.178	*Rand, Leaf, AOT40, Temp*	0.728	0.837	0.549
Leaf, Temp	0.121	0.092	0.139	*AOT40*	0.576	0.655	0.436	*Leaf, AOT40, PAR*	0.680	0.838	0.648
Leaf, PAR	0.145	0.093	0.177	*AOT40, PAR, Temp*	0.576	0.683	0.443	*Rand, Leaf, AOT40, PAR*	0.685	0.839	0.620
PAR, Temp, RH	0.066	0.094	0.151	*AOT40, PAR*	0.576	0.687	0.449	All	0.742	0.843	0.683
Temp, RH	0.041	0.098	0.139	*Rand, AOT40, PAR, Temp, RH*	0.627	0.704	0.500	*Leaf, AOT40, PAR, Temp*	0.771	0.845	0.640
Leaf, PAR, Temp	0.162	0.101	0.176	*AOT40, RH*	0.617	0.704	0.405	*Rand, Leaf, AOT40, PAR, Temp*	0.703	0.845	0.613
Rand, Leaf	0.084	0.106	0.082	*Rand, AOT40, PAR, Temp*	0.597	0.706	0.420	*Rand, Leaf, AOT40, Temp, RH*	0.736	0.846	0.645
Rand, Temp, RH	0.048	0.107	0.073	*AOT40, PAR, RH*	0.616	0.706	0.517	*Leaf, AOT40, PAR, Temp, RH*	0.747	0.847	0.673
Rand, PAR	0.163	0.114	0.069	*Rand, AOT40, PAR, RH*	0.624	0.707	0.506	*Leaf, AOT40, RH*	0.760	0.850	0.683

Overall the analysis indicated the inputs could be ranked in the order *AOT40* > *leaf* > *RH* > *temp* = *PAR* = *Rand*, the same ordering as indicated by the weightings analysis.

12.3.2 Conventional Statistical Analysis

12.3.2.1 *Simple Linear Regression Analysis*

Results of the simple linear regression analysis (Table 12.3) indicated that plotting injury against *AOT40* produced the best fit (Fig. 12.2). This analysis produced a mean r^2 value of 0.55 (all data) for the 3 leafs. The second best performance was produced using *PAR* where the mean r^2 value dropped to 0.046. Temp was slightly below this having an r^2 value of 0.044. The remaining input variables could be ranked in the order *Rand* > *RH* with r^2 values of 0.010 and 0.007 respectively. This analysis indicated that the main linear influence on the injury formation process was *AOT40* and that the remaining variables had very little influence.

Table 12.3. Results of simple regression analysis

Influence	r^2 value for *leaf*			
	Spade	First	Second	Mean
AOT40	0.709	0.659	0.264	0.544
PAR	0.017	0.014	0.106	0.046
Temp	0.036	0.016	0.080	0.044
Rand	0.0003	0.028	0.0001	0.010
RH	0.00002	0.021	0.001	0.007

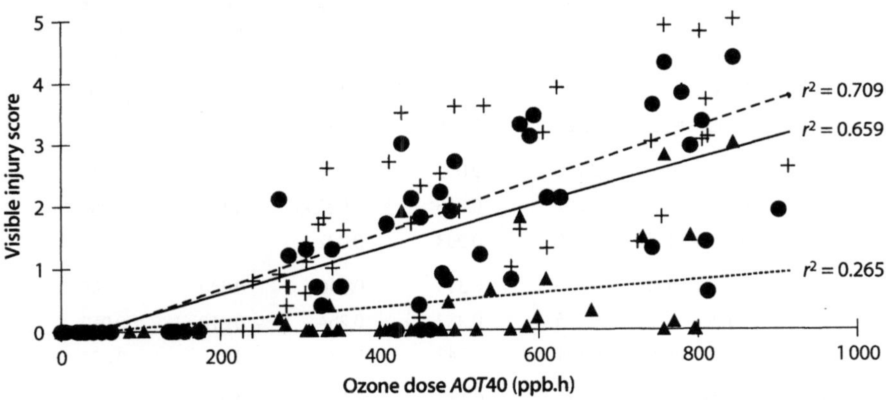

Fig. 12.2. Results of simple linear regression analysis of visible injury against *AOT40* for the spade (+), first (●) and second leafs (▲). Lines of best fit indicated as follows: spade (*dashed line*), first (*black line*) and second(*dotted line*)

12.3.2.2
Linear Regression Using Exponential, Logarithmic and Power Transformations

Regression results indicated that generally the best fit was produced using regression with a logarithmic transformation (Table 12.4). The average values for all leafs showed that the logarithmic function consistently had the strongest correlation for all of the parameters. The strongest correlation was produced with *AOT*40 using a logarithmic transformation having an average r^2 value of 0.322. This was followed by the power transformation for *AOT*40 with an average r^2 value of 0.261, then the exponential function which produced an average r^2 value of 0.255.

The next strongest correlation was produced for temperature using the logarithmic transformation which had an average r^2 value of 0.19, followed by the exponential transformation using the same variable having an average r^2 value of 0.14. The remaining variables could be ranked in the order *PAR* > *RH* > *Rand*. All of these variables had average r^2 values less than 0.14. Finally the r^2 values for *AOT*40 produced by regression using transformations never exceeded those produced by the simple linear regression analysis. A higher r^2 was however produced for *PAR*, *temp*, *RH* and *Rand* than in the simple linear regression analysis.

12.3.2.3
Multiple Regression Analysis

Further analysis of the data using multiple linear regression produced a model with an r^2 of 0.591 for the training data, only a slight improvement on the simple linear regression method.

Table 12.4. Results of nonlinear regression analysis showing r^2 values

Leaf	Regression	Parameter *AOT40*	*PAR*	*Temp*	*RH*	*Rand*
Spade	Logarithmic	0.459	0.012	0.116	0.004	0.0003
	Power	0.402	0.007	0.128	0.001	0.007
	Exponential	0.385	0.020	0.135	0.0005	0.023
First	Logarithmic	0.276	0.002	0.009	0.150	0.010
	Power	0.207	0.015	0.000006	0.173	0.003
	Exponential	0.199	0.003	0.002	0.166	0.000007
Second	Logarithmic	0.230	0.383	0.446	0.040	0.099
	Power	0.175	0.282	0.264	0.018	0.062
	Exponential	0.181	0.345	0.283	0.027	0.079
Average	Logarithmic	0.322	0.132	0.190	0.065	0.036
	Power	0.261	0.101	0.131	0.064	0.024
	Exponential	0.255	0.123	0.140	0.048	0.034

The ranking of the variables based on probability values generated by multiple regression analysis (Table 12.5) indicated that *AOT40*, *leaf* and *RH* had an important influence, whereas the influences of *PAR*, *Rand* and *temp* were weaker. These probability values indicated the probability of the hypothesis that the partial regression coefficient of the parameter was equal to zero. The regression coefficient from the multiple regression was 0.591, and the following equation was produced:

$$\text{Injury} = -0.9 - (0.07 \times Rand) - (0.36 \times leaf) + (0.0024 \times AOT40) + (0.001 \times PAR) - (0.0045 \times temp) + (0.023 \times RH)$$

When this equation was applied to the validation data, an r^2 value of 0.651 was produced for actual versus predicted data (Fig. 12.3). When this equation was applied to the test data, an r^2 value of 0.553 was produced. Clearly the method was less able to predict the mean visible injury score for the validation data and was not as good at

Table 12.5. Results of multiple regression analysis

Variable	*p*-Value
AOT40	0.000
Leaf	0.000
RH	0.002
PAR	0.169
Rand	0.538
Temp	0.807

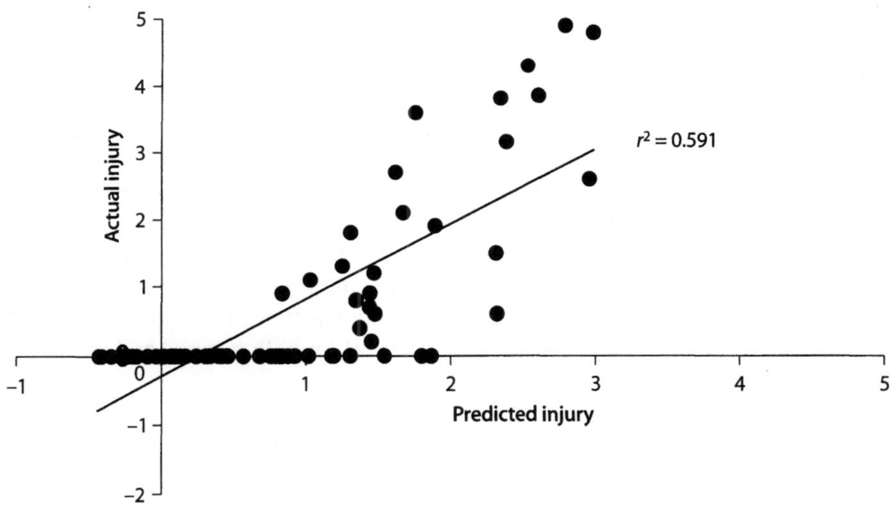

Fig. 12.3. Actual versus predicted graph for validation data applied to the multiple linear regression equation

Table 12.6. Results from forward stepwise multiple regression analysis

Variable	r^2 when included	Tolerance[a]
AOT40	0.605	0.999
Leaf	0.650	0.998
RH	0.660	0.993
Temp	not included	0.910
Rand	not included	0.735
PAR	not included	0.668

[a] Acceptable tolerance = 0.99 or greater.

Table 12.7. Results from backward stepwise multiple regression analysis

Variable	r^2 when included	Tolerance[a]
AOT40	0.605	0.999
Leaf	0.650	0.998
RH	0.660	0.993
Temp	not included	0.910
Rand	not included	0.735
PAR	not included	0.668

[a] Acceptable tolerance = 0.99 or greater.

generalizing as the ANN model. This was indicated by a large number of cases where the actual injury score was 0 but the equation predicted values from −0.49 to 0.91.

12.3.2.4
Stepwise Regression

The same results were seen from both forward and backward stepwise multiple regression analysis (Tables 12.6 and 12.7 respectively). Both methods indicated that *AOT40* had the strongest correlation, followed by *leaf*, then *RH*. The remaining variables were removed or not added to the regression analysis as their influence was not considered significant (they did not fall within the tolerance limits set within the analysis, set at 0.99). Tolerance is defined as $1 - R$ where R is the multiple correlation between the variable and all variables that are in the regression equation. The regression model had an overall r^2 value of 0.66 for the whole data set for both forward and backward stepwise regression.

12.3.2.5
Principal Components Analysis (PCA)

Eigenvalues produced from PCA and percentage variance indicated that there was a strong source from one component, which had an eigenvalue of 2.25, and a percentage variance of 37.4 in the direction of the eigenvector. Three components then had a similar, but lesser influence, with eigenvalues of 1.09, 1.00 and 0.87, and percentage variances of 18.2, 16.7 and 14.6 respectively. The next two components had little influence with eigenvalues of 0.58 and 0.2 with percentage variances of 9.7 and 3.3 respectively. So, from PCA of this data set the majority of variation arises from 4 or 5 principal linear combinations of the input variables accounting for 87 to 97 percent of the variance.

12.3.3 Combination of PCA with Least Squares Regression (LSR) and ANNs

Testing the PCA/LSR and PCA/ANN methods, using a number of different epochs, produced a range of r^2 values for the validation data (Fig. 12.4). When no hidden nodes were used, i.e. LSR with PCA, the best performance was produced within 1 000 epochs when the r^2 value for the validation data reached 0.712 (Fig. 12.5). This approach appeared to produce generalized predictions for all of the validation data that were close to the actual data. Training for 5 000 and 10 000 epochs failed to increase the r^2 value. This analysis also indicates that a low number of hidden nodes were required for optimum performance. The best performance was produced using 2 hidden nodes. However, with training for more than 1 000 epochs, a decline in performance was seen, indicating a level of overtraining. This did not occur with 1 hidden node, although the performance was lower. This would indicate that for the PCA/ANN model, between 1 and 2 hidden nodes would produce the optimum performance without overtraining. Although this is impossible, it does indicate a partial redundancy in some connections of the ANN model.

When two hidden nodes were used r^2 increased to a value of 0.834 (Fig. 12.6). This plot indicated the approach produced generalized predictions for all of the validation data that were close to the actual data. The best predictions were produced when the

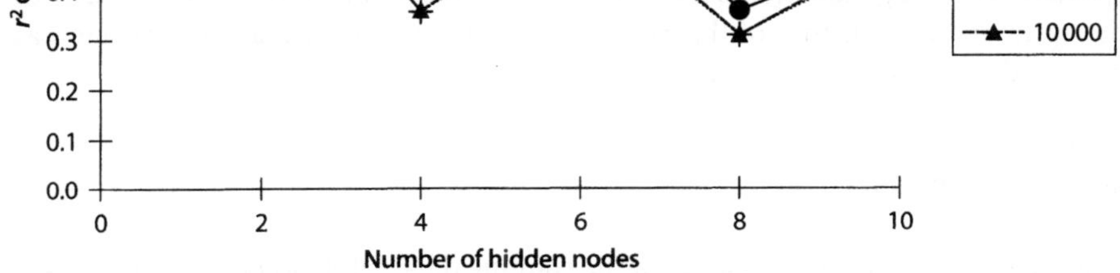

Fig. 12.4. Optimization of the model which combined PCA and ANNs based on validation data performance after 1 000, 5 000 and 10 000 epochs. Results of the models which combined PCA and LSR are also indicated by the points with no hidden nodes

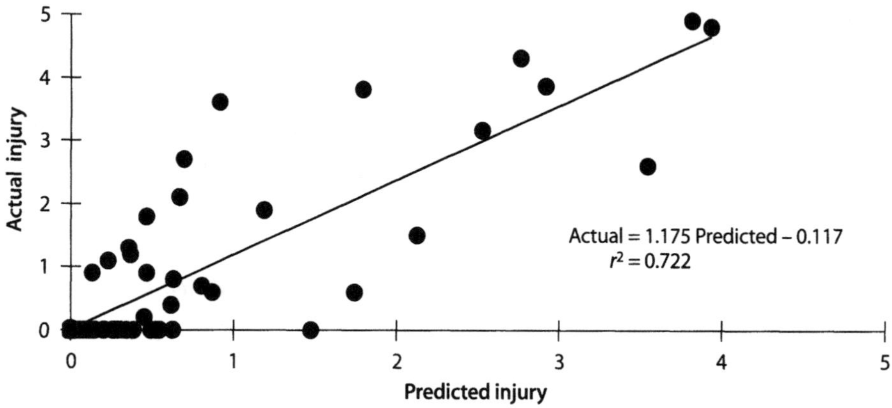

Fig. 12.5. Actual versus predicted graph for validation data applied to principal components analysis (PCA) combined with least squares regression (LSR)

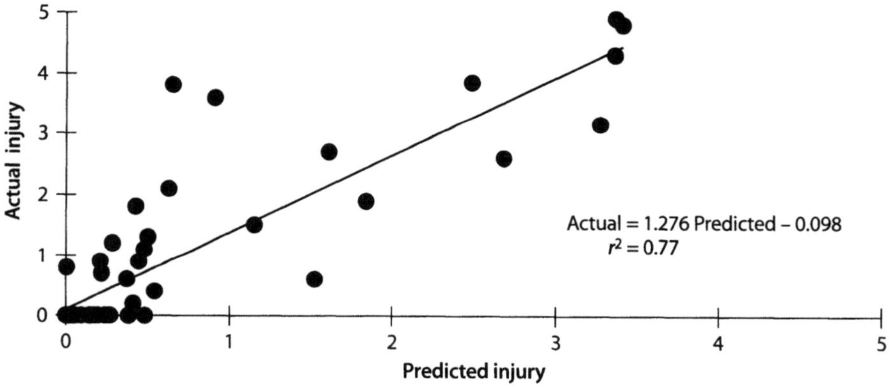

Fig. 12.6. Actual versus predicted graph for validation data applied to principal components analysis (PCA) combined with artificial neuronal networks (ANNs)

training was carried out for 1 000 epochs. When training was carried out for more epochs or using a greater number of hidden nodes the r^2 was consistently lower. Thus the optimum performance of the ANN model trained on principal components was produced after 1 000 epochs using two hidden nodes.

12.3.4 Summary of Results

The predictive performance based on the training data, validation data and test data set (where applicable) of the various methods are summarized in Table 12.8.

Table 12.8. Summary of analysis results

Method	r^2 for predictions on the training data	r^2 for predictions on the validation data	r^2 for predictions on the test data
The ANN model using all available inputs	0.742	0.843	0.683
The optimised ANN model	0.790	0.850	0.668
Simple linear regression	0.544[a]	–	–
Stepwise regression (forward and backward)	0.660	–	–
Multiple linear regression	0.591	0.651	0.553
Non–linear regression	0.459[a]	–	–
PCA with LSR (all inputs)	0.754	0.712	0.597
PCA with ANN (all inputs)	0.805	0.834	0.616

[a] Based on the mean value of the parameter with the best performance.

12.4 Discussion

12.4.1 Comparison of Modelling Performance

Analysis of the ANN model showed that an r^2 of 0.850 was produced for validation data (Table 12.6) and 0.79 for training data. Results of the simple linear regression of the training data produced r^2 values that were much lower than those produced by the ANN model. The highest r^2 produced using simple linear regression analysis was 0.513 for ozone dose *AOT*40, whereas the ANN model produced an r^2 value of 0.79. Thus, the simple linear regression analysis of this data set compare poorly with analysis using ANNs, as it can only cope with single parameters and it has no ability to model nonlinear functions.

As expected, multiple linear regression of the data produced a slightly better performance than simple linear regression, achieving an r^2 value of 0.591. When the equation produced from the analysis was applied to the validation data, an r^2 of 0.651 was produced. This result shows that the regression has successfully generalized, but only to the most general features of the injury response. As with single variable linear regression, nonlinearities are not being modelled successfully.

However, regression using transformations produced the lowest performance of all, having an r^2 value of 0.459 for *AOT*40 and the spade *leaf*. The other variables had higher r^2 values than those in the simple linear regression analysis, indicating that their influence may be less linear. The problem with regression using transformations may be that the choice of a particular transformation function is arbitrary. The multilayer

ANN approach has a greater ability to model the data in a piecewise fashion, giving a variety of gradients at different points.

PCA indicated that four main components accounted for 87% of the variation within the data. Thus, this analysis indicated that four of the linear combinations of parameters had a strong influence on visible injury. PCA is a useful tool for indicating the parameters, which act as the source of variance within the data. However, it cannot be used to produce a predictive model or a 'fit' to the data unless it is combined with another analysis method.

In this case, the PCA was combined with a least squares regression method. PCA served as a data reduction method so that the influence of variables on visible injury, and the extent to which they accounted for variation within the data, were organized into a number of principal components for the data. The PCA/LSR model produced an r^2 of 0.754 when actual results were plotted against predicted values. This was a substantial improvement on the previous conventional analysis techniques and closely matched the accuracy of the ANN approach. One possible reason for this was that the PCA part of the analysis reduced the amount of variation within the data, simplifying the modelling process, so the LSR part of the analysis can produce a more accurate model of the data.

Since PCA combined with LSR was a linear method, the high performance of the approach suggests that the data being modelled is substantially linear. However, some evidence for nonlinearity in the data is provided by the further improvement of the accuracy of the predictions was produced when ANNs were combined with the PCA, this approach having an r^2 of 0.834. The simplification of the data was indicated by a reduction in the number of hidden nodes required for optimum performance from 10 in ANN model to 2 in the PCA/ANN model. Hidden nodes detect features within the data; the data reduction by the PCA meant fewer hidden nodes were needed to detect fewer features. Also, as the ANN plus PCA approach produced a better performance than the PCA plus LSR approach, it could be deduced that the data set had a small number of nonlinear functions associated with it. Modelling of these nonlinear functions requires hidden nodes to detect and incorporate them into the model.

Clearly ANN models are better at producing accurate predictions for the data used in this study than all conventional statistical techniques examined. The performance of ANNs has been compared to conventional statistical methods and, as in the majority of cases (Wise et al. 1995; Lek et al. 1996; Timofei et al. 1997; Paruelo and Tomasel 1997), ANNs show a better performance when modelling nonlinear data.

12.4.2 Determination of the Importance of Input Variables

Analysis of the weightings of the ANN model (with random number) indicated that, as in all previous models, *AOT*40 had the strongest influence on the extent of visible ozone injury, followed by *leaf* and *RH*. Analysis of the importance of selected input combinations indicated that the inputs were ranked in the same order as in the weightings analysis. This method also indicated that *AOT*40 was the primary influence, and that all of the other input variables were secondary influences on the *AOT*40 dose response. Analysis of input combinations is more dependable than direct analysis of the weightings of the ANN model, which is essentially a linearisation of the nonlinear ANN model. By retraining, after the removal of combinations of inputs in groups, the importance of inputs which are partially bi-linearly or collinearly related can be determined. In this case, all possible combinations of inputs are used to indicate the effect (importance) of input combinations, by their simultaneous absence, as well as single variables. But in general, removals need only include groupings of variables which are removable in smaller groups or individually. In this way, it is possible to establish the key variables that are able to substitute for the ones that have been removed.

Analysis of the r^2 values for each of the influences on the visible injury score using simple linear regression indicated that as in the ANN model, *AOT40* had the strongest influence. The next strongest influences were found to be in the order of *PAR* > *temp* > *Rand* > *RH*. Leaf could not be analysed by this method, as it was an integer. Clearly with the exception of *AOT40*, there was no similarity between the ranking of parameters using simple linear regression and the ANN methods. In the analysis all of the parameters, other than *AOT40*, had very low r^2 values, indicating that they had little or no influence on the extent of visible injury.

Multiple regression analysis indicated that the influences on visible injury could be ranked in the order of *AOT40* > *leaf* > *RH* > *PAR* > *Rand* > *temp*. This agreed with the findings of weightings analysis of the ANN model with the exception that *PAR* and *temp* swapped positions in the ranking around the random number. Stepwise regression indicated that three variables should be included. These were *AOT40*, *leaf* and *RH*. The remaining variables were not ranked in the same order as ANN analysis and multiple regression, the random number having a very high tolerance value (low importance) and temperature having the lowest.

Analysis of the importance of input variables from simple regression using transformations indicated the input variables could be ranked in the order *AOT40* > *temp* > *PAR* > *RH* > *Rand*. As in simple linear regression, this differed greatly from the ANN model with the exception that *AOT40* was the most important variable.

PCA indicated that four components were influential in the data. This would appear to back up the findings of the ANN model which indicate the strongest influences came from *AOT40*, *leaf*, *temp* and *RH*. *PAR* and the random number had little or no influence.

Mechanistic studies have indicated the importance of the air humidity in the form of VPD and temperature as influences on the flux of ozone into the plant (Leuning 1995; Emberson 1996). This has also been confirmed by analysis of field measurements (Gutiérrez et al. 1994; Herbst 1995; Kallarackal and Somen 1997). Both of these components are contained in the relative humidity variable. Analysis also indicated leaf age was of high importance. Pääkkönen et al. (1995) also found this to be the case for beech leafs.

12.5 Conclusion

ANN models produced the best performance. The conventional statistical techniques produced a poor performance when modelling the data and were unable to produce accurate predictions on unseen data. The best non-ANN method of modelling was achieved by combining PCA with LSR. This approach could be improved by inclusion of the PCA with ANN techniques. In general the input combination approach is the most reliable means of determining input variable importance and the importance of input combinations. However, findings from weightings analysis and multiple linear regression indicated the same ordering of importance of input variables. Stepwise regression predicted the same variables for the three most important variables but did not put the remaining variables in the same order. Simple linear regression and regression using transformations only predicted the most important variable, which was *AOT40*. The ranking of variables produced by the ANN methods and by multiple regression was confirmed by surveys of the literature.

References

Aphalo PJ, Jarvis PG (1991) Do stomata respond to relative humidity? Plant Cell and Environment 14:127–132

Ball GR, Skarby L, Fuhrer J, Sanchez-Gimeno B, Palmer-Brown D, Mills GE (1998) Identifying factors which modify the effects of ambient ozone on white clover (*Trifolium repens* L.) in Europe. Environmental Pollution 103:7–16

Balls GR, Sanders GE, Palmer-Brown D (1995) Unravelling the complex interactions between microclimate ozone dose and ozone injury in clover (*Trifolium subterraneum* L. cv. Geraldton). Water Air and Soil Pollution 85:1467–1472

Balls GR, Sanders GE, Palmer-Brown D (1996) Investigating microclimatic influences on ozone injury in clover (*Trifolium subterraneum* L. cv. Geraldton) using artificial neuronal networks. New Phytologist 132:271–280

Benitez JM, Castro JL, Requena I (1997) Are artificial neuronal networks black boxes? IEEE Transactions on Neuronal Networks 8(5):1156–1164

Colls JJ, Sanders GE, Geissler PA, Bonte J, Galaup S, Weigel H-J, Ashmore MR, Jones M (1993) The responses of beans exposed to air pollution in open-top chambers. Air Pollution Research Report 48. The European open-top chamber project: Assessment of the effects of air pollutants on agricultural crops

Comrie AC (1997) Comparing neuronal networks and regression models for ozone forecasting. Journal of the Air & Waste Management Association 47(6):653–663

Emberson LD (1996) Defining and mapping relative potential sensitivity to ozone. Ph.D. Thesis, Centre for Environmental Technology, Imperial College of Science, Technology and Medicine, London

Fry JC (1993) Biological data analysis. A practical approach. Oxford University Press, Oxford

Grantz DA, Meinzer FC (1990) Stomatal response to humidity in a sugar cane field: Simultaneous porometric and micrometeorological measurements. Plant Cell and Environment 13:27–37

Gutiérrez MV, Meinzer FC, Grantz DA (1994) Regulation of transpiration in coffee hedgerows: Co-variation of environmental variables and apparent responses of stomata to wind and humidity. Plant Cell and Environment 17:1305–1313

Herbst M (1995) Stomatal behavior in a beech canopy – an analysis of Bowen-ratio measurements compared with porometer data. Plant Cell and Environment 18(9):1010–1018

Kallarackal J, Somen CK (1997) Water use by *Eucalyptus tereticornis* stands of differing density in Southern India. Tree Physiology 17(3):195–203

Kerstiens G, Federholzner R, Lendzian KJ (1992) Dry deposition and cuticular uptake of pollutant gasses. Agriculture, Ecosystems and Environment 42:239–253

Lek S, Delacoste M, Baran P, Dimopoulos I, Lauga J, Aulagnier S (1996) Application of neuronal networks to modelling non-linear relationships in ecology. Ecol Model 90(1):39–152

Leuning R (1995) A critical-appraisal of a combined stomatal-photosynthesis model for C-3 plants. Plant Cell and Environment 18(4):339–355

McIlraith AL, Card HC (1997) Birdsong recognition using backpropagation and multivariate statistics. IEEE Transactions on Signal Processing 45(11):2740–2748

Pääkkönen E, Metsärinne S, Holopainen T, Kärenlampi L (1995) The ozone sensitivity of birch (*Betula pendula* L.) in relation to the developmental stage of the leaves. New Phytologist 132:145–154

Paruelo JM, Tomasel F (1997) Prediction of functional characteristics of ecosystems: A comparison of artificial neuronal networks and regression models. Ecol Model 98(2–3):173–186

Roadknight CM, Balls GR, Mills GE, Palmer-Brown D (1997) Modeling complex environmental data. IEEE Transactions on Neuronal Networks 8(4):852–862

Sanders GE, Skärby L, Ashmore MR, Fuhrer J (1995) Establishing critical levels for the effects of air pollution on vegetation. Water Air and Soil Pollution 85:189–200

Simpson R, Williams R, Ellis R, Culverhouse PF (1992) Biological pattern recognition by neuronal networks. Mar Ecol Prog Ser 79:303–308

Tan SS, Smeins FE (1996) Predicting grassland community changes with an artificial neuronal-network model. Ecol Model 84(1–3):91–97

Timofei S, Kurunczi L, Suzuki T, Fabian WMF, Muresan S (1997) Multiple Linear Regression (MLR) and Neuronal Network (NN) calculations of some disazo dye adsorption on cellulose. Dyes and Pigments 34(3):181–193

Urquidimacdonald M, Macdonald DD (1994) Performance comparison between a statistical model, a deterministic model, and an artificial neuronal-network model for predicting damage from pitting corrosion. Journal of Research of The National Institute of Standards and Technology 99(4):495–504

Weiser G, Havranek WM (1995) Environmental control of ozone uptake in *Larix decidua* Mill.: a comparison between different altitudes. Tree Physiology 15:253–258

Wise BM, Holt BR, Gallagher NB, Lee S (1995) A comparison of neuronal networks, non-linear biased regression and a genetic algorithm for dynamic model identification. Chemometrics and Intelligent Laboratory Systems 30(1):81–89

Part IV

Artificial Neuronal Networks in Genetics and Evolutionary Ecology

Application of the Self-Organizing Mapping and Fuzzy Clustering to Microsatellite Data: How to Detect Genetic Structure in Brown Trout (*Salmo trutta*) Populations

J.L. Giraudel · D. Aurelle · P. Berrebi · S. Lek

13.1 Introduction

Artificial Neuronal Networks (ANNs) are now currently used for various purposes, from physical and chemical studies to biological ones. Even if they are less used in ecology and populations genetics, recent studies have shown that they can be very efficient for such problems (Cornuet et al. 1996; Foody 1997; Mastrorillo et al. 1997; Guégan et al. 1998). ANNs have several advantages: they can be applied to various data, from environmental variables to genotypes, and are usually more efficient than classical statistical techniques (FDA, for example; see Cornuet et al. 1996). In order to classify biological objects (individuals or populations, for example) using ANNs, two main types of methods can be applied: supervised and unsupervised learning. Supervised learning can be applied to the classification of individuals of unknown origin among already well-defined groups: the network will be trained to recognize these categories by using reference samples. This has been successfully applied to genetic data on bees (Cornuet et al. 1996, with some phylogenetically well separated lineages), and on trout (Aurelle et al. 1998, but with some less clearly differentiated groups).

With unsupervised learning, on the other hand, no groups are a priori defined, and the network will try to find an organization itself (and then classify individuals) in the global data set. This can be useful when the categories are previously unknown or when no pure reference samples are available for a supervised learning. The genetic analysis of French trout populations corresponds to this last situation. In southwest France, several genetic entities coexist in the same basins, or even in the same river (Aurelle and Berrebi 1998). Some so-called *wild modern* (according to Hamilton et al. 1989) *Atlantic trout* can be separated from *ancestral Atlantic trout* (also wild) using allozymes. Both are naturally present in southwest France. Moreover, stocking practices have led to the introduction of modern trout, born in hatcheries. These fish, which usually do not originate from the river where they are released, will be called *domestic modern Atlantic trout*; they cannot be separated from wild modern trout using allozymes. The main usefulness of the analysis of these populations is to classify the three types of individuals that one river can contain. In order to describe the genetic composition of the different populations belonging to these rivers, we decided to use microsatellites; they were hoped to better separate modern and ancestral trout and, within modern fish, to separate wild and domestic trout. Nevertheless, because of their high variability and particular properties (homoplasy, ancestral polymorphism, etc.), numerous alleles were shared between the different forms (Aurelle and Berrebi 1998). As each locus separately brought little information, a multilocus analysis was necessary. Indeed, the simultaneous presence of particular alleles from different loci in the same fish is a better indicator of the genetic origin of this individual than what is shown at each locus. That is why multilocus analysis was necessary to classify the sampled trout and to answer several questions:

- Is the separation between modern and ancestral trout discovered using allozymes also supported by microsatellites?
- What is the origin of sampled modern individuals: wild or domestic?
- What is the structure of the river populations: are they homogeneous or heterogeneous (because of natural or artificial propagation)?

Despite all the problems already mentioned and as shown by the study of Cornuet et al. (1996), ANNs seemed well suited to answer to such questions because of their ability to analyse microsatellites. We first used supervised learning associated with a back propagation algorithm (Aurelle et al. 1998) which gave some good results. Nevertheless, some populations or individuals were not easily classified. Several phenomena could explain this: some of these populations were probably genetically quite similar, and some samples used as references (and so assumed to be homogeneous) may in fact be heterogeneous (they could contain several different forms, domestic and wild modern Atlantic for example), which would prevent efficient learning. In order to verify the conclusions of the supervised approach, we decided to use an unsupervised network in order to get an objective image of the genetic relationships between and within our samples. The results should then not be influenced by our knowledge of the sampling points and the allozymic characteristics of the individuals; for example, a so-called modern individual sampled in a river used as a reference for wild modern type during supervised learning could in fact be a domestic fish. Moreover, unsupervised analysis is an efficient tool to reveal the genetic structure of the different populations analysed.

For this purpose we decided to use a technique which aims at representing individuals positioned in multidimensional space of the variables on a 2D map, with a minimum of distortion. Graphical representation of high-dimensional data is a difficult problem when the number of dimensions rises above three. In this work we used the Kohonen Self-Organizing Map (SOM) (Kohonen 1995), a model of Artificial Neuronal Network (ANN) for visualisation of vectors in a two-dimensional space. Through an unsupervised learning process, the SOM algorithm performs a topology-preserving projection of the data space onto a regular two-dimensional grid (topographic map). Unlike the commonly used approaches such as Principal Components Analysis (PCA), SOM is a nonlinear approach, and so formal proofs of convergence are almost impossible (Blayo and Demartines 1991). SOM has already been used successfully in ecological study (Chon et al. 1996). In our study, this analysis aims at representing the different individuals on a map. Then a cluster analysis based on fuzzy sets (Zadeh 1965; Foody 1996) is used to define the different groups which can be recognized among them; this partition into genetic groups is discussed and compared to what we already know about these samples.

13.2
Material and Methods

13.2.1
Biological Samples

Four river populations and 3 hatchery strains were analysed. The different sample origins and their sizes are given in Table 13.1. Numbers (map) in Table 13.1 refer to Fig. 13.1, and the percentages of *LDH5*90* allele give information about the genetic composition of the populations. The ancestral form is characterized by allele 100 at

Table 13.1. Origin and characteristics of the samples

No. (map)	Locality	River	Basin	Sample size	*LDH5**90 (%)
	La Canourgue	hatchery		50	95
	Brassac	hatchery		30	100
	Suech	hatchery		36	99
1	Sare	Beherekobentako	Nivelle	24	0
2	Dancharia	Nivelle	Nivelle	30	2
3	Bidarray	Bastan	Adour	29	4
4	Argeles	Luz	Adour	88	95

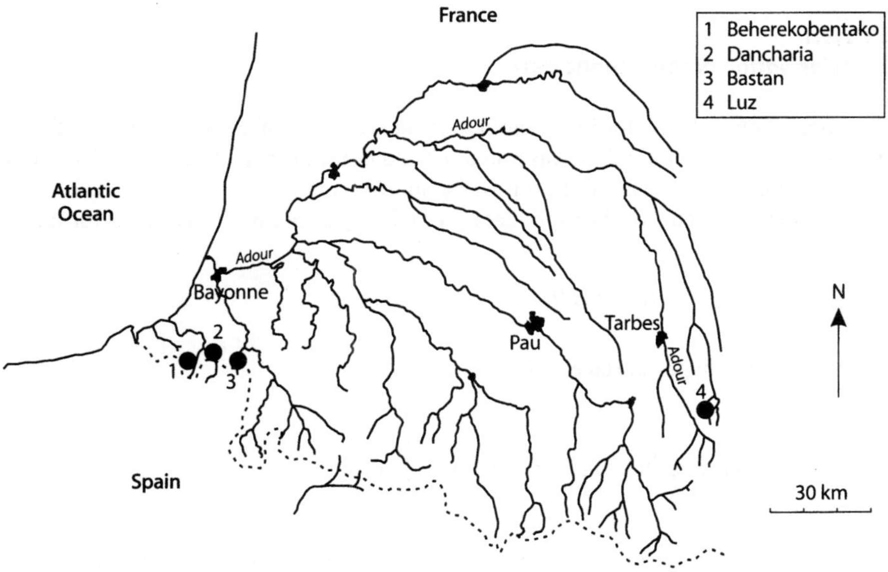

Fig. 13.1. Location of the sampling points

this locus, whereas the two modern forms possess allele 90. A population with 100% *LDH5*90* is then considered as modern, but we do not know whether these fish are wild or domestic.

Some river samples were mainly composed of modern fish (Luz), while others were almost completely ancestral (Dancharia, Béhérékobentako, Bastan). According to local managers, these populations have not been stocked for several years. Moreover, morphological characteristics would tend to show that Luz fish are mainly wild. Three hatchery strains used in southern France for stocking have also been analysed: Brassac, Canourgue and Suech.

13.2.2 Microsatellites

Among the four microsatellite loci used, two were highly variable: *Strutta 58* and MSU4, with 37 and 16 alleles respectively. The first has been cloned by Poteaux (1995). MSU4 has been published in Genbank, under accession number U43694; it was directly submitted by O'Reilly et al. (1996) and has been identified in salmon (*Salmo salar*).

Two other loci presented only a few alleles compared with usual microsatellite variability: MST 73 (6 alleles) and MST 15 (7 alleles). Both have been published by Estoup et al. (1993).

PCR and analysis procedures are described in Aurelle and Berrebi (1998).

13.2.3 Artificial Neuronal Networks

13.2.3.1 *Transformation of the Genetic Data*

The components of a vector in the input data set are alleles of a trout coded as follows: for each allele, each individual was noted 0 if it did not possess it, 0.5 if the fish was a heterozygote for this allele, and 1 if it was a homozygote.

In this way, a vector of the input data set (Table 13.2) can be seen as a vector in:

$$(\Sigma) = H^{n_1} \times H^{n_2} \times H^{n_3} \times H^{n_4}$$

where H^{n_i} is the hyper-surface of R^{n_i} defined by:

$$\forall i \in \{1;\ldots;4\}; \forall j \in \{1;\ldots;n_i\};\ x_j^i \geq 0 \text{ and } \sum_{k=1}^{n_i} x_k^i = 1$$

where n_i corresponds to the number of alleles of locus i; 4 loci have been used.

In order to choose the closest neuron to a particular genotype (see the SOM algorithm below), we define a distance in (Σ) as follows:

Table 13.2. Components of a vector in the input data set

Individual	Locus 1	Locus 2	Locus 3	Locus 4
X_1	$x^1_{11}\ x^1_{12}\ \dots\ x^1_{1n1}$	$x^2_{11}\ x^2_{12}\ \dots\ x^2_{1n2}$	$x^3_{11}\ x^3_{12}\ \dots\ x^3_{1n3}$	$x^4_{11}\ x^4_{12}\ \dots\ x^4_{1n4}$
...	...	...	...	...
...	...	...	...	...
X_p	$x^1_{p1}\ x^1_{p2}\ \dots\ x^1_{pn1}$	$x^2_{p1}\ x^2_{p2}\ \dots\ x^2_{pn2}$	$x^3_{p1}\ x^3_{p2}\ \dots\ x^3_{pn3}$	$x^4_{p1}\ x^4_{p2}\ \dots\ x^4_{pn4}$

Let be X and Y, 2 vectors in (Σ)

$$X = \left[x_1^1, \dots, x_{n_1}^1, x_1^2, \dots, x_{n_2}^2, x_1^3, \dots, x_{n_3}^3, x_1^4, \dots x_{n_4}^4\right]^T;$$

$$Y = \left[y_1^1, \dots, y_{n_1}^1, y_1^2, \dots, y_{n_2}^2, y_1^3, \dots, y_{n_3}^3, y_1^4, \dots y_{n_4}^4\right]^T$$

$$D(X;Y) = 1 - \frac{1}{4}\sum_{i=1}^{4}\left(\sum_{j=1}^{n_i}\sqrt{x_j^i y_j^i}\right) \tag{13.1}$$

This distance is derived from Nei distance D_A (Takezaki and Nei 1996), initially defined to estimate genetic distances between populations using allelic frequencies. Here it is used as an inter-individual distance which gives 0 when 2 individuals present the same genotype, and 1 when they share no alleles.

In this study, we had 245 individuals ($p = 245$); the number of alleles for each locus is: $n_1 = 37$; $n_2 = 6$; $n_3 = 7$, $n_4 = 16$ and 66 was the total number of alleles.

13.2.3.2
The Kohonen Self-Organizing Map (SOM)

13.2.3.2.1
Presentation

The Kohonen neuronal network consists of two layers: the first (input) layer is connected to a vector of the input data set (alleles coded as already explained); the second (output) layer forms a map: a rectangular grid laid out on an hexagonal lattice (Fig. 13.2).

We have n neurons ($n = n_1 + n_2 + n_3 + n_4 = 66$) in the input layer and S ($S = 80$) neurons in the output layer. The different parameters of the SOM algorithm were determined by experiment. An input vector $X(t)$ is connected to each neuron i in the output layer through an n-dimensional weight vector m_i.

$$\forall i \in \{1;\dots;S\}; m_i = \left[\mu_{i1}^1, \dots, \mu_{in_1}^1, \mu_{i1}^2, \dots, \mu_{in_2}^2, \mu_{i1}^3, \dots, \mu_{in_3}^3, \mu_{i1}^4, \dots \mu_{in_4}^4\right]^T$$

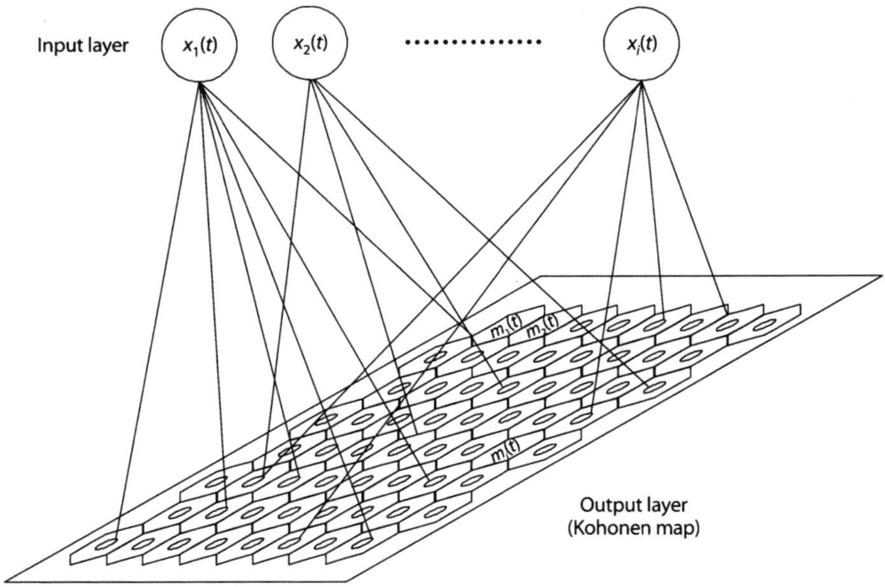

Fig. 13.2. An example of a Kohonen self-organizing map

The distribution of the weights should reflect the probability density $p(x)$ of the input data set in (Σ). Moreover, neurons which are neighbours on the grid are also expected to represent neighbouring clusters of objects. Neurons on the grid having a large distance to each other, in terms of distance in (Σ), are expected to be distant in feature space. Therefore, the SOM can be used to visualise data separability.

13.2.3.2.2
The SOM Algorithm

The SOM algorithm is unsupervised learning and can be summarized as follows:

For each neuron u, a neighbourhood $N_u(t)$ is defined. $N_u(t)$ is a decreasing function of t (Fig. 13.3); this means that as the time t (the number of iterations) increases, the neighbourhood of u, that is the number of neurons which will be modified with u (see step 5), will decrease.

Let $\alpha(t)$ be a decreasing function of time t.

During the learning phase, a data vector is randomly presented, the weights $m_i(t)$ are modified according to the algorithm below:

Step 1: $t = 0$, the weight vectors $m_i(0)$ contained in all the nodes are initialised with random samples drawn from the input data set;

Step 2: An input vector $X(t)$ is randomly selected from the input set.

Step 3: The distances between $X(t)$ and each weight vector $m_i(t)$ are computed: $d_i(t) = D(X(t); m_i(t))$, where D is defined in Eq. 13.1.

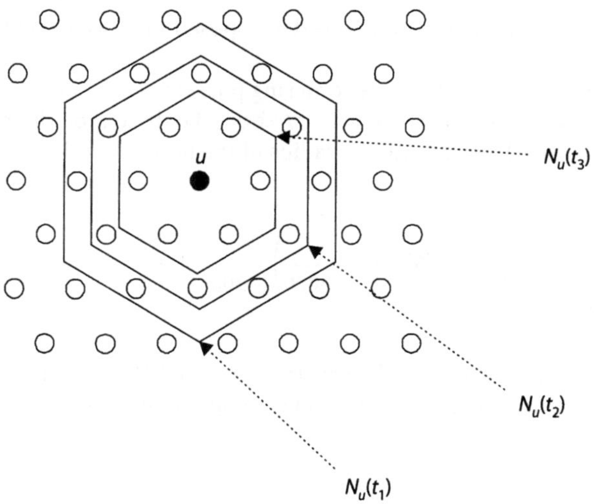

Fig. 13.3. Example of neighbourhood for neuron u ($t_1 < t_2 < t_3$)

Step 4: The Best Matching Unit (BMU) u is chosen as the winning neuron for the input vector $X(t)$: $u = \text{argmin}_i\{D(X(t); m_i(t))\}$. This chooses the neuron with the smallest distance to the input vector, or in other words, the neuron which responds maximally to this input.

Step 5: The neighbouring neurons to the winning node are updated with the rule as follows:

- If $i \in N_u(t)$, $m'_i(t+1) = m_i(t) + \alpha(t)[x(t) - m_i(t)]$, and then in order to obtain a vector in (Σ), the components of $m_i(t+1)$ are computed as follows:

$$\forall q \in \{1,\ldots,4\}; \forall j \in \{1;\ldots,n_l\}; \mu_{ij}^{q}(t+1) = \frac{\mu'^{q}_{ij}(t+1)}{\sum_{r=1}^{n_l} \mu'^{q}_{ir}(t+1)}$$

  this ensures that the sum of the new weights will be equal to 1, and so the matching and updating laws are mutually compatible with respect to the same metric (Kohonen 1995). These corrections on the winning neuron and its neighbourhood allow their weights to change in order to further reduce the distance between the weight and the input vector.

- If $i \notin N_u(t)$, $m_i(t+1) = m_i(t)$

Step 6: Increase time to $t+1$.
If $t < t_{max}$, go to step 2 or else end or stop the training.

In Step 1, the grid was chosen with 80 hexagonal lattices, so we can say that SOM is robust regarding the initialisation of the weights m_i. The choice made here accelerates algorithm convergence.

In Step 3, we consider the distance D defined in Eq. 13.1.

In Step 5, the training is decomposed into two periods:

- During the first one (ordering phase: $0 \le t \le 2\,000$), α starts with the value 0.9 then decreases: $\alpha(t) = 0.9(1 - t / 2\,000)$. Let $|N_u(t)|$ be the radius of the neighbourhood. $|N_u(t)|$ starts with the radius of the network and

$$|N_u(t)| = |N_u(0)| \times \left(1 - \frac{t}{2000}\right) + \frac{t}{2000}$$

like this, during the ordering phase $|N_u(t)|$ decreases to one unit.
- During the second period (tuning phase: $2\,001 \le t \le t_{max}$), α decreases very slowly

$$\alpha(t) = 0.02\left(1 - \frac{t}{t_{max}}\right) + 0.00001$$

and the neighbourhood $N_u(t)$ contains only the neurons closest to the BMU.

So, when time increases, and the distance between the winner nodes and the corresponding inputs decreases, the amplitude of the correction decreases.

In Step 6, t_{max} does not depend on the number of vectors in the input data set; it has been fixed at 40 000, i.e. 500 times the number of neurons in the Kohonen map.

When the training is finished, the BMU is determined for each vector of the input data set, and each sampling unit is set in the corresponding hexagon of the Kohonen map.

13.2.3.2.3
Map Measures

As the training is unsupervised, it is not easy to measure the map quality. Several criteria have been suggested (Kraaijveld et al. 1995; Der et al. 1994; Zupan et al. 1993; Hämäläinen 1994). In our work, we have computed the topographic error e which gives the proportion of sample vectors for which the two best matching vectors are not in adjacent hexagons in the map (Kiviluoto 1996):

$$\varepsilon = \frac{1}{245}\sum_{i=1}^{245} \chi(X_i)$$

where, if X_i is a sample vector, $\chi(X_i) = 1$ if the first and the second BMUs of X_i are not adjacent units, otherwise zero.

Once the SOM had been created, we tried to classify the individuals into different groups according to this map.

13.2.3.3
Clustering

The goal of cluster analysis was to subdivide a data set into different groups according to similar characteristics. For example, in our case, individuals with similar genotypes were assigned to the same clusters. The quality of the grouping would be apparent if the distances between the elements of a cluster were small and the distances between the clusters were large. For clustering, we can use SOM algorithm results as input data. Several methods have been described (Kraaijved et al. 1995; Tsao et al. 1994). In this work, we have chosen to combine SOM algorithm and Fuzzy Clustering Mean (FCM). Preprocessing data with Artificial Neuronal Network makes fuzzy clustering easier (Foody 1997).

13.2.3.3.1
The FCM Algorithm

In contrast to a hard classification (where individuals are classified to one strictly defined group), fuzzy clustering does not assign every data element to exactly one cluster; this is well suited to genetic data. Indeed, genetic data given locus by locus can be different in terms of origin. A given trout can bear ancestral alleles at one locus and modern genes at another. This method, proposed by Bezdek et al. (1984), is founded on the use of fuzzy sets. Fuzzy sets (Zadeh 1965) differ from "classical" sets in that each sample can belong to multiple sets to different degrees. The membership to a fuzzy set is characterized by a "degree of membership" which takes a value between 0 and 1 (whereas in hard classification, this function would take only two values, 0 or 1: an individual belonging to a group or not). Let $M = \{m_1; \ldots; m_S\}$ be a set of S individuals (here, the S neurons in the Kohonen map: $S = 80$). A fuzzy c-partition of M is represented by a real $c \times S$ matrix $U = (u_{ij})$ $[1 \le i \le c$ and $1 \le j \le S]$ such that:

$$u_{ij} \in [0;1] \tag{13.2a}$$

$$\forall i, 1 \le i \le c, \sum_{j=1}^{S} u_{ij} > 0 \tag{13.2b}$$

$$\forall j, 1 \le j \le S, \sum_{i=1}^{c} u_{ij} = 1 \tag{13.2c}$$

Consequently, for each fuzzy set, the membership value of at least one individual is not 0 and for each individual, the sum of its membership values is 1. A membership value of 0 would indicate that it does not belong to the fuzzy set, while a membership value of 1 denotes that it fully belonging to the fuzzy set.

The FCM algorithm is based on the minimization of a generalized least-squares objective function

$$J_v = \sum_{j=1}^{S}\sum_{i=1}^{c}\left(u_{ij}\right)^{v} D\left(m_j;e_i\right)$$

where e_i is the centre of cluster i.

The determination of the exponent v, and of the number of clusters c (cluster validity) is a difficult problem. Experimental strategy was used to select v and, following Bezdek et al. (1984), we chose c which optimizes the normalized entropy:

$$H_c(U) = -\sum_{j=1}^{S}\sum_{i=1}^{c}\frac{u_{ij}\ln(u_{ij})}{S}$$

13.2.3.3.2
The FCM Algorithm Applied to the Kohonen Map

We consider the set $M=\{m_1; \ldots; m_S\}$ where m_i are the weight vectors computed at the end of the SOM training. We determine the c-fuzzy clusters in M. The distance used will be D (defined in Eq. 13.1). The FCM algorithm is applied with

$$e_i = \left[\varepsilon_{i1}^{1},\ldots,\varepsilon_{in_1}^{1},\varepsilon_{i1}^{2},\ldots,\varepsilon_{in_2}^{2},\varepsilon_{i1}^{3},\ldots,\varepsilon_{in_3}^{3},\varepsilon_{i1}^{4},\ldots\varepsilon_{in_4}^{4}\right]^T$$

Moreover, the centres of the clusters have to be in (Σ) and are modified at each iteration as follows:

$$\forall q \in \{1,\ldots,4\}; \forall j \in \{1;\ldots,n_q\}; \varepsilon_{ij}^{q} = \frac{\varepsilon'^{q}_{ij}}{\sum_{r=1}^{n_q}\varepsilon'^{q}_{ir}}$$

After which, a surface motif is chosen for each hexagon according to the cluster to which it belongs (Fig. 13.5). We used mixed surface for weakly separated hexagons.

13.3
Results and Discussion

13.3.1
The Self-Organizing Map

After the end of training, in order to evaluate the quality of the SOM, the topographic error ε was calculated and we found $\varepsilon = 0.0041$. According to this result, for only one sample vector, the first and the second BMUs are not adjacent hexagons. So, the smoothness of the Kohonen map is can be considered as good.

The results of the distribution of the individuals belonging to the different populations on the Kohonen map is given in Fig. 13.4. On the 80 hexagons, the size of the label is proportional to the number of individuals of each population in the cluster. To interpret of this map, it should be noted that two neighbouring hexagons contain more closely related individuals than distant hexagons, but the relationship is not regular. For example, two contiguous hexagons can in fact contain more distant individuals than another pair of hexagons; This is related to the fact that it is a two-dimensional representation of a multidimensional space, and there are some distortions: two close hexagons can in fact be separated by a kind of "valley" indicating that going from one to another, there is some great distance.

Nevertheless, inspection of this map shows several interesting groupings. The individuals coming from ancestral populations (according to allozymes: Béhérékobentako (*Be*), Dancharia (*Da*) and Bastan (*Ba*)) are mainly in the right side of the map and among them, Dancharia individuals are mainly on the upper part and Bastan and Béhérékobentako individuals in the lower part. Nevertheless, these distributions overlap and a few hatchery individuals are found on this part of the map. On the contrary,

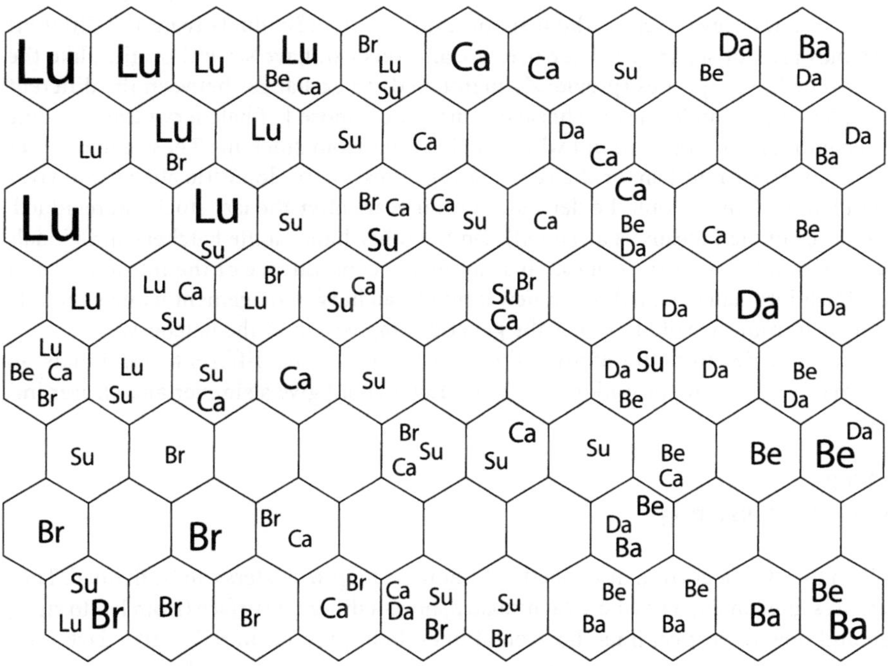

Xx = 1 Individual in population Xx

Xx = 10 Individuals in population Xx

Xx = 20 Individuals in population Xx

Fig. 13.4. Distribution of individuals on the Kohonen map. *Ba*: Bastan, *Be*: Beherekobentako, *Br*: Brassac, *Ca*: Canourgue, *Da*: Dancharia, *Lu*: Luz, *Su*: Suech. In each hexagon, the size of the print is proportional to the number of individuals in each locality only very few individuals from ancestral populations can be found on the left part of the map (see hexagons *41* and *4*).

The left and middle parts correspond to modern populations, with assumed wild modern individuals (Luz river, *Lu*) on the left upper part and domestic ones (hatchery samples: Brassac (*Br*), Canourgue (*Ca*), and Suech (*Su*)) on the left lower and the middle part of the map. Luz individuals seem to be well grouped and separated from other modern individuals. The distribution of the different hatchery strains is less clear and several hexagons contain individuals of various origins.

These observations can be interpreted from a biological point of view. First, they confirm the separation between ancestral and modern individuals using a new kind of marker (microsatellites) and without predefining some samples or populations. In the same way, among modern individuals we can see a separation between river individuals (assumed to be predominantly wild) and domestic individuals; This agrees well with the morphological characteristics of these fish and with the preliminary results obtained with classical population genetics analyses (Aurelle et al. 1998). These distinctions show that unsupervised artificial neuronal networks can be successfully applied to genetic data, and even to microsatellite data where there is quite a lot of "noise" (due to homoplasy and ancestral polymorphism; see Jarne and Lagoda (1996) for a review). Nevertheless, the distributions of the individuals from the different samples are not separated. They overlap, and this can have several origins, like the microsatellite properties (previously mentioned) or exchanges between the different samples. For example, in some hexagons mainly composed of hatchery samples, some river individuals can be found which could result from stocking. These could either be recently released fish, or the result of introgression of domestic alleles into river populations (which could be detected several years after the last stocks were added; see for example Poteaux et al. (1998)). On the other hand, some hatchery individuals can be found among river ones. This can be the consequence of the introduction of wild fish into hatcheries: this is a new kind of hatchery management using generally male fish from the wild to renew the genetic composition of the hatchery.

But, in order to better analyse the distribution of the different individuals on the map, a clustering procedure can be useful: it could give a simpler and clearer image.

13.3.2
The Fuzzy Clustering

When the FCM algorithm is applied to Kohonen's map, 6 clusters can be defined. They are presented in Fig. 13.5 where each cluster shows a different surface (from left to right, up and down): vertical lines, horizontal lines, light grey, hachured, white, dark grey. Hexagons with several surfaces correspond to "mixed classifications," with some membership functions less than 0.8. These "mixed" hexagons are not numerous, and, for example, there is no mixing between the white and dark grey clusters. Mixing is only encountered between vertical line and horizontal line clusters, light grey and horizontal line, light grey and hachured. One hexagon is the result of mixing between light grey, vertical line and horizontal line clusters.

For each of these clusters, we can relate the composition to the origin of the fish. The distribution of the samples among the clusters is given in Table 13.3. It should be

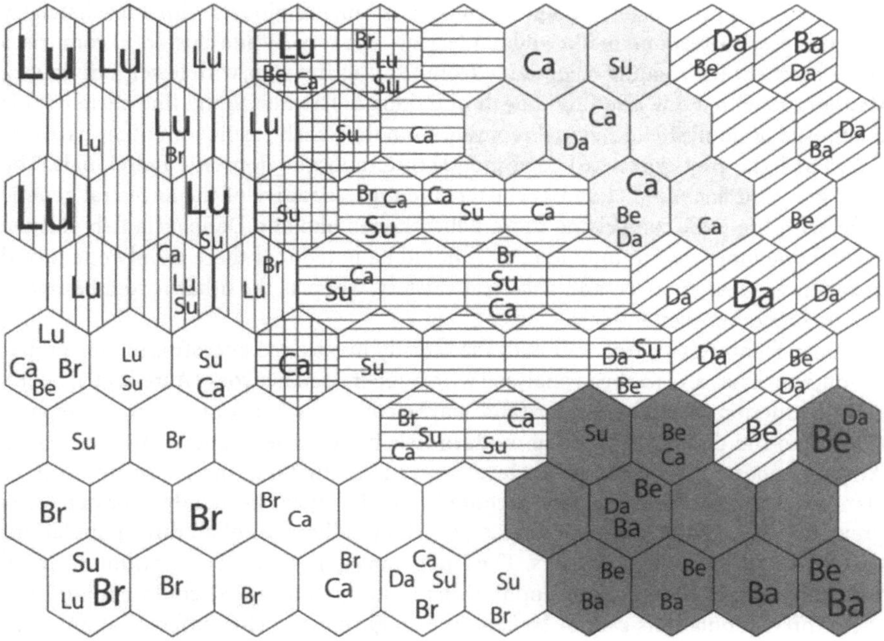

Fig. 13.5. Clustering of trout deduced from Fuzzy Clustering Mean. Each of the 6 clusters has a different surface. The mixed-surface hexagons correspond to values of membership functions of less than 0.8

Table 13.3. Proportions of individuals of the different samples found among the 6 clusters created by the FCM algorithm

Samples	**Cluster 1** (*dark grey*)	**Cluster 2** (*vertical lines*)	**Cluster 3** (*horizontal lines*)	**Cluster 4** (*white*)	**Cluster 5** (*hachured*)	**Cluster 6** (*light grey*)
Canourgue	0	0.17	0.32	0.1	0.05	0.36
Brassac	0	0.14	0.14	0.72	0	0
Suech	0	0.22	0.5	0.19	0.06	0.03
Behereko.	0.29	0.08	0	0	0.59	0.04
Dancharia	0.04	0	0	0.04	0.85	0.07
Bastan	0.74	0	0	0	0.26	0
Luz	0	0.98	0.01	0.01	0	0

noted that in order to calculate the percentages of individuals of a given population found in one cluster, we assigned each individual to a cluster, whereas in a fuzzy logic analysis, there is no such hard classification. Nevertheless, these percentages can give an estimation of the cluster composition.

The vertical line cluster (2) mainly comprises Luz individuals with a few hatchery fish; this may correspond to the wild modern genetic type. White cluster (4) comprises some hatchery fish, mainly originating from Brassac and also, with fewer individuals, Canourgue with a few river and Suech individuals. The horizontal line cluster (3) is composed of domestic individuals coming from the Suech strain with other hatchery fish. The light grey cluster (6) corresponds to Canourgue domestic individuals. The dark grey and hachured (1 and 5) clusters are representative of the ancestral genetic type, with the pink comprising more Béhérékobentako and Dancharia fish and the second one more Bastan individuals. It should be noted that there is only one mixed hexagon between modern and ancestral clusters, whereas the other mixed hexagons can be found between modern clusters.

These separations agree well with the already mentioned ancestral/modern separation, which has already been analysed with a supervised network (Aurelle et al. 1998). This distinction would be clearer (and maybe correspond to a more ancient separation) than the distinction among modern fish, between wild and domestic strains. Among ancestral fish, the procedure found two clusters, one corresponding to Dancharia and Béhérékobentako (*hachured*), and the other one mainly to Bastan (*dark grey*). But this separation is not clear, and certain individuals of all three populations can be found in the two clusters. The important genetic differentiation observed between natural brown trout populations (Estoup et al. 1998) could explain why these three populations cannot be kept in a single cluster. This differentiation probably indicates quite a long divergence time among these populations and/or strong genetic modifications because of recurrent bottlenecks or colonization events for example.

For the three hatchery strains three clusters have been defined, and they can be attributed to the three strains. Brassac is the only hatchery mainly belonging to one cluster (72% in Cluster 4, whereas Suech and Canourgue are more equally distributed). This can be related to the characteristics of this hatchery which comprises only one strain and is small compared to the others, and would then look genetically more "particular." But in each of the three clusters some individuals can be found originating from the three different domestic samples, and this "overlap" is greater than between the two ancestral clusters. This can be related to the management practices of domestic strains, with a lot of exchanges between hatcheries which reduces the genetic diversity between them, but increases the diversity inside each strain (Guyomard 1989). But despite this, the FCM algorithm found three clusters among the strains: the genetic diversity within these samples is probably high enough to identify several tendencies in each one; and because of the low inter-sample differences, these tendencies could also be related in other strains. Nevertheless, Clusters 3, 4 and 6 probably correspond to domestic fish different from the river ones.

The last cluster, corresponding to Luz (*vertical line*) is easy to interpret: it corresponds to an apparently wild modern sample well separated from hatchery clusters, with little overlapping (the meaning of these overlaps has already been discussed for the SOM interpretation). It should be noted that the overlap between the river populations (modern or ancestral) and the domestic ones is low, which confirms the clear genetic separation of these forms and also indicates low impact of stocking for these populations (as already noted for Mediterranean rivers: Poteaux and Berrebi 1997). Further analyses with several different modern samples will be interesting in order to analyze the natural genetic differentiation within this group and compare it to ancestral populations.

13.4 Conclusion

These results show that unsupervised network and fuzzy clustering algorithms can be successfully applied to complex genetic data, such as microsatellites which have been used for the description of at least 3 groups and a lot of subgroups of trout. These techniques can give an image of the genetic structure of populations without using a priori knowledge about their composition. In our case, this gave us some useful information in different ways. From a theoretical point of view, it confirmed the existence of several wild forms in southwest France. From a more practical point of view, it allowed us to evaluate the genetic impact of stocking on several river populations, which is useful for the management of genetic diversity of this species. This new approach can then be considered as complementary to more classical techniques.

Acknowledgements

This research was supported by the Bureau des Ressources Génétiques (grant no. 95011), the Conseil Supérieur de la Pêche (grant no. 96027) and the Club Halieutique Interdépartemental. The field captures were made by the local Fédérations de Pêche kindly assisted by students from ENSAT (Toulouse) and volunteers from Montpellier II University.

References

Aurelle D, Berrebi P (1998) Microsatellite markers and management of brown trout *Salmo trutta fario* populations in south-western France. Génétique, sélection. Evolution 30:75–90
Aurelle D, Giraudel JL, Lek S, Berrebi P (1998) Utilisation des réseaux de neurones multicouches pour classifier des populations de truites à partir des données génétiques. In: E.N.S.A. (ed) 6ièmes rencontres de la Société Francophone de Classification. Montpellier, pp 11–14
Bezdek JC, Ehrlich R, Full W (1984) The Fuzzy *c*-Means clustering algorithm. Computers and Geosciences 10:191–203
Blayo F, Demartines P (1991) Data analysis: How to compare Kohonen neuronal networks to other techniques ? In: Prieto A (ed) Artificial neuronal networks. Springer-Verlag, Berlin, pp 469–475
Chon T-S, Park YS, Moon KH, Cha E, Pa Y (1996) Patternizing communities by using an artificial neuronal network. Ecol Model 90:69–78
Cornuet JM, Aulagnier S, Lek S, Franck P, Solignac M (1996) Classifying individuals among infra-specific taxa using microsatellites data and neuronal networks. C R Acad Sci Paris, Life sciences 319:1167–1177
Der R, Villmann Th, Martinetz Th (1994) New quatitative measure of topology preservation in Kohonen's feature maps. In: Proc. ICNN'94, IEEE Service Center, Piscataway, pp 645–648
Estoup A, Presa P, Krieg F, Vaiman D, Guyomard R (1993) (CT)n and (GT)n microsatellites: A new class of genetic markers for *Salmo trutta* L. (brown trout). Journal of the Genetical Society of Great Britain 71:488–496
Estoup A, Rousset F, Michalakis Y, Cornuet JM, Adriamanga M (1998) Comparative analysis of microsatellite and allozyme markers: A case study investigating microgeographic differentiation in brown trout (*Salmo trutta*). Molecular Ecology 7:339–353
Foody GM (1996) Fuzzy modeling of vegetation from remotely sensed imagery. Ecol Model 85:3–12
Foody GM (1997) Fully fuzzy supervised classification of land cover from remotely sensed imagery with an artificial neuronal network. Neuronal Computing & Applications 5(4):238–247
Guégan JF, Lek S, Oberdorff T (1998) Energy availability and habitat heterogeneity predict global riverine fish diversity. Nature 391:382–384

Guyomard R (1989) Diversité génétique de la truite commune. Bull Fr Pêche Piscic 314:118–135
Hämäläinen A (1994) A measure of disorder for the self-organizing map. In: Proc. ICNN'94, IEEE Service Center, Piscataway, pp 659–664
Hamilton KE, Ferguson A, Taggart JB, Tomasson T, Walker A (1989) Post-glacial colonisation of brown trout, *Salmo trutta* L.: Ldh-5 as a phylogeographic marker locus. J Fish Biol 35:651–664
Jarne P, Lagoda PJL (1996) Microsatellites, from molecules to populations and back. Tree 11:424–428
Kiviluoto K (1996) Topology preservation in self-organizing maps. The 1996 IEEE International Conference on Neuronal Networks (Cat. No. 96CH35907) 1:294–299, New York
Kohonen T (1995) Self-organizing maps. Springer-Verlag, Heidelberg (Series in Information Sciences, 30)
Kraaijveld MA, Mao J, Jain AK (1995) A non-linear projection method based on Kohonen's topology preserving map. IEEE Transactions on Neuronal Networks 6:548–559
Mastrorillo S, Lek S, Dauba F, Belaud A (1997) The use of artificial neuronal networks to predict the presence of small-bodied fish in river. Freshwat Biol 38:237–246
O'Reilly PT, Hamilton LC, McConnell SK, Wright JM (1996) Rapid analysis of genetic variation in Atlantic salmon (*Salmo salar*) by PCR multiplexing of dinucleotide and tetranucleotide microsatelitte. Can J Fish Aquat Sci 53:2292–2298
Poteaux C (1995) Interactions génétiques entre formes sauvages et formes domestiques chez la truite commune (*Salmo trutta fario* L.). Thesis, University Montpellier II, Montpellier
Poteaux C, Berrebi P (1997) Intégrité génomique et repeuplements chez la truite commune du versant méditerranéen. Bull Fr Pêche Piscic 344/345:309–322
Poteaux C, Bonhomme F, Berrebi P (1998) Differences between nuclear and mitochondrial introgressions of brown trout populations from a restocked main river and its unrestocked tributary. Biological Journal of the Linnean Society 63:379–392
Takezaki N, Nei M (1996) Genetic distances and reconstruction of phylogenetic trees from microsatellite DNA. Genetics 144:389–399
Tsao C-K, Bezdek E, Pal NR JC (1994) Fuzzy Kohonen clustering networks. Pattern Recognition 27(5):757–764
Zadeh LA (1965) Fuzzy sets. Information and Control 8:338–353
Zupan J, Li X, Gasteiger J (1993) On the topology distortion in self-organizing maps. Biological Cybernetics 70:189–198

Chapter 14

The Macroepidemiology of Parasitic and Infectious Diseases: A Comparative Study Using Artificial Neuronal Nets and Logistic Regressions

J.F. Guégan · F. Thomas · T. de Meeüs · S. Lek · F. Renaud

14.1 Introduction

Of the about 270 species of helminths, protozoa and arthropods which may permanently or occasionally infect human populations, less than 45 species, or about 16 percent, are strictly dependent on humans for their survival (Ashford 1991; Petney and Andrews 1998). For the few West European countries providing reasonably reliable demographic data before the nineteenth century, epidemics, famines and wars are favoured as the three critical controlling mechanisms in human demographic crisis (Jones 1990). For instance, bubonic plague, one of the most dreadful epidemic killers, dominated the pattern of mortality variation from 1340 to its disappearance after 1670 in Europe, when smallpox epidemics may well have assumed a similar determining effect. Undoubtedly, mankind has experienced such disease effects along its evolution, leaving each time relatively resistant populations (Anderson and May 1991; Ewald 1994). Adopting a wider perspective, the "health" of man is determined essentially by his behaviour, his food and the nature of the world around him, and as such he is directly or indirectly influenced by different forms of parasitic and infectious diseases (Combes 1995). It may appear obvious that human conditions represent foci for a wide range of diseases (Anderson and May 1991). Unfortunately, the intimate interactions between different forms of diseases and human life-history traits have been virtually neglected (Immerman 1986). Life-history theory predicts that faced with virulent parasites, hosts should adjust their reproductive biology by increasing reproductive output and/or reducing age at maturity (Stearns 1992; Michalakis and Hochberg 1994; McNamara and Houston 1996; Sorci et al. 1996; Reeson et al. 1998; Kris and Lively 1998; Brooke et al. 1998). Intuitively, variation in parasite species composition across countries might be sensitive to human life-history traits, and vice versa as predicted by theory. In such a perspective, the determination of the exact relationships between abiotic and biotic characteristics and both presence/absence and spatial occupancy of diseases may appear crucial in that they might probably help to improve our understanding of the underlying processes that generate them. However, the nature of factors affecting the presence/absence of a given disease and its spatial distribution has been derived from a combination of expert opinion, limited data and the use of geographical and climate descriptors. This is largely due to a traditional individual-centred medical preoccupation in understanding these diseases (Jones 1990), and the disciplinary gap that exists between biomedical scientists, ecologists and evolutionary biologists (Petney and Andrews 1998). As recently pointed out by Craig et al. (1999), none has a clear and reproducible numerical definition of Malaria distribution in Africa, for instance; consequently, its comparative value is rather limited. Interestingly, large global data sets including environmental and human population data are now available which make them suitable for comparative studies.

Given these above remarks, our primary focus in this contribution was to model, and then to predict the spatial representation of different human infectious and parasitic diseases, some of which with very deleterious effects on populations across countries. The traditional method for depicting distribution and temporal patterns of major diseases used by epidemiologists and geographers has been to map rates of change in spatial distribution, or to provide a series of static "snapshot" maps (Pedersen 1995). In practice, this facilitates a form of prediction in which we can calculate declines in site occupancy or colonizations of new sites by diseases. Even though these models are entirely relevant since they permit one to present the results to a wide audience (Martin 1996), they do not authorise crude predictions for risk assessment of disease re-emergence or colonization of new sites. Here, we conduct a comparative analysis on a global scale using two multivariate methods, i.e. logistic regressions and artificial neuronal networks, to precisely predict the spatial distribution of diseases. We then compare the performance of both logistic and artificial neuronal network models in predicting the actual spatial distribution of the infectious and parasitic diseases under study. Finally, we explore the utility of such predictive models in epidemiology, and notably how variations in time, say climate change, may affect the actual disease distribution.

14.2 Materials and Methods

14.2.1 Materials

We compiled data for a total set of 168 different countries located all over the world and for which all population, geographical and epidemiological information was available. Large global data sets are now available which make the modelling of disease spatial distribution entirely relevant. In doing so, one should make the maximum use of the available data, as accurate as possible, and be able to appreciate the potential inaccuracy in the results. Epidemiological data were obtained from two main sources, the World Health Organization (W.H.O.) and the Center for Disease Control and Prevention (C.D.C.), a quick and convenient method entirely reliable for such a purpose. Our models assume that all variables, e.g. presence of a given disease, have a homogeneous distribution across each source country. From the total data set of 168 countries, we considered a subset of 153 countries for a phylogeny-based comparison analysis (see below).

14.2.1.1 *Spatial Patterns*

Since geographical and ecological factors might strongly influence the variation of parasite species distribution across countries, we considered five ecogeographical variables for each country. These spatial descriptors are those which are probably the most usually invoked for explaining free-living species occupancy and distribution on largest scale (Brown 1995; Rosenzweig 1995; Whittaker 1975). They are: *(1)* total surface area of a given country (in km^2), since larger land masses may harbour higher species diversity than smaller masses do, and thus the likelihood it incorporates a given parasite species is higher; *(2)* mean latitude (in degree and minutes which refers to the value taken at the geographical centre of each country), since higher species diversity is generally found under tropical areas, and many human infectious diseases are primarily concentrated in those regions, when compared to more septentrional provinces; *(3)* mean longitude (in degree and minutes, measured as previously) which takes into account the fact that parasite species might have dispersed along an east-west gradient from their centre of dispersion. These three environmental parameters were log-transformed in logistic regression models in order to minimize effects of nonnormality on statistics (Zar 1996). They have been kept unchanged in artificial neuronal net procedures. Furthermore, we considered whether or not a country was located *(4)* on the northern or southern hemisphere, since countries are more numerous in the northern part of the world which represents a statistical artefact, and *(5)* on a land mass or an island since island populations may present frequent fade-outs of infection (Rhodes and Anderson 1996) or a given disease may be extinguished there (Esch et al. 1990). These two variables were coded as categorical variables (0/1). Initially, the three continuous variables were incorporated into principal component analysis to reduce dimensionality and eliminate collinearity between these source variables (Sheldon and Meffe 1995; Oberdorff et al. 1998). Only one synthetic output principal component (PC*GEO*1, eigenvalue = 2 107.91) explained 99.58% of the total inertia, 89.98% of which depending on the latitudinal effect. Because of no effect of multicollinearity on final models, we decided to use raw variables. All this spatial information can be accepted as synthetic as possible reflecting other potential physical variables influencing parasite species distribution, e.g. temperature or ecosystem productivity as well. All this available data may contribute to the actual distribution of environmentally determined diseases. Spatial data were compiled from World Atlas v. 2.1.0 ©, on a Macintosh personal computer.

14.2.1.2
Economic, Social and Demographic Patterns

Because the human disease characteristics might differ so much across countries having more or fewer inhabitants, more or less urbanization, or more or fewer financial supports for health care campaigns, we also compiled demographic and economic data for the 153 countries. Data for population geography were essentially obtained from the 1992 world population data sheet (Jones 1990). Five demographic or economic parameters were retained for each country: *(1)* total population (in number of people per country), which represents the potential colonizing pool for any disease; *(2)* total population growth (per 1 000 people), which gives an estimate of the reproductive ability in growing populations; *(3)* population density (number of people per km^2), which permits one to separate countries on a continuum of populations with high aggregate behaviour (as for high urbanization areas) to lower level (in rural areas), which can strongly influence the likelihood of disease successful transmission; *(4)* death rate (per 1 000 people), which gives an estimate of differential mortality in the area, and which can be attributed in part to the deleterious effect some diseases may actually have on human health; *(5)* per capita gross national product (GNP in US$ a year) to evaluate the resource, or income effect, on disease spatial distribution through financial supports granted by local politicians and governments in health care campaigns. The three variables, i.e. total population, population density and GNP, were log-transformed and the two parameters, i.e. total population growth and death rate, were arcsine-transformed to deal with nonlinearity before introduction into logistic regression models (Zar 1996). These variables were kept unchanged into artificial neuronal net methods. To avoid multicollinearity between all these variables, we proceeded as previously in using principal components analysis. Since final models did not change between using raw variables or principal components, we kept predictive variables unchanged. Conceptually, all these factors may reflect the probability of transmission occurring or not occurring from high (e.g. countries with high population density and low incomes) to low transmission intensity (e.g. countries with low population density and large incomes).

14.2.1.3
Historical Patterns

Generally, closely related taxa are more likely to exhibit similar traits than distant taxa since they have been subject to similar evolutionary constraints inherited from a common ancestor (Harvey and Pagel 1991). Intuitively, two groups of relative human ethnic groups might share similar traits. For instance, they might harbour the same infectious disease, or group of co-occurring diseases, which both represent a result of phylogenetic history. Immunodeficiency and reduced antibody levels to both related tribes may be invoked to explain this historical component, but such assumptions have seldom been elucidated completely. Although such aspects remain to be investigated, there is evidence indicating that two populations cannot be treated as statistically independent points (Martins 1996; Martins and Hansen 1997). To deal with the confounding effect of common history on parasitism, we used the human group phylogeny based on the findings of Cavalli-Sforza (1997). Unfortunately, it was impossible to directly use the entire phylogeny, mainly due to the difficulty of crossing this phylogenetic information based on the existence of more or less well-recognized ethnological groups with our data concerning political nations. From the entire database of 168 countries, we retained a subset of 153 countries for performing a phylogeny-based correction analysis. We decided to consider only the eight large divisions of ethnological groups as defined hereafter. Then, we used only countries for which at least 50% of inhabitants belong to one majority ethnological group. We omitted some countries, e.g. Brazil, *f*SU, South Africa, Chile, with a high human polymorphism. To control for the confundant effect exerted by common history on calculation, different comparative methods have been developed to take this nonindependence into account (see Martins 1996 for review). For the purpose of this work, we used a General Least-Square Model, which permits one to remove the variance due to common history using categorical codes (Grafen 1992). The choice of this phylogenetic method was based on its better robustness toward misspecification of our models (Martins and Hansen 1997). Phylogenetic coding variables, coded 0 or 1, were introduced into predictive models as dummy categorical variables. The eight main divisions of human groups considered in this work are:

I. Africans and Nilotics (except native people from the Maghreb);
II. Europeans (including people from the Middle-East);
III. Indians;
IV. Mongoloids, Japanese and Koreans;
V. Amerindians;
VI. New Guineans-Papous;
VII. Melanesians;
VIII. Mhongs, Khmers, Thais, Filipinos, Indonesians and related tribes.

The first coding variable separates the tribe division I from all other tribes, and it removes the differences between the means of these two hierarchical groups. Next, and nested within it, we separate the group formed by tribes II to V from the group including tribes VI to VIII, and so on down to the last bifurcating branch of the phylogeny.

14.2.1.4
Human Life-History Trait Patterns

Life-history theory assumes that reproduction is costly, and that strategic decisions should be selected by organisms over their lifetime in contrasting ecological and social environments (Alexander 1987; Shykoff et al. 1996; Lively 1987; Hochberg et al. 1992; Forbes 1993; Lafferty 1993). Parasitism may be such an underlying environmental condition, which could interfere with survival and reproduction capabilities in humans across distinct regions of the world. Since parasites use resources from their human hosts for their maintenance and own reproduction, costs of reproduction in humans can be predicted to increase in the presence of parasitism (see Møller 1997; Teriokhin 1998 for a theoretical viewpoint). In addition, differences in male and female life histories such as life span variability across sexes, i.e. constant shorter life span for man than for his congener world-wide (Teriokhin and Budilova 2000), may possibly interfere with varying levels of parasitism. In this study, three different human traits, susceptible to interfere with infection, were available for predicting disease spatial distribution: *(1)* fertility rate, which indicates the number of children that would be born to 1 000 women during their lifetime passing through the child-bearing ages; *(2)* female life expectancy at birth (in years); and *(3)* male life expectancy at birth (in years). Life expectancies at birth are substantially higher for females than males, and thus sex ratios are directly affected, and in turn both sexes may represent differential hosts for parasites. This refers to the three most current parameters used in evolutionary ecology to estimate the degree of organism fitness (see Teriokhin and Budilova 2000).

14.2.1.5
Parasite Patterns

Disease occurrences in the 153 different countries were compiled from information available on mainly two different web sites, the World Health Organization (Geneva, Switzerland at http://www.WHO.int/) and the Center for Disease Control and Prevention (Atlanta, USA at http://www.CDC.gov/) sites. Many of the investigations of human parasites have been based on the microscopic examination of the patients' stool

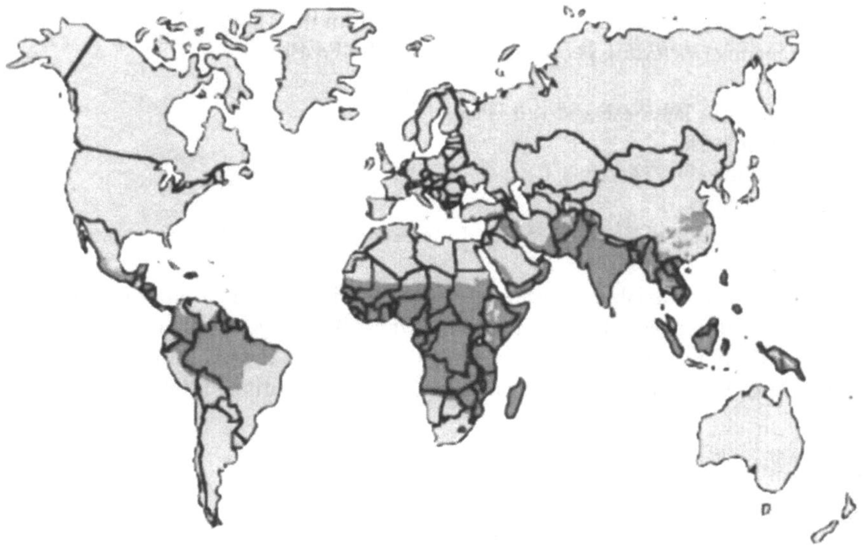

Fig. 14.1. Actual distribution of malaria in the world. A total of 111 different countries (66.1%) on 168 present in our total database are affected by paludism (after W.H.O./C.T.D. 1997)

for helminth eggs. Thus, the estimates of the overall human parasitofauna are certainly underevaluated. Therefore, we were able to collect data on presence/absence for a set of 15 categories of human diseases known to have a more or less large impact on human health. When information at the species level was not available, we decided to pool these data by category of diseases. Disease categories are as follows: Hepatitis A, Hepatitis B, Malaria (see Fig. 14.1 for illustration), Schistosomiasis, Filariosis, Meningococcosis, Yellow fever, Dengue fever, Cholera, Trypanosomiasis, Dracunculosis, Chagas, Lyme, cutaneous Leishmaniosis and visceral Leishmaniosis. Other diseases were available, e.g. Typhoid fever, but they were widespread species with a range size of 153 countries. We are absolutely conscious that these values are subject to some sources of error, e.g. some parasite species may have been recently introduced into countries and thus they do not yet appear into the available check-lists, or they have been annihilated for a couple of decades, but their potential effect on human populations is still effective, or the presence/absence of diseases have not been declared to the two international organizations, or this refers to only a sub-sample of what really exists, but these data are what is really available to us today! In many ways, many of these infectious and parasitic diseases represent the actual most dreadful killers occurring on earth.

14.2.2 Methods

From the variety of multivariate statistics that can be used to predict a presence/absence event, we opted for two distinct techniques for estimating the probability that an event of presence (or absence) of a parasitic or infectious disease occurs across countries. There are logistic regressions (Jongman et al. 1995; Norusis 1997) and artificial neuronal networks (Rumelhart et al. 1986; Edwards and Morse 1995). A third potential method, i.e. multiple discriminant analysis, was not relevant here since the data incorporate categorical variables as independent parameters (Jongman et al. 1995).

14.2.2.1
Logistic Regression Models

Presence, or absence, of a given infectious disease across the 153 countries was fitted to the 20 independent variables listed above (see Section 14.2.1) using logistic regression procedure. The general linear model can be written as follows:

$$prob(1) = 1 / (1 + e^{-z})$$

$$prob(0) = 1 - prob(1)$$

with e is the base of the natural logarithms, $prob(1)$ the associated probability that a given disease occurs in a country, $prob(0)$ the associated probability that a given disease does not occur, and z the linear combination of the independent variables of the form

$$z = b_0 + b_1(\text{surface area}) + b_2(\text{mean latitude}) + \ldots + b_{20}(\text{ethnyVIII})$$

The logistic model in terms of the log of odds, or log*it*, can be written as follows:

$$\log it\,[prob(1) / prob(0)] = z \text{ or } prob(1) / prob(0) = e^{z}$$

The parameters of the logistic regression model were estimated using the maximum likelihood method. The nine categorical variables, i.e. hemisphere, landmass/island, and the seven ethnic groups, were entered into regressions as indicator-variable coding. The other variables were considered as continuous variables. The Wald's statistic and its associative significance level were used to detect significant independent variables within the logistic model, using a significance level of 0.10. R statistics were used for determination of partial correlation between the disease occurrence dependent variable and each of the independent parameters. Accuracy of fit of the logistic models was tested using −2 times of the likelihood ($-2LL$) with a model perfectly fitting data having a score of 0. The proportion of total explained variation in logistic regression was given by the Nagelkerke R-square statistics.

We then compare predictions obtained from logistic models to the observed outcomes using contingency tables with cut values of 0.50. In the case of a cut value of 0.50, this indicates whether the estimated probability is greater or less than one-half.

All the independent variables were first entered into logistic regression models which permits one to control for the effect of other independent variables on a given descriptor variable (general model). Second, we proceeded to backward elimination procedure in order to identify minimal models with a subset of independent variables as good predictors as the total set of independent variables entered into general models.

14.2.2.2
Artificial Neuronal Networks

Artificial neuronal nets are known for their capacity to process nonlinear relationships (Rumelhart et al. 1986; Freeman and Skapura 1992). In the present work, we used one of the general principles of artificial nets, i.e. the back propagation algorithm (Gallant 1993) for training the database with a typical three-layer feed-forward 20-3-2 network (Fig. 14.2), that is, 20 input neurons corresponding to the 20 independent parameters introduced into the model, 3 hidden neurons determined as the optimal configuration to obtain a best compromise between bias and variance and 2 output neurons for disease occurrence, i.e. one for presence of the disease and one for its absence. The performance of the artificial neuronal nets was analysed using two different techniques of partitioning the total data set into a first subset for training the neuronal model and a second set for testing its real predictive power. Since there is

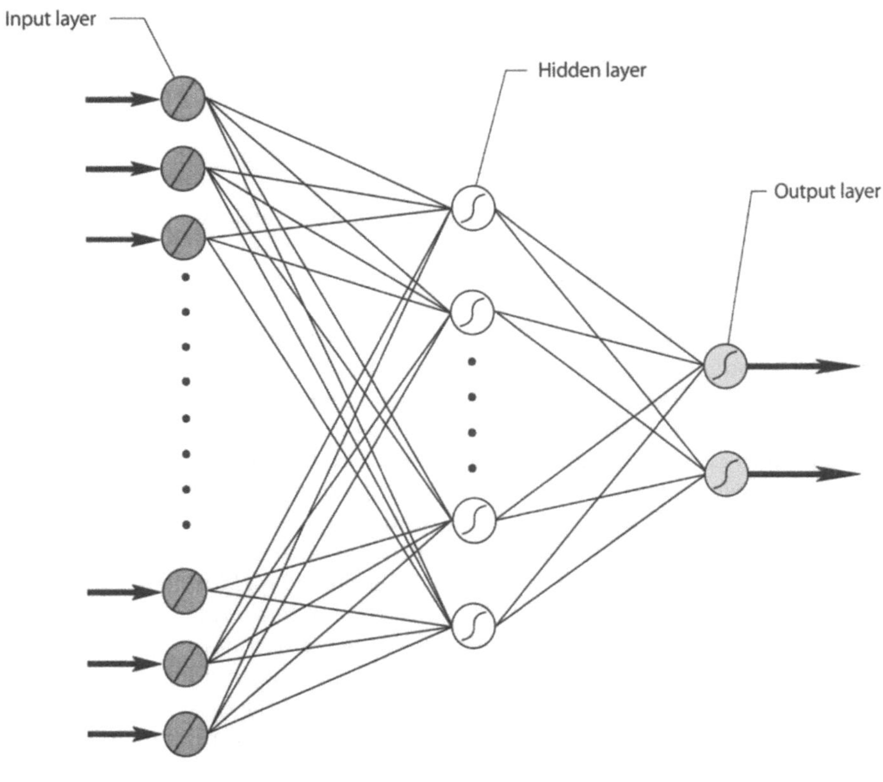

Fig. 14.2. Schematic representation of a three-layered feed-forward neuronal network, with one input layer, one hidden layer and one output layer as used in the present work. The left side shows the input parameters used in back propagation network models, and the right side illustrates the output parameters, i.e. presence or absence of a given disease

actually a large debate on the adequacy of partitioning methods and its effect on model error rates (Fielding and Bell 1997) without any special consensus, we opted to develop both "leave-one-out" (Efron 1983; Kohavi 1995) and "hold-out" (Efron 1983; Kohavi 1995; Friedman 1997) cross-validations for our data-set.

The "leave-one-out", or jack-knife procedure, leaves out a test set (one country × 20 inputs) from the training set (152 countries × 20 inputs), and this is repeated for each country. Then, the model run with the training set may be used to predict the presence/absence of a given disease in the test set. This was repeated with a maximum of 1 000 iterations for each country.

In the "hold-out" procedure also called *k*-fold cross-validation, two random sets are extracted from the total data set: a trained set (¾, i.e. 115 countries) and a test set (¼, i.e. 38 countries). Similar to the jack-knife procedure, the model is first adjusted with the training set, and then it is used for prediction of presence/absence in the remaining test set (Kohavi 1995; Friedman 1997). This procedure was repeated 10 times to provide a better compromise prediction on the random test sets.

As for logistic regressions, we compared predictions obtained from neuronal net models to the observed outcomes using classification tables with cut values of 0.50.

Readers will be able to find further details on the different neuronal network procedures in the first chapter of this book.

All statistical analyses were performed with SPSS 7.5 and MatLab 5.0 for a personal computer

14.2.2.3
Comparative Analysis

We evaluated the efficacy of both multivariate models for classifying the presence, or absence, of the 15 infectious and parasitic diseases by plotting the ratios of sensibility values (i.e. the true positive fraction) versus 1-specificity values (i.e. false positive fraction) against the different levels of occurrence frequency, i.e. prevalence observed across the total set of diseases. In fact, as pointed out by Fielding and Bell (1997) and more recently by Manel et al. (1999), a decreasing frequency of occurrence may be responsible for an exaggerated inflation of positive prediction errors. It is well accepted that logistic regressions are sensitive to such biases (Norusis 1997), but unfortunately we do not have any specific ideas how neuronal networks may be affected by the frequency of events occurring.

Additionally, the use of an arbitrary threshold probability, or cut-off value, which discriminates between predicted probabilities, of saying 0.50 and greater to be classified as having positive nodes, and of 0.50 and smaller as having negative ones, may be strongly influenced by prevalence values, i.e. the number of positive occurrences in the total data-set. To deal with the effect of selecting a specific cut-off value on prediction error, we compared the predictive performance of both logistic regression and artificial neuronal net models using ROC (Received Operating Characteristics) curves across different levels of threshold probabilities as recognized by Zweig and Campbell (1993) and Manel et al. (1999).

14.3 Results

14.3.1 Logistic Regressions and Artificial Neuronal Networks Face to Face

14.3.1.1 *Logistic Regressions*

Table 14.1 illustrates the classification results obtained for the 15 infectious diseases using logistic regression methods. The percentages of good classification for the 15 diseases were strongly high, varying from 90.2 to 100% (mean 97.3%, *SD* = ±3.5) of countries well classified when all independent variables were kept in regressions. True absence scores, which determine the negative predictive power of data, varied from 90.6 to 100% (mean 91.9%, *SD* = ±2.9), and true presence scores, which represent the positive predictive power, ranged from 85.2 to 100% (mean 95.8%, *SD* = ±5.2). Interestingly, these results show that there is no substantial difference between classification performance of positive and negative cases.

In a backward stepwise selection procedure generating a minimal logistic model for each disease, we obtained comparable scores of classification for countries (data not illustrated for clarity of the manuscript): overall good classification between 49.4 and 100% (mean 90.8%, *SD* = ±13.8); true absence scores between 46.9 and 100% (mean 86.7%, *SD* = ±27.3); true presence scores between 33.3 and 99% (mean 71.6%, *SD* = ±33.7). Two diseases, Chagas and Hepatitis A, strongly affected overall performance through lower scores obtained for these diseases, i.e. 49.4 and 65.5% respectively, which contributed to the relatively high standard deviations observed across all different minimal models. For illustration, Table 14.2 shows the results of both general and minimal logistic models for Schistosomiasis with the contribution of the different significant factors for explaining the presence or absence of this disease across countries.

14.3.1.2 *Artificial Neuronal Nets*

Table 14.1 shows results obtained from artificial neuronal networks with the total set of 20 inputs for the 15 diseases using jack-knife procedure. As previously observed for logistic regression (see above), the percentages of good classification scores were very high, ranging from 88.9 to 100% (mean 96.3%, *SD* = ±3.6) when all input variables were kept in models. True absence and true presence scores varied from 0 to 100% (mean 90.3%, *SD* = ±25.3) and from 0 to 100% (mean 80.6%, *SD* = ±29.7), respectively. Three diseases (Cutaneous and Visceral Leishmaniosis and Hepatitis A) strongly affected classification scores in that true positive performances for the two Leishmaniosis, i.e. 20% and 0% respectively, and true negative performances for Hepatitis A, i.e. 0%, were extremely low. Moreover, predictions obtained with minimal neuronal networks formed with only 3 input parameters, i.e. the 3 human life-history traits, performed nearly as well as global nets to model the disease occupancy per country: 79.2 to 98.2% (mean 92.7%, *SD* = ±5.5) of total good prediction scores, 0 to 100% (mean 82.8%,

Table 14.1. Results obtained from both logistic regressions and neuronal nets for predicting the presence/absence of 15 infectious and parasitic diseases across a set of 153 countries when all independent parameters are kept into models. True positive and true negative estimates with their respective percentage values and overall good classification percentage values were derived from leave-one-out procedure. Data show that both methods converge in efficiently predicting the occurrence of the different diseases across countries, with a slight higher performance of good achievement of prediction for logistic regression procedure (but see Tables 14.2 and 14.3). See text for further explanation on statistics. C. Leishmaniosis refers to Cutaneous Leishmaniosis and V. Leishmaniosis to Visceral Leishmaniosis.

Model	Logistic								Neuronal Net				
Disease	True Negatives	%	True Positives	%	Overall Performance	−2*LL*	*R*	*p*	True Negatives	%	True Positives	%	Overall Performance
Chagas	136	100	17	100	100	0.0	1.00	<0.0001	136	100	16	94.1	99.3
Cholera	87	90.6	51	89.5	90.2	77.6	0.762	<0.0001	87	90.6	49	85.9	88.9
C. Leishmaniosis	148	100	5	100	100	>100	1.00	<0.0001	148	100	1	20.0	97.4
Dengue fever	87	94.6	52	85.2	90.8	57.1	0.840	<0.0001	86	93.5	51	83.6	89.5
Dracunculosis	133	99.2	18	94.7	98.7	13.5	0.917	<0.0001	134	100	17	80.5	98.7
Filariosis	90	94.7	53	91.4	93.5	58.6	0.832	<0.0001	90	94.7	53	91.4	93.5
Hepatitis A	1	100	152	100	100	0.0	1.00	ns	0	0.0	152	100	99.3
Hepatitis B	3	100	150	100	100	0.0	1.00	ns	3	100	150	100	100
Lyme	142	100	11	100	100	0.0	1.00	<0.0001	142	100	9	81.8	98.7
Malaria	49	96.1	101	99.0	98.0	26.4	0.927	<0.0001	49	96.1	99	97.1	96.7
Meningococcosis	107	98.2	39	88.6	95.4	31.3	0.902	<0.0001	106	97.2	37	84.1	93.5
Schistosomiasis	91	95.8	53	91.4	94.1	43.2	0.882	<0.0001	92	96.8	54	93.0	94.1
Trypanosomiasis	81	100	72	100	100	0.0	1.00	<0.0001	78	96.3	72	100	98.0
V. Leishmaniosis	150	100	3	100	100	0.0	1.00	ns	150	100	0	0.0	98.1
Yellow fever	114	99.1	37	97.4	98.7	20.5	0.931	<0.0001	114	99.1	37	97.4	98.7
Mean		91.9		95.8	97.3					90.3		80.6	96.3
SD		2.9		5.2	3.5					25.3		29.7	3.6

Table 14.2. Results of logistic regression procedure for modelling the effect of continuous and discrete independent variables on one discrete dependent variable such as the Schistosomiasis occurrence (presence or absence) across the 153 countries. Table illustrates the results obtained in a global linear model (**a**) and after a backward stepwise elimination procedure (**b**). Both models converge in that the same set of explanatory variables is retained. Estimation parameters for both models are highly significant (Model a: $-2LL = 43.169$, Goodness of fit = 49.917, Nagelkerke $R^2 = 0.882$, Chi-square model = 159.897, $d.f. = 20$, $p = 0.0000$, Overall classification = 94.12%, true negative score = 95.79%, true positive score = 91.38% with a cut-off value of 0.50; Model b: $-2LL = 52.007$, Goodness of fit = 70.815, Nagelkerke $R^2 = 0.854$, Chi-square model = 151.059, $d.f. = 7$, $p = 0.0000$, Overall classification = 93.46%, true negative score = 96.84%, true positive score = 87.83% with a cut-off value of 0.50). Details on statistics are given in the text

a **Independent factors**	***B***	**Wald's**	***d.f.***	***p***	***R***
Spatial characteristics					
Surface area	<0.0001	1.3049	1	ns	0.0000
Mean latitude	−0.1447	3.8996	1	0.0483	−0.0967
Mean longitude	−0.0590	2.8664	1	0.0904	−0.0653
Land mass vs. island	−0.2281	0.0162	1	ns	0.0000
North/South	1.2638	0.6156	1	ns	0.0000
Human traits					
Fertility rate	0.8045	3.0667	1	0.0799	0.0725
Female life span	0.6544	2.8330	1	0.0923	0.0640
Male life span	−0.8534	3.2046	1	0.0734	−0.0770
Demographic characteristics					
Population growth	0.3337	0.9190	1	ns	0.0000
Population density	−0.0005	0.0725	1	ns	0.0000
Death rate	−2.1056	0.7520	1	ns	0.0000
GNP	−0.0003	1.5130	1	ns	0.0000
Total population	<0.0001	0.0234	1	ns	0.0000
Historical patterns					
Dummy Group I	−0.8507	0.2568	1	ns	0.0000
Dummy Group II	0.0389	0.0002	1	ns	0.0000
Dummy Group III	8.0851	0.0009	1	ns	0.0000
Dummy Group IV	5.8435	0.0044	1	ns	0.0000
Dummy Group V	7.1893	0.0071	1	ns	0.0000
Dummy Group VI	13.8095	0.0141	1	ns	0.0000
Dummy Group VII	12.2174	0.0308	1	ns	0.0000
Constant	−34.3221	0.0110	1	ns	

$SD = \pm 30.5$) of true negative scores and 16.7 to 100% (mean 73.1%, $SD = \pm 28.8$) of true positive scores (Table 14.3). Except for the lowest negative scores obtained for the two Hepatitis A and B, and the lowest positive scores obtained for the Chagas disease and

Table 14.2. *Continued*

b Variables in the equation	*B*	Wald's	*d.f.*	*p*	*R*
Spatial characteristics					
Surface area	<0.0001	4.5876	1	0.0322	0.1129
Mean latitude	–0.1917	14.1435	1	0.0002	–0.2445
Mean longitude	–0.0547	11.4365	1	0.0007	–0.2156
Human traits					
Fertility rate	0.5692	5.2436	1	0.0220	0.1264
Demographic characteristics					
Population growth	0.6483	7.3891	1	0.0066	0.1629
Historical patterns					
Dummy Group VI	11.9461	0.0257	1	ns	0.0000
Dummy Group VII	11.3130	0.0756	1	ns	0.0000
Constant	–26.5672	0.0972	1	ns	

Model if constant removed	Log likelihood	2 log *LR*	*d.f.*	*p*
Spatial characteristics				
Surface area	28.316	4.625	1	0.0315
Mean latitude	41.055	30.103	1	0.0000
Mean longitude	38.099	24.191	1	0.0000
Human traits				
Fertility rate	–29.077	6.148	1	0.0132
Demographic characteristics				
Population growth	–31.112	10.216	1	0.0014
Historical patterns				
Dummy Group VI	30.977	9.947	1	0.0016
Dummy Group VII	34.949	17.890	1	0.0000

the two Cutaneous and Visceral Leishmaniosis, we obtained similar prediction scores compared to neuronal net models with the 20 input parameters (see Table 14.1). These findings particularly demonstrate that minimal neuronal nets perform as well as general models to predict and model the distribution of those infectious diseases. In addition, all these results corroborate those obtained using logistic regressions. Again, artificial neuronal nets, as well as logistic regressions, were not influenced, or slightly perturbed, by a presence/absence effect for prediction of true negative plots versus true positive plots.

Table 14.3. Results obtained with artificial neuronal network modelling for predicting the presence/absence of the 15 infectious and parasitic diseases across the total set of 168 countries with only the three human-life history traits, i.e. fertility and life spans for both sexes. Since historical information was excluded from this analysis, it was possible to use the total number of countries available in our database. True positive and negative scores with their respective percentage values and overall good classification percentage values were derived from leave-one-out procedure (see text). Data show that modelling efficiently predicts the occurrence of most of diseases solely using human characteristics as input parameters

Disease	True negatives	(%)	True positives	(%)	Prediction performance
Chagas	147	100	5	23.8	94.4
Cholera	97	91.5	48	77.4	86.3
Cut. Leishmaniosis	162	100	1	16.7	97.0
Dengue fever	90	90.9	60	87.0	89.3
Dracunculosis	149	100	16	84.2	98.2
Filariosis	90	91.5	46	74.2	79.2
Hepatitis A	0	0.0	165	100	98.2
Hepatitis B	1	20.0	163	100	97.6
Lyme	153	98.1	7	58.3	95.2
Malaria	43	75.4	107	96.4	89.3
Meningococcosis	120	97.6	35	77.8	92.3
Schistosomiasis	102	95.3	51	83.6	91.1
Trypanosomiasis	128	96.2	33	94.3	95.8
Visc. Leishmaniosis	164	100	1	25.0	98.2
Yellow fever	107	84.9	41	97.6	88.1
Mean		82.8		73.1	92.7
±*SD*		30.5		28.8	5.5

Hold-out modelling gave similar results of good recognition patterns. Table 14.4 shows results of the ten tests for two diseases, Schistosomiasis and Yellow fever. Scores of overall correctly classified countries were between 81.6 and 94.7% (mean 87.6%, $SD = \pm 3.73$) for Yellow fever disease, and 78.9 to 92.1% (mean 85.3%, $SD = \pm 4.86$) for Schistosomiasis.

14.3.2
Occurrence and Threshold Effects

14.3.2.1
Occurrence Effects

As illustrated in Fig. 14.3, we did not observe any occurrence effect (range values from 25 to 93% for the 15 parasite species across the 153 countries) on sensitivity/1–specificity ratios for both neuronal network models (Fig. 14.3a) and logistic regressions (Fig. 14.3b). This means that, at least in this study, infectious and parasitic diseases

Table 14.4. Results of artificial neuronal network modelling after the partitioning of the data set into a training set (¾ of countries) and a test set (¼ of countries), or hold-out procedure. Table shows results of the ten tests for two diseases, i.e. Schistosomiasis (**a**) and Yellow fever (**b**). Also given are mean values of the ten trials and their corresponding standard deviations

a	True negatives	(%)	True positives	(%)	Prediction performance
Trial 1	17	80.9	15	88.2	84.2
Trial 2	21	91.3	10	66.7	81.6
Trial 3	20	87.0	10	66.7	78.9
Trial 4	22	91.7	13	92.8	92.1
Trial 5	21	87.5	12	85.7	86.8
Trial 6	19	86.3	14	87.5	86.8
Trial 7	22	95.6	13	86.7	92.1
Trial 8	18	78.3	12	80.0	78.9
Trial 9	19	86.4	15	93.7	89.5
Trial 10	21	87.5	11	78.6	84.2
Mean	20	87.3	12	82.7	85.3
SD	1.70	5.03	1.84	9.65	4.86

b	True negatives	(%)	True positives	(%)	Prediction performance
Trial 1	21	87.5	11	78.6	84.2
Trial 2	19	79.2	12	85.7	81.6
Trial 3	21	87.5	12	85.7	86.8
Trial 4	23	92.0	11	84.6	89.5
Trial 5	20	83.3	13	92.9	86.8
Trial 6	22	91.7	12	85.7	89.5
Trial 7	23	95.8	13	92.9	94.7
Trial 8	19	82.6	13	86.7	84.2
Trial 9	20	90.9	14	87.5	89.5
Trial 10	22	88.0	12	92.3	89.5
Mean	21	87.9	12	87.3	87.6
SD	1.49	5.03	0.95	4.46	3.73

having a very low spatial occupancy on earth, i.e. endemic species with a very limited area (e.g. Chagas disease restricted to some countries of Southern America), may be as best predicted as widespread diseases since the occurrence effect plays a poor or a very slight role when considering all the predictive variables (inputs) into modelling. Nevertheless, minimal models built with the 3 human life-history traits only were more sensitive to variation in prevalence values (see Table 14.3). Considering minimal models, both logistic regressions and neuronal nets were affected by those variations in

Fig. 14.3. Sensitivity/1-specificity ratios against occurrence values for the 15 infectious and parasitic diseases: **a** scatter diagram obtained using artificial neuronal network procedure; **b** scatter diagram obtained after logistic regression method. Both statistical techniques converge in that sensitivity/1-specificity profiles are not or very slightly influenced by occurrence effect (see text for further details)

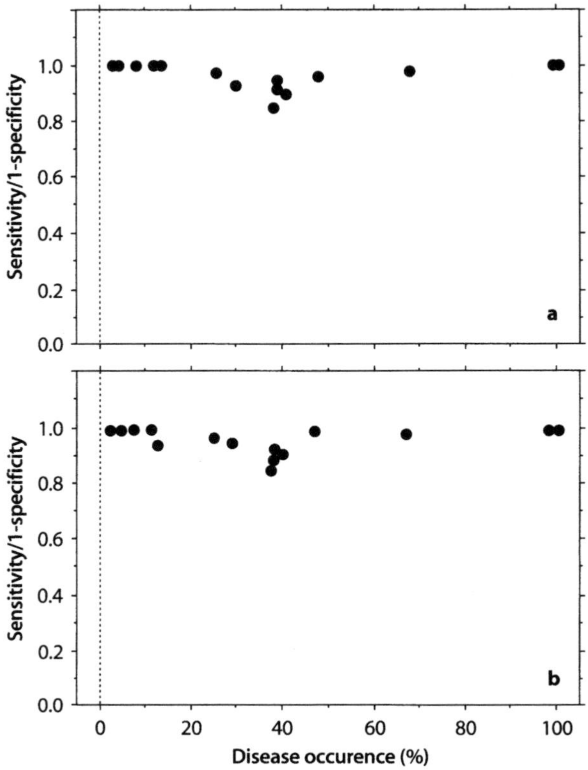

prevalence values, and we found both low prevalence and high prevalence effects on classification scores. This can be observed in Table 14.3 for the 5 diseases, i.e. the Hepatitis A and B (high prevalence effect observed on true negative scores), the Cutaneous and Visceral Leishmaniosis and the Lyme disease (low prevalence effect on true positive scores). However, the lowest sensitivity/1–specificity ratios were observed for the average values of prevalence, i.e. diseases having a prevalence value of around 35–40%, i.e. the Cholera and Dengue fevers and the two helminthic Schistosomiasis and Filariosis diseases. We did not find any special explanations to these findings. These results conflict in part with previous studies showing low prevalence values of an event which might be responsible for better prediction of true absences than true presences (see Manel et al. 1999).

14.3.2.2
Threshold Effects

Figure 14.4 plots the relationship between sensitivity values (i.e. the true positive fraction) against 1-specificity values (i.e. the false positive fraction) across different thresholds of cut-off-values obtained by jack-knife procedures for both classifying methods. The procedure was repeated for each of the 15 infectious diseases. These results show that a sufficiently high rate of true positives and a low rate of false positives may

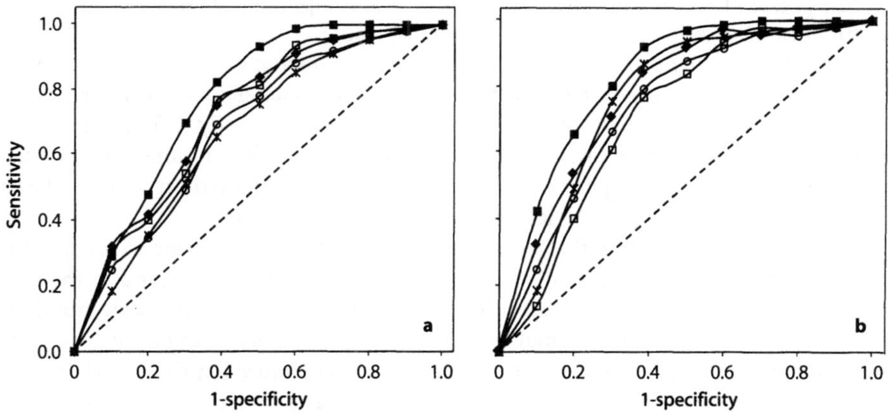

Fig. 14.4. Received operating characteristic curves; **a** artificial neuronal networks; **b** logistic regressions illustrating the relationship between sensitivity values (i.e. the true positive fraction) against 1-specificity values (i.e. the false positive fraction) across different thresholds of cutoff-values obtained by jack-knife procedures. These results show that a sufficiently high rate of true positives and a low rate of false positives may be achieved at a threshold probability of 0.5 for both logistic regression and neuronal net methods. Curve behaviours illustrate the general tendency observed across the 15 different diseases. The *dotted line* corresponds to *x* equals *y*

be achieved by both neuronal net (Fig. 14.4a) and logistic regression (Fig. 14.4b) methods. However, sensitivity estimates increase sharply to larger values for logistic regression than for neuronal nets, which means that the former method more correctly classifies new cases irrespective of the threshold value of probability to accept presence when compared to the latter. Artificial neuronal network procedure was more sensitive in that it classified false presence with a high probability of event. Our present findings correspond to those obtained by Manel et al. (1999) in that we observed a general but slight tendency for better performance in logistic regressions for classifying countries across distinct threshold probabilities than neuronal nets did.

14.4 Discussion

There have been very scarce comparison studies of the ability of logistic regressions and neuronal networks to discriminate between the presence and the absence of an event (de Garine-Wichatitsky et al. 1999). We provide in this work an attempt to compare the ability of these two methods to make accurate predictions of the spatial occupancy and occurrence for 15 human diseases, some of which can be considered as endemic on a continent, e.g. Chagas disease, and some others have a widespread distribution world-wide, e.g. Hepatitis A and B. Logistic regressions have been widely used and proposed for several applications relevant to ecological and evolutionary investigations (Schoener and Adler 1991; Trexler and Travis 1993; Veltman et al. 1996; Gaston 1998; Sorci et al. 1998). Several texts provide informative discussions on logistic regressions (McCullagh and Nelder 1989; Norusis 1997). Unfortunately, very little attention has been paid to comparing the efficiency of these two methods, and we wish to aid in this present investigation the development of a comparative analysis of statistical techniques when an event is expected to occur or not.

First, defining the precise spatial distribution of parasitic and infectious diseases is of prime interest, but it remains difficult. On smallest spatial scales ecological variability and temporal changes make disease distributions not easily definable in space (Sutherst 1998). Many infectious diseases have their distribution in time, which waxes or wanes with the natural periodicity of events. This is particularly true for Malaria distribution in Sub-Saharan Africa, for instance (Craig et al. 1999).

We have demonstrated in this work that simple models may be used to predict the actual distribution of different diseases on a global scale. The data sets and the two methodological approaches we used are entirely relevant at this largest scale, but they do not consider small-scale anomalies that evidently affect distribution, such as local extinction of vectors, arid zones, deforestation, etc. As preconised by Craig et al. (1999), the modelling of disease distribution may be viewed as a four-tier approach: *(1)* the first level, at a large scale, defines the broad distribution of diseases (this work); *(2)* the second level, at a southcontinental scale, takes into account differences between ecological zones; *(3)* the third level, at a regional or national scale, defines the transmission intensity within a given zone of transmission ecology, such as perennial, seasonal or bi-seasonal transmission; *(4)* the fourth level, at a local scale of say several km-squares, which allows more detailed precisions. On largest scales, prediction of a disease in terms of presence according to very global environmental and biological descriptors is highly appreciable. Interestingly, such models are entirely relevant at smallest scales, since the inclusion of other smaller-scale data sets should permit more detailed predictions.

Second, logistic regression procedure tends to be the method of choice for classifying an event of presence or absence, simply because it is used widely and is generally understood. It is worth noting here that logistic regressions gave similar results to artificial neuronal nets. In addition, logistic regressions were less time-consuming than neuronal nets. On the practical side, logistic regressions generally provide an information about the importance of predicting factors on the likelihood of an event. Astonishingly, we did not find any major effect of occurrence, i.e. the proportion of absence scores on presence scores, on classification using general models. Contrastingly, minimal models for both methods were more sensitive to the occurrence effect. One possible explanation is that in both statistical techniques, data are assumed to implicitly contain the information necessary to establish the relation, an assumption more probable when performing a general model than in a minimal model. The very high scores of classification across the 15 infectious and parasitic diseases on a global scale we obtained with general models probably reveal the fact that there is good reason for using these factors to indicate the probable presence of a given disease, at least at this scale of investigation. In addition, the examples we show confirm that human life-history traits allow very good predictive scores of presence of a disease. This may be particularly true when faced with virulent parasites, hosts should adjust their reproductive biology by increasing reproductive output and/or reducing age at maturity as suggested by theory. Thus, these findings would tend to show that some diseases, e.g. schistosomiasis, responsible for high incidences of morbidity and mortality in human populations, might be associated with their host life history characters, and vice versa. Although it is well known that some diseases cause severe impacts on human populations, they do not necessary imply a subtle response in the adjustment of human life-history traits. Prediction does not demonstrate causation between the presence of a given disease and shifts in life-history parameters. According to Møller (1998), parasites should impose stronger selection pressures on their hosts in the tropics compared to nontropical climatic zones, and thus parasite constraints on man should be stronger around the Equator line. Three alternative hypotheses could explain our findings: *(1)* both parasite species occurrence and human characteristics may be related to a third surrogate variable, or group of surrogate variables, not encountered into our models, which would act simultaneously on both these parameters, *(2)* parasite impacts should select for optimal human responses, or *(3)* the presence of one disease, or one group of co-occurring diseases, might be determined by humankind. As of present, we cannot evaluate the relative importance of these three mechanisms, but just apprehend them by modelling. Nevertheless, we demonstrate that human-parasite biological systems are probably characterized by nonlinear spatially extended relationships, which might reflect the connectedness that would exist in real communities between man and his infectious and parasitic diseases.

14.5 Conclusion

It is remarkable to see how such simple predictive models approximate the actual distribution of the different infectious and parasitic diseases across the world so well. Adopting a wider perspective, i.e. macroepidemiology, we show here that this modelling can be easily repeated and manipulated in combination with other new available data sets and combining data from multiple sources. Collations of data for infectious and parasitic diseases on the field are numerous, but the modelling of diseases is still in its infancy. Probably, more global data sets including various abiotic and biotic parameters are needed in conjunction with geographical information systems. Thus, we view the modelling of infectious and parasitic diseases as a promising avenue of research.

Our results suggest the existence of nonlinear laws for the spatial distribution of some diseases and human life-history traits. Many factors determine why an area in a particular region harbours a disease (or not). Classical theories of biogeographers explain the spatial distribution of diseases by cultural, geographical and socio-economical factors, with southtropical areas harbouring the bulk of parasitic and infectious diseases on earth (Anderson and May 1991). Because a wide array of factors may contribute to the actual variation in space of human diseases, it is actually difficult to disentangle the respective effect of the different variables involved. However, our findings tend to show the existence of a correlation between diseases and human characteristics, with fertility being the most important factor.

In free-living organisms, there is evidence of rare-common species differences (see Kunin and Gaston 1993; Gaston et al. 1998; Blackburn et al. 1998). Widespread species tend, on average, to be locally more abundant than rare restricted species at large spatial scales. A positive interspecific (in space) and intraspecific (in time) relationship arises because common species are abundant at some sites, whereas rare species are rare at all sites. Unfortunately, little is known about parasitic and infectious diseases. Restricted parasite endemics, e.g. Chagas disease, may probably differ in its ecological interactions with hosts, and/or in its population genetics to a more similar widespread disease. Such differences might result from evolved adaptations to the condition of rarity, or, on the contrary, not all species might be equally armed to become rare. Extensive data sets are now available on parasitic and infectious diseases to be analysed in this way. Combining different information about genetics, physiology, life history, epidemiology etc., may allow us to address some unresolved issues in public health. We are developing in our laboratory such methodological approaches to integrate a vast amount of data (see Rapport et al. 1998) on some core research projects, i.e. Chagas disease, trypanosomiasis, paludism and liver-fluke.

Changes in attribute data entered into models are very easy to deal with. In some ways, it is possible to modify a variable, or a group of values within a variable, allowing comparison of trends before and after modification, and to observe effects on prediction of disease occurrences across countries. For instance, this would be possible with population density, death rate, GNP data or fertility rate values to detect declines in site occupancy for a disease, synchronous declines for a group of co-occurring diseases, or on the contrary dispersion of some others. In addition, climate change, deforestation and desertification scenarios, which are highly probable in the near future, may provide an immediate response on some spatial disease occupancies. Lattice illustrations using Kohonen (1984) mapping may be ideal for representing human disease distributions in space and time, for detecting regions or areas at risk of new invasions by infective agents (work in progress). We feel that such analysis methods are not yet particularly well developed, and they are completely general to answer some urgent problems the earth is facing within the very near future, such as emerging and resurging diseases (Gratz 1999)!

Acknowledgements

This work was funded by I.R.D. (J.F.G.) and C.N.R.S. (F.T., T.d.M., S.L. and F.R.). We thank the staff of C.E.P.M. at Montpellier (France), two anonymous referees for comments on an earlier draft of this work, and the thousands of field epidemiologists who have carried out the disease census-taking.

References

Alexander HM (1987) Pollinator limitation in a population of *Silene alba* infected by the anter-smut fungus *Ustilago violacea*. J Ecol 75:771–780

Anderson RM, May RM (1991) Infectious diseases of humans: Dynamics and control. Oxford University Press, Oxford

Ashford RW (1991) The human parasite fauna: towards an analysis and interpretation. Ann Trop Med Parasitol 85:189–198

Blackburn TM, Gaston KJ, Greenwood JJD, Gregory RD (1998) The anatomy of the interspecific abundance-range size relationship for the British avifauna: II. Temporal dynamics. Ecology Lett 1:47–55

Brooke M de L, Davies NB, Noble DG (1998) Rapid decline of host defences in response to reduced cuckoo parasitism: Behavioural flexibility of reed warblers in a changing world. Proc R Soc Lond B 265:1277–1282

Brown JH (1995) Macroecology. University of Chicago Press, Chicago

Cavalli-Sforza LL (1997) Genes, peoples, and languages. Proc Natl Acad Sci USA 94:7719–7724

Combes C (1995) Interactions durables: Ecologie et evolution du parasitisme. Masson Ed., Paris

Craig MH, Snow RW, le Sueur D (1999) A climate-based distribution model of malaria transmission in Sub-Saharan Africa. Parasitology Today 15:105–111

Edwards M, Morse DR (1995) The potential for computer-aided identification in biodiversity research. Tree 10:153–158

Efron B (1983) Estimating the error rate of a prediction rule: Improvement on cross-validation. J Am Stat Ass 78:316–330
Esch GW, Bush AO, Aho JM (1990) Parasite communities: Patterns and processes. Chapman and Hall Ltd., London
Ewald PW (1994) Evolution of infectious diseases. Oxford University Press, Oxford
Fielding AH, Bell JF (1997) A review of methods for assessment of prediction errors in conservation presence/absence models. Env Cons 24:38–49
Forbes MRL (1993) Parasitism and host reproductive effort. Oikos 67:444–450
Freeman JA, Skapura DM (1992) Neuronal networks: Algorithms, applications and programming techniques. Addison-Wesley Publishing Company, Reading
Friedman JH (1997) On bias, variance, 0/1-loss and the curse-of-dimensionality. Data Mining and Knowledge Discovery 1:55–77
Gallant SI (1993) Neuronal network learning and expert systems. The MIT Press, Cambridge
Garine-Wichatitsky M de, Meeûs T de, Guégan JF, Renaud F (1999) Spatial and temporal distributions of parasites: Can wild and domestic ungulates avoid African tick larvae? Parasitology 119:455–466
Gaston KJ (1998) Species-range size distributions, extinction and transformation. Philos Trans R Soc Lond B 353:219–230
Gaston KJ, Blackburn TM, Gregory RD, Greenwood JJD (1998) The anatomy of the interspecific abundance-range size relationship for the British avifauna: I. Spatial patterns. Ecology Lett 1:38–46
Grafen A (1992) The uniqueness of the phylogenetic regression. J Theor Biol 156:405–423
Gratz NG (1999) Emerging and resurging vector-borne diseases. Annu Rev Entomol 44:51–75
Harvey PH, Pagel MD (1991) The comparative method in evolutionary biology. Oxford University Press, Oxford
Hochberg ME, Michalakis Y, de Meeüs T (1992) Parasitism as a constraint on the rate of life-history evolution. J Evol Biol 5:491–504
Immerman RS (1986) Sexually transmitted disease and human evolution: survival of the ugliest. Hum Ethol Newsletter 4:6–7
Jones H (1990) Population geography. Paul Chapman Publishing Ltd., London
Jongman RHG, ter Braak CJF, van Tongeren OFR (1995) Data analysis in community and landscape ecology. Cambridge University Press, Cambridge
Kohavi R (1995) A study of cross-validation and bootstrap for estimation and model selection. Proceedings of the 14th International Joint Conference on Artificial Intelligence, Morgan Kaufman Publishers Inc., pp 1137–1143
Kohonen T (1984) Self-organization and associative memory. Springer-Verlag, Berlin
Kris AC, Lively CM (1998) Experimental exposure of juvenile snails (*Potamopyrgus antipodarum*) to infection by trematode larvae (*Microphallus* sp.): Infectivity, fecundity compensation and growth. Oecologia 116:575–582
Kunin WM, Gaston KJ (1993) The biology of rarity: Patterns, causes and consequences. Tree 8:298–301
Lafferty KD (1993) The marine snail, *Cerithidea californica*, matures at smaller sizes where parasitism is high. Oikos 68:3–11
Lively CM (1987) Evidence from a New Zealand snail for the maintenance of sex by parasitism. Nature 328:519–521
Manel S, Dias JM, Ormerod SJ (1999) Comparing discriminant analysis, neuronal networks and logistic regression for predicting species distributions: a case study with a Himalayan river bird. Ecol Model 120:337–347
Martin D (1996) Depicting changing distributions through surface estimations. In: Longley P, Batty M (eds) Spatial analysis: Modelling in a GIS environment. GeoInformation International, Cambridge
Martins EP (1996) Phylogenies and the comparative method in animal behavior. Oxford University Press, Oxford
Martins EP, Hansen TF (1997) Phylogenies and the comparative method: A general approach to incorporating phylogenetic information. Am Nat 149:646–667
McCullagh P, Nelder JA (1989). Generalized linear models. Chapman and Hall Ltd., London
McNamara JM, Houston AI (1996) State-dependent life histories. Nature 380:215–221
Michalakis Y, Hochberg ME (1994) Parasitic effects on host life-history traits: A review of recent studies. Parasite 1:291–294
Møller AP (1997) Parasitism and the evolution of host life history. In: Clayton DH, Moore J (eds) Host-parasite evolution: General principles and avian models. Oxford University Press, Oxford
Møller AP (1998) Evidence of larger impact of parasites on hosts in the tropics: Investment in immune function within and outside the tropics. Oikos 82:265–270
Norusis MJ (1997) SPSS for Windows. Advanced Statistics 7.5., Chicago

Oberdorff T, Hugueny B, Compin A, Belkesam D (1998) Non-interactive fish communities in coastal streams of the North-Western France. J Anim Ecol 67:472–484
Pedersen M (1995) Interactive and animated cartography. Prentice Hall Int., Upper Saddle River, N.J.
Petney TN, Andrews RH (1998) Multiparasite communities in animals and humans: Frequency, structure and pathogenic significance. Int J Parasitol 28:377–393
Rapport DJ, Costanza R, McMichael AJ (1998) Assessing ecosystem health. Tree 13:397–402
Reeson AF, Wilson K, Gunn A, Hails RS, Goulson D (1998) Baculovirus resistance in the noctuid *Spodoptera exempta* is phenotypically plastic and responds to population density. Proc R Soc Lond B 265:1787–1791
Rhodes CJ, Anderson RM (1996) Power laws governing epidemics in isolated populations. Nature 381:600–602
Rosenzweig ML (1995) Species diversity in space and time. Cambridge University Press, Cambridge
Rumelhart DE, Hinton GE, Williams RJ (1986) Learning representations by back-propagating errors. Nature 323:533–536
Schoener TW, Adler GH (1991) Greater resolution of distributional complementarities by controlling for habitat affinities: A study with Bahamian lizards and birds. Am. Nat. 137:669–692
Sheldon AL, Meffe GK (1995) Path analysis of collective properties and habitat relationships of fish assemblages in coastal plain streams. Can J Fish Aquat Sci 52:23–33
Shykoff JA, Bucheli E, Kaltz O (1996) Flower lifespan and disease risk. Nature 379:779
Sorci G, Clobert J, Michalakis Y (1996) Cost of reproduction and cost of parasitism in the common lizard, *Lacerta vivipara*. Oikos 76:121–130
Sorci G, Møller AP, Clobert J (1998) Plumage dichromatism of birds predicts introduction sucess in New Zealand. J Anim Ecol 67:263–269
Stearns SC (1992) The evolution of life-histories. Oxford University Press, Oxford
Sutherst RW (1998) Implications of global change and climate variability for vector-borne diseases: generic approaches to impact assessments. Int J Parasitol 28:935–945
Teriokhin AT (1998) Evolutionary optimal age of repair: Computer modelling of energy partition between current and future survival and reproduction. Evolutionary Ecology 12:291–307
Teriokhin AT, Budilova EV (2000) Evolutionary optimal networks for controlling energy allocation to reproduction and repair in men and women. (see Chapter 15 of this book)
Trexler JC, Travis J (1993) Nontraditional regression analyses. Ecology 74:1629–1637
Veltman CJ, Nee S, Crawley MJ (1996) Correlates of introduction success in exotic New Zealand birds. Am Nat 147:542–557
Whittaker RH (1975) Communities and ecosystems. Macmillan, New York
Zar JH (1996) Biostatistical analysis. Prentice Hall Int., Upper Saddle River, N.J.
Zweig MH, Campbell G (1993) Receiver-operating characteristic (ROC) plots: A fundamental tool in clinical medicine. Clin Chem 39:561–577

Evolutionarily Optimal Networks for Controlling Energy Allocation to Growth, Reproduction and Repair in Men and Women

A.T. Teriokhin · E.V. Budilova

15.1 Introduction

This paper may be considered as a continuation of the approach of some recent works (Abrams and Ludwig 1995; Cichon 1997; Teriokhin 1998) which treat quantitatively, using an evolutionary optimization approach, the so-called disposable soma theory of ageing (Kirkwood 1981). This theory affirms that the senescence of an organism with age is due to insufficient repair caused by evolutionarily profitable diversion of energy to the organism's other needs, mainly to reproduction.

Evolutionary optimization methodology is based on the assumption that during the evolution of a species some criterion of Darwinian fitness is maximized. Usually, the Malthusian parameter, which is defined implicitly by the Euler-Lotka equation, is used as such a criterion.

This equation relates the Malthusian parameter, ρ, which characterizes the rate

$$1 = \int_0^{\infty} e^{-\rho t} f_t l_t dt \qquad (15.1)$$

of increase of a population, to individual characteristics such as f_t, fertility at age t, and l_t, probability to survive to age t. If the population number is stationary then the maximization of ρ is equivalent to the maximization of F,

$$F = \int_0^{\infty} f_t l_t dt \qquad (15.2)$$

the life time reproductive success of individual (Taylor et al. 1974).

It is evident that in the absence of any constraints the problem of evolutionary optimization becomes trivial: the greater f_t and l_t the greater F. It is therefore necessary to add some constraints is to make the problem meaningful. We recognize three levels of introducing constraints in the evolutionary optimization problem statement.

The first is the level of the so-called trade-off curves (e.g. Stearns 1992). In this case f_t and l_t are directly related by some equation in a sort that f_t decreases with increasing l_t, and vice versa.

The second level (e.g. Perrin and Sibly 1993) is when f_t and l_t are related implicitly by a model for allocation of an organism's resources (energy, for brevity) among its different needs (growth, reproduction, repair, maintenance etc.). This approach can be adequately formulated in terms of the theory of optimal control (e.g. Bellman 1957; Pontryagin et al. 1962). The main concepts of this theory include a criterion of optimality (F, in our case), control variables (e.g. fractions of energy allocated to different organism's needs), state variables (variables characterizing the physiological state of organism and the state of environment), and state equations (differential equations describing the dynamics of state variables). The final goal of the theory is to find an optimal strategy, i.e. a rule optimally matching a set of admissible values of control variables to any set of admissible values of state variables.

The third level (e.g. Mangel 1990; Budilova et al. 1995) is characterized by a still more detailed consideration of the physiology. In this case, one tries not simply to find an optimal strategy of energy allocation but also to define the structure and parameters of the system which realizes such a strategy. The neuronal network approach (e.g. Rumelhart and McClelland 1986) provides an adequate mean to pose and solve this problem. State and control variables of the optimal control approach become here input and output nodes of a cognitive control network, and the rules describing interactions between the nodes play the role of state equations.

In this paper we consider, on the second and third levels, a particular problem of modelling the evolution of life histories of men and women. Special attention is paid to explaining sex distinctions in their life histories. For example, it is well known from demographic data that men, as compared with women, have later age of maturity, greater body size, shorter life span, and have no menopause. We will show that all these distinctions immediately emerge in evolutionarily optimal life histories if only one assumption is made about the difference between the physiology of men and women, namely, if we assume that women, but not men, can accumulate reproductive energy in their offspring.

15.2 Optimal Control Model

Let us specify the assumptions of the model of evolutionary optimization of men's and women's life histories. We will consider the problem of allocating energy to only three needs of the organism: growth, repair, and reproduction. Correspondingly, three age dependent control variables, u_{wt}, u_{qt}, and u_{rt} (with $u_{wt} + u_{qt} + u_{rt} = 1$), denoting the fractions of energy allocated to growth, repair, and reproduction are considered. Our state variables are w_t, body size, q_t, organism vulnerability, and r_t, energy accumulated in offspring (only for women).

The state equation for the size is

$$\frac{dw}{dt} = u_{wt}e_t, \quad w(0) = w_0 \tag{15.3}$$

where e_t is the rate of energy production at age t which itself depends on the body size, w_t, in the following way

$$e_t = M_e \frac{(w_t / M_w)^d}{c + (w_t / M_w)^d} \tag{15.4}$$

so that for not very great sizes e_t is proportional to w_t to the power d. The constants M_e, M_w, c, and d in Eq.15.4 are unknown parameters to be fitted to demographic data (M_e and M_w denote maximum possible values of e_t and w_t).

The state equation for vulnerability (which at the population level is better interpreted as mortality rate) is

$$\frac{dq}{dt} = a(1 / u_{qt} - u_{qt})^b, \; q(0) = q_0 \tag{15.5}$$

That means that the vulnerability does not increase with age if all the energy is spent on repair ($u_{qt} = 1$), and increases with infinite rate if no energy is allocated for repair ($u_{qt} = 0$), a and b being parameters which should be fitted to demographic data. The vulnerability normally grows with age, according to such a state equation, because the organism cannot spent all its energy on repair. Knowing q_t, we can find the survival l_t, i.e. the probability of surviving to age t, using the equation

$$l_t = e^{-\int_0^t q_s ds} \tag{15.6}$$

We assume in addition that the reproductive effect depends nonlinearly on the energy invested in reproduction

$$f_t = M_f \frac{(r_t / M_r)^h}{g + (r / M_r)^h} \tag{15.7}$$

where g, h, M_r, M_f are parameters to be fitted to demographic data (M_r and M_f denote maximum possible values of r_t and f_t), and r_t is the energy invested in offspring since the moment of previous offspring releasing. We, however, suppose that only women, but not men, can accumulate energy according to the state equation

$$\frac{dr}{dt} = u_{rt} e_t, \; r(t_0) = 0 \tag{15.8}$$

where t_0 is the beginning of a recurrent period of offspring carrying and bringing up.

15.3 Computing Optimal Strategies

We use the dynamic programming method (Bellman 1957; Mangel and Clark 1988) to compute evolutionarily optimal strategies of energy allocation. For this, we set lower and upper limits for the lifetime and state variables and divide these intervals into sufficiently small parts.

Namely, we set all the lower limits for t, w, q, r equal to 0, the upper limits to T, M_w, M_q, M_r, and the numbers of division to N_t, N_w, N_q, N_r. We also divide the interval from 0 to 1 for energy fractions u_{wt}, u_{qt}, and u_{rt} into N_u parts and assume that energy can be allocated either to growth or reproduction, but not to both. In addition, we introduce a control variable u_{ft} which is allowed to take only three values: 0 for growing, 1 for accumulating reproductive energy, and 2 for releasing offspring (u_{ft} cannot be 1 for men).

The dynamic programming computes the lifetime reproductive success of

$$F(w_t, q_t, r_t, t) = \max_{u_{wt}, u_{qt}, u_{rt}, u_{ft}} \{[F(w_{t+1}, q_{t+1}, r_{t+1}, t+1) + f_t]\exp(-q_t)\} \tag{15.9}$$

individual F starting with $F(w_t, q_t, r_t, t) = 0$ at $t = T$ and proceeding backwards up to t=0 according to recurrent equation where (assuming that $dt = T / N_t = 1$), we set

$$w_{t+1} = w_t + u_{wt}e_t, \text{ if } u_{ft} = 0; \quad w_{t+1} = w_t, \text{ if } u_{ft} \neq 0 \tag{15.10}$$

$$q_{t+1} = q_t + a(1/u_{qt} - u_{qt})^b \tag{15.11}$$

$$r_{t+1} = r_{t+1} + u_{rt}e_t, \text{ if } u_{ft} = 1; \quad r_{t+1} = 0, \text{ if } u_{ft} \neq 1 \tag{15.12}$$

$$f_t = M_f \frac{(r_t / M_e)^h}{g + (r_t / M_e)^h}, \text{ if } u_{ft} = 2; \quad f_t = 0, \text{ if } u_{ft} \neq 2 \tag{15.13}$$

As a result of this backward procedure we obtain, for all sets of state variables, values at each time step, optimal values of control variables. That allows us, now proceeding forward from some initial values of state variables, to find the optimal life history strategy of investing energy to growth, reproduction, and repair and corresponding dependencies of size, offspring, and vulnerability on age.

15.4 Fitting the Optimal Control Model

The above dynamic programming procedure can be performed only if all the parameters q_0, w_0, a, b, c, d, g, h, M_e, M_w, M_r, and M_f in the model are defined. We do not know these parameters, but we can try to choose them in such a way that the resulting optimal life history characteristics (age and size at maturity, mean life span for men and women, age of menopause for women) should be close to real demographic data, and, in particular, demonstrate differences in men's and women's life histories. We have tried a number of sets of parameters and found that the following set of parameter values (the same for men and women) satisfies, to some extent, these requirements: $q_0 = 0.005$, $w_0 = 3$, $a = 0.003$, $b = 2$, $c = 0.5$, $d = 0.5$, $g = 0.2$, $h = 1.7$, $M_e = 45$, $M_w = 80$, $M_r = 30$, $M_f = 35$.

Figures 15.1 and 15.2 illustrate the dependencies of e_t on w_t and f_t on r_t defined by Eqs. 15.4 and 15.7 for these parameter values. The dependence of e_t on w_t shows an approximate proportionality of e_t to w_t to the power 0.5, and that agrees with widely accepted approximations for this dependence (e.g. Bertalanffy 1957). As regards the

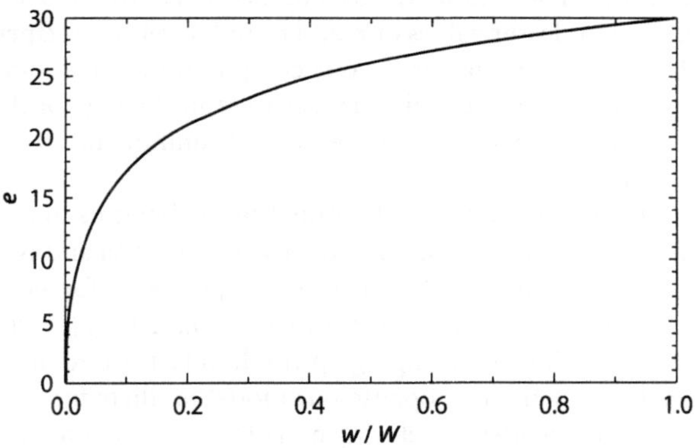

Fig. 15.1. Energy production rate (e) as a function of size (w)

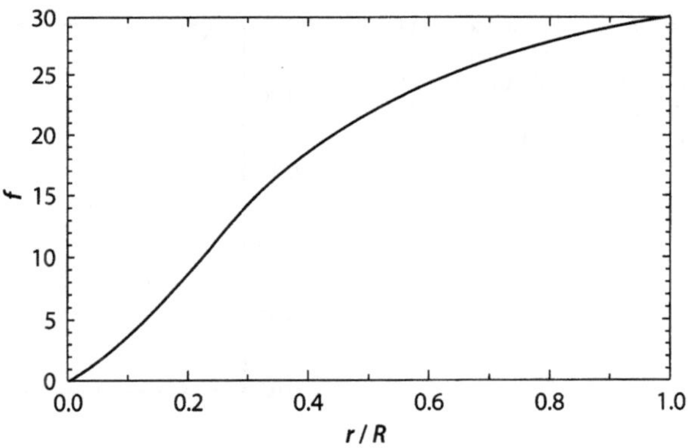

Fig. 15.2. Reproductive output (f) as a function of amount of energy invested in reproduction (r)

dependence of f_t on r_t, it can be seen from Fig. 15.2 that it is sigmoidal. This implies that it is advantageous to invest more (to some extent) energy into offspring and this can be done in two ways. First, an individual can accumulate energy in offspring before releasing this energy (we will see that this is the women's strategy) and, second, an individual can release a great amount of energy instantaneously if it has a big body size (it is the life strategy of men).

Using the above set of parameter values (with $N_t = 140$, $N_w = N_q = N_u = 20$, $N_r = 5$), we computed the evolutionarily optimal (i.e. maximizing lifetime reproductive success) strategies for men and women (the only difference between men and women was that u_{ft} could not take the value 1 for men, i.e. men could not accumulate reproductive energy).

Figures 15.3 and 15.4 present, for men and women correspondingly, the dependencies of the control variable u_{ft} (taking values 0 and 2 for men and 0, 1, and 2 for women) on state variables (on w_t and q_t for men and on w_t, q_t, and r_t for women).

In Fig. 15.3 we see that it is evolutionarily optimal for men to grow ($u_{ft} = 0$) when w_t and q_t are not very large and to reproduce ($u_{ft} = 2$) when w_t or q_t become sufficiently great. In Fig. 15.4 we see also that it is optimal for women to grow ($u_{ft} = 0$) for lesser values of w_t, q_t, and r_t and to reproduce ($u_t = 1$ or 2) for greater values of these state variables. In addition, we observe that it is optimal to accumulate energy in offspring ($u_{ft} = 1$) for lesser values of r_t and q_t and to release offspring ($u_{ft} = 2$) for greater values of these state variables. In our computations we assumed in advance that growing abruptly and irreversibly changes to reproducing, for it may be shown that this is so because of linear dependence of the Hamiltonian for our model on u_{wt} (Ziolko and Kozlowski 1983).

Applying the described optimal rules allows us to compute optimal dependencies of state and control variables on age. In Fig. 15.5a and 15.5b we show the dependencies of u_{ft} on t for men and women. The dependence for women differs from that for men in the presence of a period, from age of maturity $t_{mat} = 14$ to age of menopause $t_{mp} = 50$, when carrying and bringing up children lasts more than one time step (2 years, to be exact). After the menopause our model predicts for women cessation of accumulating energy in offspring during more than one time step. Assuming that it is physiologi-

Fig. 15.3. Dependence of optimal decision rule u_f on size (w) and vulnerability (q) for men ($u_f = 0$ corresponds to growing and $u_f = 2$ to reproducing)

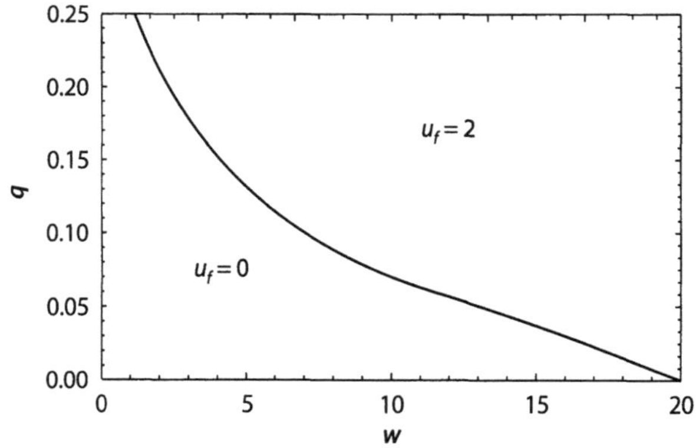

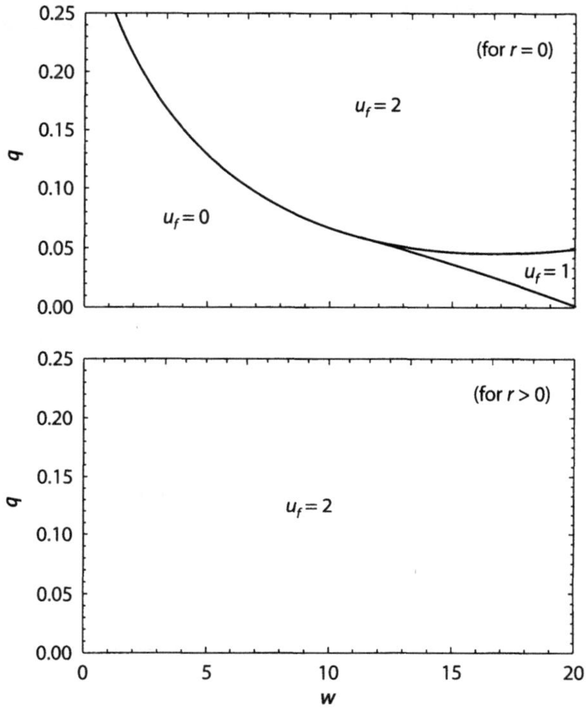

Fig. 15.4. Dependence of optimal decision rule u_f on size (w), vulnerability (q), and accumulated reproductive energy (r) for women ($u_f = 0$ corresponds to growing, $u_f = 1$ to accumulating reproductive energy, and $u_f = 2$ to reproducing)

cally impossible to reproduce full-value human offspring during only one year, we may suppose that women, during their evolution, changed carrying and bringing up their own children in older age to, say, bringing up grandchildren.

In Fig. 15.6 we show the optimal dependencies of size w_t on age t for men and women. We see that the size of maturity for men, $w_{mat} = 16$, is greater than that for women, $w_{mat} = 14$ (we put, both for men and women, $w_0 = 3$).

In Fig. 15.7 are shown, again for men and women, the optimal dependencies of fraction of energy allocated to repair, u_{qt}, on age. We see that with age this fraction steadily decreases both for men and women, though for women this decreasing is slower. As a consequence, the vulnerability, q_t, grows with age at an increasing rate as can be seen in Fig. 15.8. Again, this growing is slower for women as compared with men, which results in greater mean life span for women, $t_{mls} = 61$, as compared with that for men, $t_{mls} = 54$.

Thus we do see that the only assumption about impossibility for men, as opposed to women, to accumulate energy in offspring leads to well known distinctions in men's evolutionarily optimal life histories: later maturation, bigger size, shorter life span, absence of menopause. Note that an analogous assumption (higher promiscuity in men as opposed to women) was already successfully used, though in a different model, for an evolutionary explanation of slower ageing in women (Rossler et al. 1995).

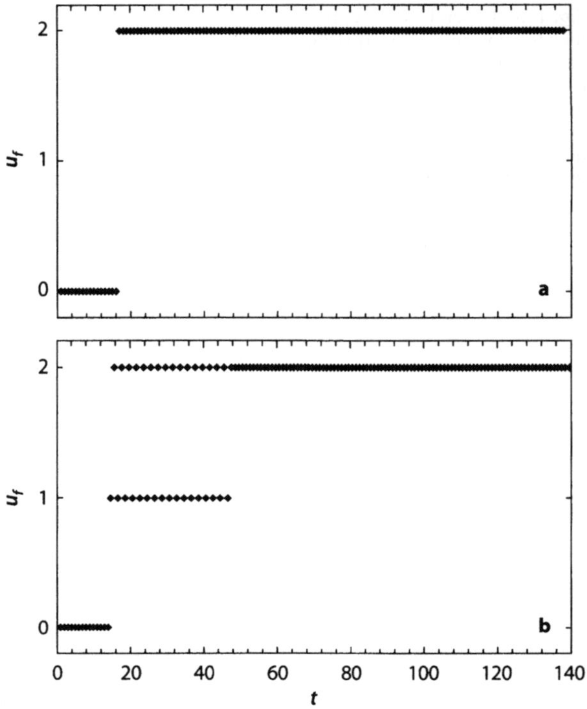

Fig. 15.5. Life history dynamics of optimal decision making; **a** men; **b** women

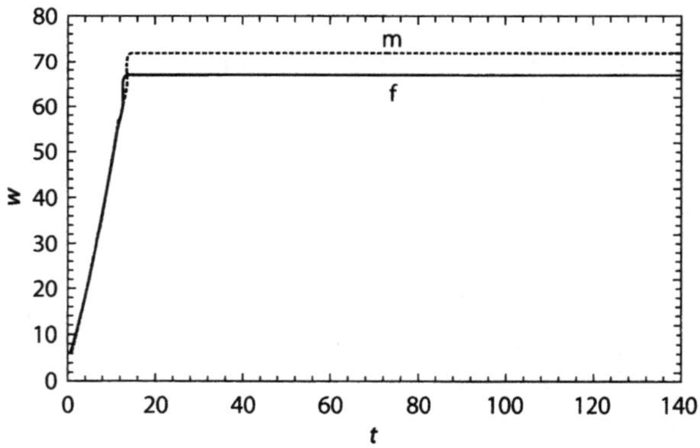

Fig. 15.6. Optimal life history dynamics of growing (increase in size w_t with age t) for men (m) and women (f)

15.5 Network Control Model

So far, we have not touched on physiological mechanisms capable of realizing evolutionarily optimal age-dependent strategies of energy allocation. Now we will try to do this, partly relying on the results of previous paragraphs.

In Fig. 15.9 we present a network scheme for simulating the processes of allocating energy in a human organism during its life history. Like traditional neuronal networks

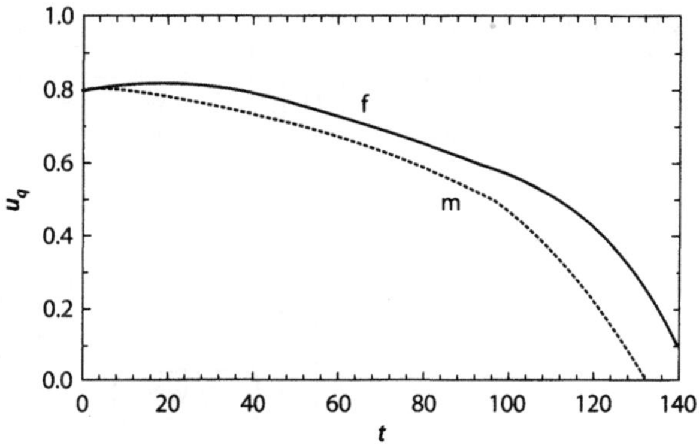

Fig. 15.7. Optimal life history dynamics of the fraction of energy allocated for repair (u_{qt}) for men (m) and women (f)

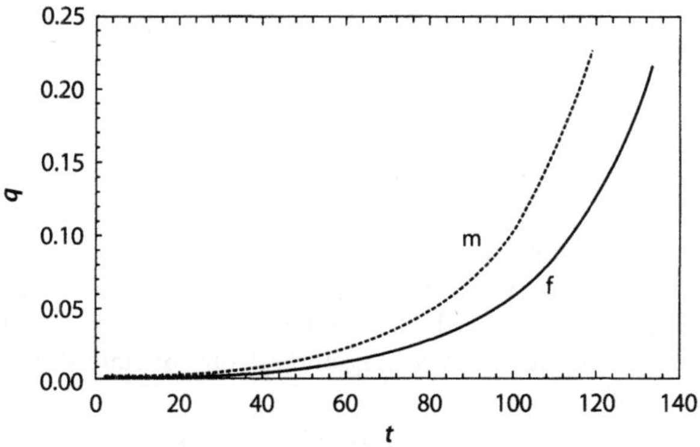

Fig. 15.8. Optimal life history dynamics of senescing (increase in vulnerability q_t with age t) for men (m) and women (f)

(e.g. Rumelhart and McClelland 1986), our network, at least partly, consists of units joined by weighted links. However, there are essential distinctions. Not all the units are as simple as classical formal neurons (e.g. Hopfield 1982). And not all links and units deal with information: some of them operate with energy flows. We would better describe this network not as truly neuronal but rather as somatoneuroendocrine.

The network proceeds as follows. At each time step t unit E produces some amount of energy e_t defined by the size of individual, w_t, in accordance with Eq. 15.4. This amount of energy is divided between growth (unit W), repair (unit Q), and reproduction (unit R) proportionally to the values of control signals u_{wt}, u_{qt}, and u_{rt}. The changes in size, w_t, vulnerability, q_t, and reproductive energy, r_t, depend on the fractions of energy directed to growth, repair, and reproduction in the same way as in the above optimal control model, i.e. according to Eqs. 15.3, 15.5, and 15.8. One more control variable, u_{ft}, taking only two values (1 or 2) prevents ($u_{ft} = 1$) or stimulates ($u_{ft} = 2$) releasing accumulated reproductive energy r_t. Reproductive output f_t (directed to output F), depends on accumulated reproductive energy, r_t, as it stated above by Eq. 15.7.

In their turn, the values of control signals u_{wt}, u_{qt}, u_{rt}, and u_{ft} themselves depend on the values of state signals w_t, q_t, and r_t. This dependence is realized through units Q-WR, W-R, and R-F. Unit Q-WR gets as input the values of state signals with weights s_1, s_2, and s_3 and generates two output signals: u_{qt} and $1 - u_{qt}$. The first output is used

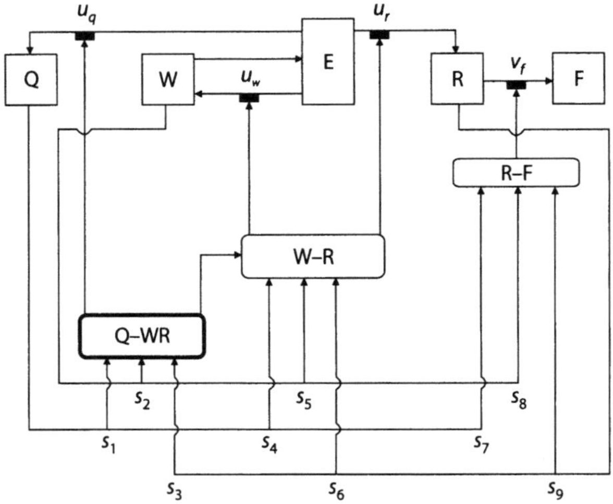

Fig. 15.9. Network scheme modelling somato-neuroendocrine processes of energy allocation for growth, repair, and reproduction in a human organism

to control the fraction of energy allocated to repair, and the second signal is sent to unit W-R which redirects it either to growth or reproduction, that is, generates either $u_{wt} = 1 - u_{qt}$ and $u_{rt} = 0$ or $u_{wt} = 0$ and $u_{rt} = 1 - u_{qt}$. The choice is defined by the values of state signals weighted by s_4, s_5, and s_6. The value of control signal u_{ft} (1 or 2) is generated by unit R-F, which functions as a simple formal neuron: the output is 2 if the weighted (with weights s_7, s_8, and s_9) sum of state signals is greater than a threshold, and 1 otherwise.

It is assumed that the network can die at any time (age) step with probability $\exp(-q_t)$ and the amount of reproductive output, accumulated in F by this time, defines the lifetime reproductive success of this network.

15.6 Optimizing the Network Model

The network described above contains two sets of unknown parameters. The first set of unknown parameters consists of the weights of links: s_1, s_2, s_3, s_4, s_5, s_6, s_7, s_8, and s_9. To find them we will use the so called genetic optimization algorithm (e.g. Bounds 1987). Finding weights s_1–s_9 in the neuronal network model corresponds to finding optimal strategies in the optimal control problem statement. The second set consists of parameters already used in the optimal control model for describing processes of energy transformation: q_0, w_0, a, b, c, d, g, h, M_e, M_w, M_r, and M_f. We will simplify our task and use for these parameters the values already found when fitting the optimal control model.

The genetic algorithm proceeds in our case as follows. We generate randomly N sets of weights and simulate the life history of each of N corresponding networks. Depending on the obtained random values of weights and on randomness of death age, different networks will have different lifetime reproductive successes. After the death of all individuals, we form a new generation that consists of copies of networks from the previous generation, the number of copies being proportional to the lifetime reproductive successes of prototypes. Additionally, the weights in the new generation networks are subject to mutating and crossing. The procedure is repeated to produce newer and newer generations, and as a result, a greater and greater fraction of networks in newer generations will have higher lifetime reproductive success. When the average increase in reproductive success becomes too slow, we stop the procedure and take the values of weights for the best networks in the last generation as optimal.

Performing genetic optimization, we have made some additional simplifications to reduce computer time. Taking into account some properties of optimal strategies, calculated by the method of dynamic programming, we set several weights, namely s_1, s_3, s_6, s_7, and s_9, equal to 0.

Finally, the operations performed by units Q-WR, W-R, and R-F are summarized as follows. Unit Q-WR generates a control signal according to a sigmoid formula $u_{qt} = A / [1 + s_1(q_t)^B]$ (A and B are additional optimized parameters). Unit W-R allocates the fraction $1 - u_{qt}$ of energy to growth if $s_5 w_t + s_4 q_t < 1$, and to reproduction otherwise. Unit R-F stimulates accumulation energy in offspring (i.e. generates $u_{ft} = 1$) if $s_7 q_t < 1$, and stimulates releasing offspring (i.e. generates $u_{ft} = 2$) otherwise (note that u_{ft} is always 2 for men).

The genetic optimization with $N = 1500$ and number of generations equal to 500 resulted in the following values for the unknown parameters: $A = 0.76$, $s_1 = 29\,863$, $B = 4.4$, $s_5 = 0.013$, $s_4 = 7.1$ for men and $A = 0.78$, $s_1 = 46\,563$, $B = 4.8$, $s_5 = 0.015$, $s_4 = 14$, $s_7 = 40$ for women.

In Fig. 15.10 we present (for women) the resulting optimal strategy. As we see, it roughly approximates that in Fig. 15.4 obtained with dynamic programming. Figure 15.11 shows the dependence of u_{qt} on q. We see that the fraction of energy allocated to repair decreases when vulnerability increases (faster for men than for women), and this tendency especially accelerates for higher values of vulnerability ($q = 0.05$ and greater). The main life history characteristics of networks are: $t_{mat} = 15$, $t_{mls} = 67$, $w_{mat} = 77$ for men and $t_{mat} = 13$, $t_{mls} = 77$, $w_{mat} = 62$, $t_{mp} = 60$ for women. They are close to the optimal ones found by dynamic programming and, in particular, also demonstrate differences between men's and women's life history traits.

So we may conclude that indeed the network presented in Fig. 15.9 exhibits a roughly optimal life history strategy of energy allocation for growth, repair, and reproduction after having been subject to genetic optimization.

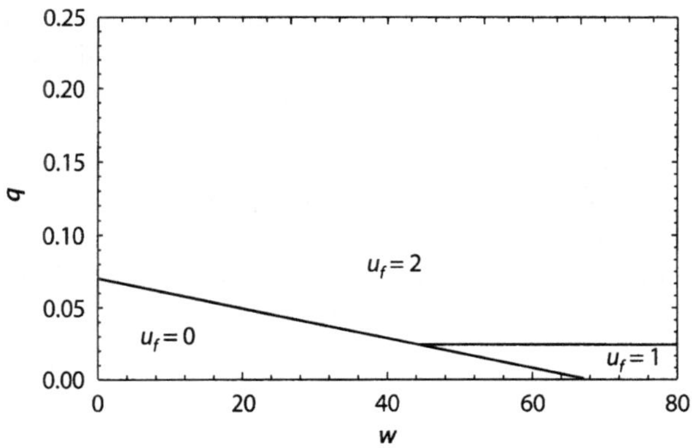

Fig. 15.10. Dependence of decision rule u_f on size (w) and vulnerability (q) for women (for $r = 0$ at the beginning of time step) obtained for a genetically optimized network

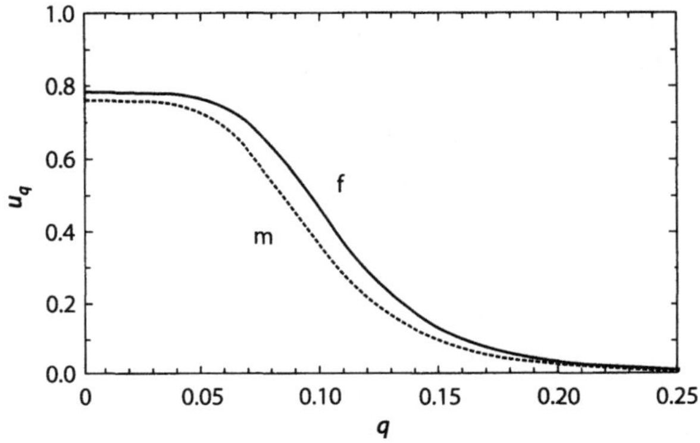

Fig. 15.11. Dependence of the fraction of energy allocated for repair (u_q) on vulnerability (q) for men (m) and women (f) obtained for a genetically optimized network

15.7 Discussion

Thus we have shown that many differences in men's and women's life histories may be explained as evolutionarily advantageous consequences of the only distinction in women's physiology: the necessity of carrying and bringing up their children over a rather prolonged period of time. We also have shown that a simple evolutionarily optimal network scheme can describe the process of distribution of energy in a human organism among its different needs.

We deliberately did not refine the structure of the network. This should be done in parallel with indication of at least hypothetical body prototypes of each unit of the model, but this much more complicated task is out of the scope of the present work. Nevertheless, we can formulate some principal points, following from our approach, which should be first of all taken into account when relating the model to a real organism. Both in the optimal control and network models the decisions concerning energy allocation are made on the basis of information about the state variables. Hence the system must know the size and vulnerability of the organism as well as the stage of development of offspring (for women). That is, some sensors for the state variables (size, vulnerability, offspring status) must exist in the organism and we should look for such sensors.

Another problem concerns the signals controlling the process of allocating energy. We might suppose that these signals are realized in form of concentrations of hormones. The hormones which modify growth and sexual behaviour are well known. We are obliged to suppose that there exist also hormones which stimulate repair, say, endorphins.

Surely, a more realistic model should include, as a minimum, the hostility of environment. It was supposed in our model to be a constant (q_0) during the entire life as well as during the evolution, but in reality it changes. Supposedly, the level of concentration of stress hormones reflects the influence of environment and this should be taken into account in more elaborate models.

Acknowledgements

The authors are grateful to P. Abrams, M. Cichon, and P. Kloeden for their helpful comments. This paper was funded by the Russian Foundation for Basic Research, grant no. 98-04-49140.

References

Abrams PA, Ludwig D (1995) Optimality theory, Gompertz law, and the disposable soma theory of senescence. Evolution 49:1055–1066
Bellman R (1957) Dynamic programming. Princeton Univ. Press, Princeton
Bertalanffy L (1957) Quantitative laws in metabolism and growth. The Quarterly Review of Biology 32:217–231
Bounds DG (1987) New optimization methods from physics and biology. Nature 329:215–219
Budilova EV, Kozlowski J, Teriokhin AT (1995) Neuronal network models of life history energy allocation. In: Proc. of the First Nat. Conf. on Application of Mathematics to Biology and Medicine, Univ. of Mines and Jagiellonean Univ., Krakov, pp 13–18
Cichon M (1997) Evolution of longevity through optimal resource allocation. Proc R Soc Lond B 49:1383–1388
Hopfield JJ (1982) Neuronal networks and physical systems with emergent collective computational abilities. Proc Nat Acad Sci USA 79:2554–3558
Kirkwood TBL (1981) Repair and its evolution: Survival versus reproduction. In: Townsend CR, Calow P (eds) An evolutionary approach to resource use. Blackwell Scientific Publications, Oxford, pp 165–189
Mangel M (1990) Evolutionary optimization and neuronal network models of behavior. J Math Biol 28:237–256
Mangel M, Clark C (1988) Dynamical modeling in behavioral ecology. Princeton Univ. Press, Princeton
Perrin N, Sibly RM (1993) Dynamic models of energy allocation and investment. Ann Rev of Ecology and Systematics 24:379–410
Pontryagin LS, Boltyanskii VG, Gamkrelidze RV, Mishchenko EF (1962) Mathematical theory of optimal processes. Wiley, New York
Rossler R, Kloeden PE, Rossler OE (1995) Slower aging in women: A proposed evolutionary explanation. BioSystems 36:179–185
Rumelhart DE, McClelland JL (eds) (1986) Parallel distributed processing. MIT Press, Cambridge
Stearns SC (1992) The evolution of life histories. Oxford Univ. Press, Oxford
Taylor HM, Gourley RS, Lawrence CE, Kaplan RS (1974) Natural selection of life history attributes: An analytical approach. Theor Pop Biol 5:104–122
Teriokhin AT (1998) Evolutionarily optimal age schedule of repair: Computer modelling of energy partition between current and future survival and repair. Evolutionary Ecology 12:291–307
Ziolko M, Kozlowski J (1983) Evolution of body size: An optimization model. Math Biosci 64:127–143

Part V

Perspectives

Chapter 16

Can Neuronal Networks be Used in Data-Poor Situations?

W. Silvert · M. Baptist

16.1 Introduction

Perhaps the greatest problem that is faced in most attempts to use artificial neuronal networks for ecological applications is that the quantity of data is often very limited. Although there are a few cases where large amounts of data are available, as in the case of remote sensing or observations based on automatic telemetry, it is far more common to have to deal with limited and irregularly spaced data, and the data may not always be strictly comparable due to variations in environmental conditions between sampling periods. In most situations the collection of field data is both time-consuming and expensive. Since the training and testing of neuronal networks is very data-intensive, this poses serious obstacles to the development of neuronal network applications in ecology.

The field of marine benthic ecology is one in which data are difficult to obtain, and large databases are almost unknown. Most benthic data are obtained by laborious analysis of individual samples, usually cores or grabs, although automated sediment traps and other computerised instrumentation are beginning to be used. Because of these data restrictions, the prospects for using artificial neuronal networks in the analysis of benthic data are not promising. On the other hand, analysis of benthic data often requires highly specialised expertise that is not commonly available, so the incentive to use artificial neuronal networks is strong.

The material in this paper is based on an effort to facilitate the analysis of geochemical cores by developing an artificial neuronal network with a very limited data set. After careful review of the data and elimination of unreliable samples, we were left with a very small number of cores, and the prospects for development of a neuronal network seemed very dubious. Still, we feel that we made substantial progress by suitable preprocessing of the data, and we feel that this approach may be useful to other applications in ecology and similar data-poor fields.

16.2 Neuronal Network Training and its Limitations

The most common neuronal networks are feed-forward neuronal networks that are trained using error back propagation. This is a training method in which the network is supplied with input values and also with the desired output values. The weights in the network are adjusted based upon the error between the expected output and the computed network output until this error is minimized. In a reasonably complex network the number of weights is large and, unless there are many data pairs, the minimization process may not give meaningful results. For this reason, the application of neuronal networks to ecological data does not always lead to reliable models.

As with most empirical (e.g. statistical) modelling approaches, one divides the total data set into two parts, one of which is used for fitting the model (which is referred to as "training" in the terminology of artificial neuronal networks), and the other of which is used for testing it. When the data set is very large, typically involving thousands or even tens of thousands of data pairs, this is relatively straightforward. With smaller data sets it can be difficult to decide how much of the data can be used for training the model while still leaving enough data to test it adequately. One approach is to repeat the process several times, selecting a training set at random, then fitting the model and testing it with the remaining data.

One must, however, be careful not to "overtrain" the network, meaning that after too many iterations one stumbles upon a model which fits the training data almost perfectly, without actually providing a good representation of reality. If the test set is too small, the network may give adequate test results, leading to acceptance of a model which is not very good. This is especially true if the model has too many degrees of freedom, which is the result of including too many hidden neurons – this is analogous to the statistical problem of fitting a model with too many parameters, such as a polynomial of too high order.

Another issue related to the overtraining problem is the bias-variance dilemma (Geman et al. 1992). It can be demonstrated that the mean square value of the estimation error between the function to be modelled and the neuronal network consists of the sum of the (squared) bias and variance. With a neuronal network using a training set of fixed size, a small bias can only be achieved with a large variance (Haykin 1994). This dilemma can be circumvented if the training set is made very large, but if the total amount of data is limited, this may not be possible.

As pointed out in the introduction, ecological data sets are usually small. Unfortunately ecological processes are often nonlinear and poorly understood, and the limited data sets are barely adequate. Since the processes may not be adequately understood, the use of neuronal networks to generate empirical models of complex behaviour seems a promising technique for a better replication and understanding of ecosystem behaviour. On the other hand, the small size of the data sets makes it very difficult to train and test these models well enough to have confidence in the results.

The situation is not as hopeless as it may appear. The field of artificial neuronal networks is developing very rapidly, and new approaches resolve many formerly intractable problems. For example, general regression neuronal networks (Specht 1991) offer a promising way to deal with small data sets, and have been used successfully to construct a neuronal network for a data set not much larger than the one used in this project. There are some other neuronal network configurations that are capable of handling smaller data sets, such as Kohonen networks, but the use of sophisticated second-generation approaches is not the issue we are trying to address in this paper. Instead we have tried to explore ways of dealing with small data sets when one uses the most popular configuration of neuronal networks, feed-forward error back propagation networks.

When only a relatively small data set is available for the application of a neuronal network, a number of drawbacks arise. First, one has to split the already small data set into a training set and a testing set. Second, overtraining is more likely to occur with small data sets, because the degrees of freedom in the network rapidly increase with the number of neurons. For a good model, the number of data pairs should exceed the number of weights in the neuronal network.

16.3 The Geochemical Data Problem

Geochemical profiles sampled underneath fish farms provide valuable data on the benthic impacts of these farms, but the interpretation of these data is a complex process requiring scientific sophistication and understanding of benthic processes. It is difficult to avoid a degree of ambiguity and subjectivity in the interpretations, although this is a more general problem in the analysis of scientific data than is commonly admitted (Silvert 1997). Some typical geochemical profiles are shown in Fig. 16.1. These types of profiles are commonly found for organic carbon and organic matter, and, in the case of hyperenriched sediments, porewater concentrations of nutrients and hydrogen sulphide. The first profile, Fig. 16.1a, shows a continuous decrease in carbon levels, presumably reflecting constant deposition and degradation. The second profile, Fig. 16.1b, shows a flat plateau suggesting that the sediments have been mixed by bioturbation. Since bioturbation is indicative of a viable benthic community, the second of these profiles would normally be interpreted as showing less of an environmental impact than the first.

Angel et al. (1998) used data from dive logs to classify environmental impacts of a fish farm in the Red Sea in terms of four fuzzy sets: Nil, Moderate, Severe and Extreme impact. This paper used only visual data recorded by divers, but on some of the sampling dates the divers also took sediment cores, and these can also be used as indica-

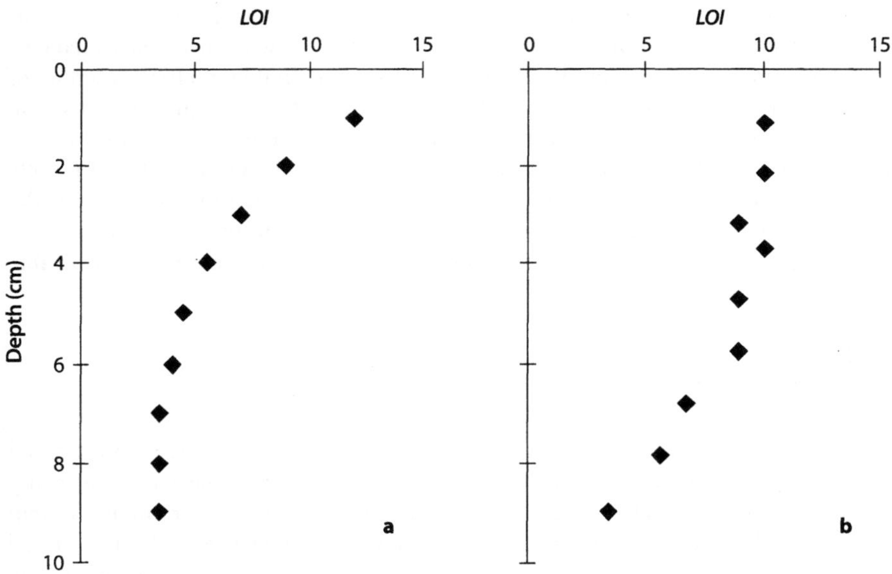

Fig. 16.1. Two typical sediment profiles; **a** simple deposition and degradation; **b** probable bioturbation

tors of environmental impact. The number of useful cores is substantially less than the number of dive logs, but because it is widely felt that chemical analysis of such cores provides a more objective measure of environmental impact than the dive logs, which involve subjective elements, research is continuing on the use of these cores as environmental indicators. In particular, sediment profiles for loss on ignition (*LOI*) data are being investigated. *LOI* is a measure of carbon content, and since carbon deposition is one of the major forms of environmental effect from fish farms (due to faecal settlement and waste feed), it is a critical indicator of both current and past deposition rates, as well as of bioturbation.

Although preliminary results indicate good agreement between the classifications based on the dive logs and those obtained from the *LOI* data, the analysis of *LOI* data is extremely difficult and requires a great deal of expertise. This makes it a prime candidate for neuronal network analysis, since the value of *LOI* data for monitoring purposes would be greatly enhanced if an expert system could be used for the analysis, and there did not seem to be any reliable way of developing a rule-based system.

We subsequently undertook the development of an artificial neuronal network to simulate the impact-assignment process in an effort to develop an expert system that does the same analysis, as described in Baptist et al. (1998).

The sediment profiles contain eleven sampling depths. The values for loss on ignition (*LOI*) at each depth were used for the input neurons, so there were eleven input neurons. The fuzzy membership values for the four impact classes were used as outputs – these memberships sum to one, so there were three independent output neurons. The best results with the feed-forward error back propagation neuronal network were obtained with a hidden layer of eleven neurons. Configurations with fewer hidden neurons, even down to three, were also tried. Although the fit to the training set was reasonable, the fit to the test set worsened with a decreasing number of hidden neurons. The total network therefore had 165 connections. The data set of *LOI* profiles measured over a four-year period was small and contained only nineteen suitable profiles. Baptist et al. (1998) were faced with a typical problem when using neuronal networks for ecological problems: the data set was too small in relation to the number of degrees of freedom in the neuronal network. Their results were typical for this kind of problem, the training set (fourteen profiles) showed an almost perfect fit, whereas the test set (five profiles) had considerable discrepancies. Figure 16.2 gives a sample of the results that were obtained by applying the artificial neuronal network to the scores for "moderate-impact." Even though the test results were still acceptable, this cannot be considered very significant in light of the small amount of data available for the test procedure.

16.4 Preprocessing Data

One of the reasons why neuronal networks are inefficient is that they are usually trained with raw data that ignores the understanding and insight that a human expert can bring to bear. For example, in the classic pattern recognition problem of training a system to recognize human faces, it is considered an achievement when an artificial neuronal network "learns" to ignore whether the subject is wearing eyeglasses. Edelman et al. (1998) discuss the differences one encounters training a neuronal network to recog-

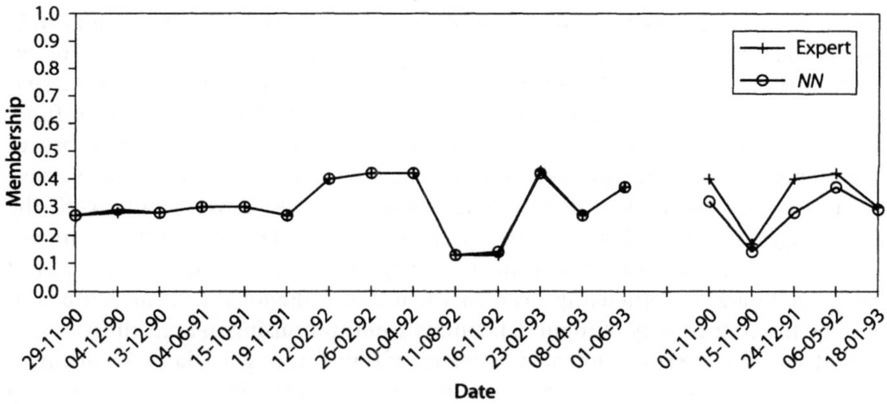

Fig. 16.2. Sample results obtained with an artificial neuronal network fit to the raw "moderate-impact" scores

nize faces when the full face is visible and when the hair is covered by a bathing cap. But we all know that eyeglasses are almost useless for identification purposes, and hair shape and colour can change rapidly. If we could teach the neuronal network in advance to ignore eyeglasses and hair style, and to concentrate on facial structure, it should be possible to make the training process much more efficient.

One way to interact with the training process is to preprocess the data before it is fed into a neuronal network. Preprocessing has been applied to pattern recognition problems, although more for the purpose of simplifying the vast amount of data contained in a matrix of pixel values than for the present purpose of extracting information as efficiently as possible from a limited data set (Huntsberger et al. 1998). We might process the images through a graphics program that would erase eyeglasses and facial hair, and generate just a three-dimensional profile that reflected the underlying bone structure. We are not aware of any effort to investigate this approach, although the underlying idea of looking for basic geometric structure has been discussed by Edelman (1998). We did however look at ways of transforming the data for the *LOI* profiles in such a way as to simplify the work of the neuronal network, and we feel that the approach led to promising results.

When an expert on benthic ecology and geochemistry assigns classifications based on different *LOI* profiles, he makes use of subjective factors, even though the data themselves are objective numerical values. An expert uses his knowledge of natural processes to analyse the shape and overall magnitude of the overall profile or parts of the profile (for example the upper 5 centimetres) to arrive at an interpretation of the degree of impact. However, since a neuronal network only sees the numbers in the profile as independent variables, it does not even have the valuable information that the depths of the samples are contiguous. It may therefore improve the neuronal network performance if the data were preprocessed and expert-based discriminants were used as input to the neuronal network. This can also simplify the neuronal network layout, speed the training process and result in a network with a smaller number of degrees of freedom.

One way in which we can link this type of human pattern recognition with artificial neuronal networks is to transform the data using orthogonal functions. For example, Edelman et al. (1998) use principal components analysis to represent the spatial data constituting a pixel image of a face in terms of a small number of eigen-functions (i.e. orthogonal functions) which are then used as input variables. This procedure seems relevant to the problem of analysing the *LOI* profiles, except that we have to keep in mind that each of these profiles contains at most eleven points, while the facial images consisted of over 15 000 pixels. The use of data transformations in pattern analysis is more a way of reducing unmanageably large data bases than a way of using small data sets efficiently. Even so, the use of orthogonal transformations to describe the connectivity of input data points seems promising in both situations.

Baptist et al. (1998) experimented with ways of transforming the data with orthogonal functions, such as Fourier transforms, to generate input variables that more closely correspond to the patterns that experts see when they analyse these kinds of data. Fourier transforms distinguish particular patterns, and at least for the lower-frequency transforms, each Fourier component emphasises a specific part of the profile and might serve as a discriminant. For example, the lowest frequency components describe the total carbon content and the gradient of the profile. Since Fourier components are based on sine and cosine functions, they have a close correspondence to the wavelets used by Huntsberger et al. (1998). The first five Fourier components were chosen as input neurons. This way, each Fourier component distinguishes a particular pattern in the *LOI* profiles. Consequently the number of input neurons were reduced to five in a network with five hidden neurons and four output neurons. This brings back the number of weights to 45. A sample of the results of this neuronal network for the "moderate-impact" scores are shown in Fig. 16.3 and were comparable with those of the raw *LOI* profiles. We found that the Fourier components were very sensitive to variations in the *LOI* values. Furthermore, the number of data pairs was too small to develop a good neuronal network.

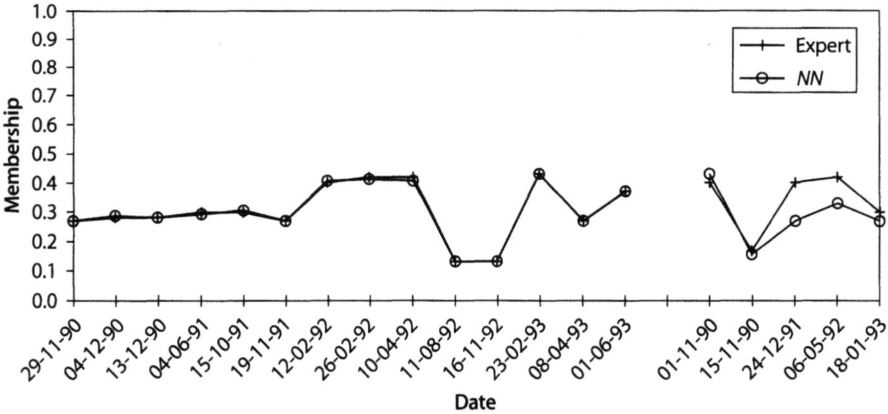

Fig. 16.3. Sample results obtained with an artificial neuronal network fit to the Fourier transforms of "moderate-impact" scores using the first five Fourier components (o, c1, s1, c2, s2)

16.5
Results

A comprehensive discussion of the results of this study would be complicated by the fact that the expert assessments that were used to train and test the neuronal network were expressed in terms of fuzzy membership functions. Because of the additional complexity that this introduced to the analysis, we refer the reader to Baptist et al. (1998) for a full discussion of the theoretical background of the study and of the detailed results. In this paper we want to focus on the problems of the small data set and the issue of whether transforming the data added to the efficiency and effectiveness of the process. We feel that these results are strongly suggestive, but not conclusive.

The analysis of the data was carried out with both the raw input data (i.e. *LOI* values as a function of depth) and with transformed values, using the first four or five Fourier components only. We were surprised to see that in both cases the neuronal networks worked quite well – it is not surprising that they fit the training sets, of course, since there were so few data and so many parameters, but the test sets were reasonable. Given the small size of the test sets, we cannot attach very much significance to this result, but they do suggest that the use of neuronal networks for even very small data sets may prove practicable.

We were, however, also surprised to see that the use of a smaller number of transformed variables as inputs did not appreciably improve the fitting of the network. On more careful examination of the network we discovered that when the raw data were used, the output was dominated by the first input neuron, corresponding to the *LOI* value just below the surface (0.5 cm), so from this point of view the raw data was already in a form that might not be improved by transformation.

16.6
Discussion and Conclusion

When data are scarce it might be useful to preprocess (transform) the data before they are used to train a neuronal network. The transformation technique must be chosen with care and should be based on ecological knowledge of the system. A preprocessing of data can reduce the number of input neurons and therefore reduce the problem of overtraining. It may also help to understand the process handling inside a neuronal network, making it less of a black box then it is when blindly applied to incoherent data sets.

However, preprocessing based on understanding requires that the understanding be reliable, and we don't always know that this is the case. Huntsberger et al. (1998) discuss the distinction between feature-based systems and image-based systems, and we have not been able to identify clearly the extent to which the experts who analysed the *LOI* profiles relied on features (such as the near-surface values) as opposed to evaluating the entire profile. They were also surprised to learn that the top part of the profile played such a major role in determining their evaluations, but in retrospect this may not be unreasonable even if we accept their assertions that they looked at the total shape of the profile. For example, a core that displays evidence of bioturbation is likely to have lower levels of surface carbon because it has been mixed deeper into the sediments. Although we feel that discussions with the experts can help resolve these issues, it is difficult when dealing with such a small data set as the one that was available for this study.

We are not convinced that Fourier transforms were the best choice of transform functions for this project, since they are best for analysing data that are distributed over a uniform interval. Given the shape of the *LOI* profiles, it might have been preferable to use Bessel or Laplace transforms, which offer a better representation of functions which are large near the origin and taper off as one moves away from it, but neither the amount of data nor the time available permitted exploration of this idea.

The use of preprocessing data to refine the inputs to artificial neuronal networks has been studied only in a few cases, mainly in the context of reducing large quantities of graphic data to manageable scale, and whether the same approach can be successfully applied to very small data sets is not clear. Even in the relatively advanced field of image processing and facial recognition, it has not been shown conclusively that preprocessing with orthogonal transformations is advantageous (Edelman et al. 1998). However we feel that the approach is promising and deserves further investigation. Certainly if neuronal network theory is to be applied to small and highly variable ecological data sets, the focus must be on how to use all the data as efficiently as possible, and we should not be using the data to establish patterns that are already known.

References

Angel D, Krost P, Silvert W (1998) Describing benthic impacts of fish farming with fuzzy sets: Theoretical background and analytical methods. J Appl Ichthyol 14:1–8

Baptist MJ, Silvert W, Krost P, Angel DL (1998) Assessing benthic impacts of fish farming with an expert system based on neuronal networks. Bull Can Soc Theo Biol, Fall 1998. [The CSTB Bulletin is an electronic journal accessible at URL http://www.mar.dfo-mpo.gc.ca/science/mesd/he/science/cstb/, and the paper itself is located at URL http://www.mar.dfo-mpo.gc.ca/science/mesd/he/staff/silvert/bska/]

Edelman B (1998) Spanning the face space. J Biol Systems 6:265–280

Edelman B, Valentin D, Abdi H (1998) Sex classification of face areas: How well can a linear neuronal network predict human performance? J Biol Systems 6:219–240

Geman S, Bienenstock E, Doursat R (1992) Neuronal networks and the bias/variance dilemma. Neuronal Computation 4:1–58

Haykin S (1994) Neuronal networks: A comprehensive foundation. Macmillan College Publishing Co., New York

Huntsberger T, Rose J, Ramaka S (1998) Fuzzy-face: A hybrid wavelet/fuzzy self-organizing feature map system for face processing. J Biol Systems 6:281–298

Silvert W (1997) Ecological impact classification with fuzzy sets. Ecol Model 96:1–10

Specht D (1991) A general regression neuronal network. IEEE Trans. on Neuronal Networks 2:568–576

Index

A

B

C

N

O

P

Q

R

T

Y

Z

MIX
Papier aus verantwortungsvollen Quellen
Paper from responsible sources
FSC® C105338
FSC
www.fsc.org

If you have any concerns about our products,
you can contact us on
ProductSafety@springernature.com

In case Publisher is established outside the EU,
the EU authorized representative is:
Springer Nature Customer Service Center GmbH
Europaplatz 3, 69115 Heidelberg, Germany

Printed by Libri Plureos GmbH
in Hamburg, Germany